HARCOURT
Math

CALIFORNIA EDITION

Harcourt School Publishers

Orlando • Boston • Dallas • Chicago • San Diego
www.harcourtschool.com

Mathematics Advisor

Richard Askey
Professor of Mathematics
University of Wisconsin
Madison, Wisconsin

Senior Author

Evan M. Maletsky
Professor of Mathematics
Montclair State University
Upper Montclair, New Jersey

Authors

Angela Giglio Andrews
Math Teacher, Scott School
Naperville District #203
Naperville, Illinois

Grace M. Burton
Chair, Department of Curricular Studies
Professor, School of Education
University of North Carolina
 at Wilmington
Wilmington, North Carolina

Howard C. Johnson
Dean of the Graduate School
Associate Vice Chancellor for
 Academic Affairs
Professor, Mathematics and
 Mathematics Education
Syracuse University
Syracuse, New York

Lynda A. Luckie
Administrator/Math Specialist
Gwinnett County Public Schools
Lawrenceville, Georgia

Joyce C. McLeod
Visiting Professor
Rollins College
Winter Park, Florida

Vicki Newman
Classroom Teacher
McGaugh Elementary School
Los Alamitos Unified School District
Seal Beach, California

Janet K. Scheer
Executive Director
Create A Vision
Foster City, California

Karen A. Schultz
College of Education
Georgia State University
Atlanta, Georgia

Program Consultants and Specialists

Janet S. Abbott
Mathematics Consultant
California

Arax Miller
Curriculum Coordinator and
 English Department Chairperson
Chamlian School
Glendale, California

Lois Harrison-Jones
Education and Management
 Consultant
Dallas, Texas

Rebecca Valbuena
Language Development Specialist
Stanton Elementary School
Glendora, California

iii

Understand Numbers and Operations

1 PLACE VALUE AND NUMBER SENSE

✓ **Check What You Know** 1
1 **Hands On:** Patterns on a Hundred Chart . . . 2
2 Understand Place Value 4
3 Understand Numbers to 10,000 • *Activity* . . . 6
  Thinker's Corner • Ducky Digits
4 Understand 10,000 • *Activity* 10
5 **Problem Solving Strategy:**
  Use Logical Reasoning 12
  Chapter 1 Review/Test 14
  Cumulative Review 15

2 COMPARE, ORDER, AND ROUND NUMBERS 16

✓ **Check What You Know** 17
1 Size of Numbers 18
2 Compare Numbers 20
  Linkup • Reading: Compare
3 Order Numbers 24
4 **Problem Solving Skill:**
  Identify Relationships 26
5 Round to Nearest 10 and 100 28
6 Round to Nearest 1,000 30
  Chapter 2 Review/Test 32
  Cumulative Review 33

Daily Review and Practice
  Quick Review
  Mixed Review and Test Prep
Intervention
  Troubleshooting, pp. H2–H7
Extra Practice pp. H32–H35

Technology Resources

Harcourt Math Newsroom Video:
Chapter 2, p. 21

E-Lab:
Chapter 1, p. 3
Chapter 3, p. 40
Chapter 4, p. 57

Mighty Math Calculating Crew:
Chapter 3, p. 47
Chapter 4, p. 63

Multimedia Glossary:
The Learning Site at
www.harcourtschool.com/
glossary math

3 ADDITION 34

✓ **Check What You Know** 35

1 Column Addition 36

2 Estimate Sums 38

3 **Hands On:** Add 3-Digit Numbers 40

4 Add 3-Digit Numbers 42

5 **Problem Solving Strategy:**
Predict and Test 44

6 Add Greater Numbers 46
     Thinker's Corner • Sum It Up

     Chapter 3 Review/Test 50

     Cumulative Review 51

4 SUBTRACTION 52

✓ **Check What You Know** 53

1 Estimate Differences 54

2 **Hands On:** Subtract 3-Digit Numbers 56

3 Subtract 3-Digit Numbers 58
     Linkup • **Science**

4 Subtract Greater Numbers 62
     Thinker's Corner • Solve It!

5 **Problem Solving Skill:**
Estimate or Exact Answer 66

6 Algebra: Expressions and
Number Sentences 68

     Chapter 4 Review/Test 70

     Cumulative Review 71

UNIT WRAPUP

Math Detective: Putting It Together 72

Challenge: Understand 100,000 73

Study Guide and Review 74

California Connections: Squaw Valley USA
     and Skiing in the Sierra Nevada 76

UNIT 2
CHAPTERS 5–6

Money and Time

5 USE MONEY 78

✓ **Check What You Know** 79
1 **Hands On:** Make Equivalent Sets 80
2 **Problem Solving Strategy:**
  Make a Table 82
3 Compare Amounts of Money 84
4 **Hands On:** Make Change 86
5 Add and Subtract Money 88
  Chapter 5 Review/Test 90
  Cumulative Review 91

Daily Review and Practice
  Quick Review
  Mixed Review and
  Test Prep
Intervention
  Troubleshooting,
  pp. H7–H8
Extra Practice
  pp. H36–H37

Technology Resources

Harcourt Math Newsroom Video:
Chapter 6, p. 100

E-Lab:
Chapter 5, p. 81
Chapter 6, p. 95
Chapter 6, p. 98

Mighty Math Calculating Crew:
Chapter 5, p. 87

Multimedia Glossary:
The Learning Site at
www.harcourtschool.com/
glossary/math

6 UNDERSTAND TIME 92

✔ **Check What You Know** 93
1 Hands On: Time to the Minute 94
2 A.M. and P.M. 96
3 Hands On: Elapsed Time 98
4 Use a Schedule 100
5 Use a Calendar 102
6 Problem Solving Skill:
Sequence Events 104
Chapter 6 Review/Test 106
Cumulative Review 107

UNIT WRAPUP

Math Detective: Making Cents of It 108
Challenge: Time, Year, Decade, Century 109
Study Guide and Review 110
California Connections: Monterey Bay
and The Aquarium Gift Store 112

Multiplication Concepts and Facts

7 UNDERSTAND MULTIPLICATION **114**

✓ **Check What You Know** 115

1 Algebra: Connect Addition and Multiplication 116

2 Multiply with 2 and 5 118

3 **Hands On:** Arrays 120

4 Multiply with 3 122

   Thinker's Corner • Solve the Riddle!

5 **Problem Solving Skill:** Too Much/ Too Little Information 126

   Chapter 7 Review/Test 128

   Cumulative Review 129

8 MULTIPLICATION FACTS THROUGH 5 . . . . **130**

✓ **Check What You Know** 131

1 Multiply with 0 and 1 132

2 Multiply with 4 134

3 **Problem Solving Strategy:** Find a Pattern 136

4 Practice Multiplication 138

   Thinker's Corner • Use Data

5 Algebra: Find Missing Factors 142

   Chapter 8 Review/Test 144

   Cumulative Review 145

Daily Review and Practice
   Quick Review
   Mixed Review and Test Prep
Intervention
   Troubleshooting, pp. H9–H13
Extra Practice
   pp. H38–H41

Technology Resources

Harcourt Math Newsroom Video:
Chapter 7, p. 165

E-Lab:
Chapter 7, p. 116
Chapter 9, p. 150

Mighty Math Carnival Countdown:
Chapter 7, p. 124
Chapter 8, p. 139

Multimedia Glossary:
The Learning Site at www.harcourtschool.com/ glossary/math

9 MULTIPLICATION FACTS AND STRATEGIES 146

✓ **Check What You Know** 147
1 Multiply with 6 148
2 Multiply with 7 150
3 Multiply with 8 152
4 **Problem Solving Strategy:**
Draw a Picture 154
5 Algebra: Practice the Facts 156
    Thinker's Corner • All Squared Off

    Chapter 9 Review/Test 160
    Cumulative Review 161

10 MULTIPLICATION FACTS AND PATTERNS 162

✓ **Check What You Know** 163
1 Multiply with 9 and 10 164
    Linkup • Reading: Analyze Information
2 Algebra: Find a Rule 168
3 Algebra: Multiply with 3 Factors 170
4 **Problem Solving Skill:**
Multistep Problems 172
    Chapter 10 Review/Test 174
    Cumulative Review 175

UNIT WRAPUP

Math Detective: Follow the Leader 176
Challenge: Square Numbers 177
Study Guide and Review 178
California Connections: Central Valley
   and Open-Air Markets 180

UNIT 4
CHAPTERS 11–13

Division Concepts and Facts

11 UNDERSTAND DIVISION 182

✓ **Check What You Know** 183
1 **Hands On:** The Meaning of Division 184
2 Relate Subtraction and Division 186
3 Algebra: Relate Multiplication
   and Division 188
4 Algebra: Fact Families • *Activity* 190
   Thinker's Corner • Tricky Triangles
5 **Problem Solving Strategy:**
   Write a Number Sentence 194
   Chapter 11 Review/Test 196
   Cumulative Review 197

12 DIVISION FACTS THROUGH 5 198

✓ **Check What You Know** 199
1 Divide by 2 and 5 200
2 Divide by 3 and 4 202
3 Divide with 0 and 1 204
4 Algebra: Write Expressions 206
5 **Problem Solving Skill:**
   Choose the Operation 208
   Chapter 12 Review/Test 210
   Cumulative Review 211

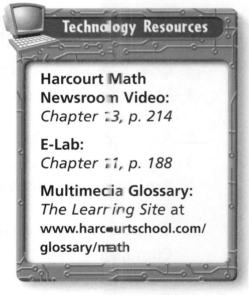

Daily Review and Practice
  Quick Review
  Mixed Review and
    Test Prep
Intervention
  Troubleshooting,
    pp. H14–H16
Extra Practice
    pp. H42–H44

Technology Resources

Harcourt Math Newsroom Video:
Chapter 13, p. 214

E-Lab:
Chapter 11, p. 188

Multimedia Glossary:
The Learning Site at
**www.harcourtschool.com/
glossary/math**

13 DIVISION FACTS THROUGH 10 212

✔ **Check What You Know** 213

1 Divide by 6, 7, and 8 214
  Linkup • Reading: Choose Important Information

2 Divide by 9 and 10 218

3 Practice Division Facts Through 10 220
  Linkup • Social Studies

4 Algebra: Find the Cost 224

5 **Problem Solving Strategy:**
  Work Backward 226

  Chapter 13 Review/Test 228

  Cumulative Review 229

UNIT WRAPUP

Math Detective: Missing Parts 230

Challenge: Divide by 11 and 12 231

Study Guide and Review 232

California Connections: Sacramento
  and Sutter's Fort 234

Data, Graphing, and Probability

14 COLLECT AND RECORD DATA 236

✓ **Check What You Know** 237
1 **Hands On:** Collect and Organize Data . . . 238
2 Understand Data 240
3 Classify Data 242
4 **Problem Solving Strategy:**
Make a Table 244
Chapter 14 Review/Test 246
Cumulative Review 247

15 ANALYZE AND GRAPH DATA 248

✓ **Check What You Know** 249
1 **Problem Solving Strategy:**
Make a Graph 250
2 Read Bar Graphs 252
3 **Hands On:** Make Bar Graphs 254
4 Line Plots • *Activity* 256
   Linkup • Reading: Use Graphic Aids
5 Locate Points on a Grid 260
6 Read Line Graphs 262
Chapter 15 Review/Test 264
Cumulative Review 265

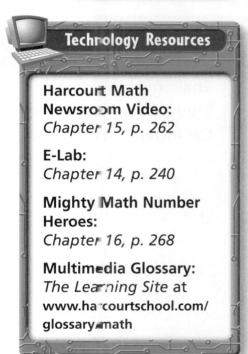

16 PROBABILITY 266

✓ **Check What You Know** 267
1 Certain and Impossible 268
2 Likely and Unlikely 270
3 **Hands On:** Possible Outcomes 272
4 Experiments • *Activities* 274
   Linkup • Science
5 Predict Outcomes • *Activity* 278
6 **Problem Solving Skill:**
   Draw Conclusions 280
   Chapter 16 Review/Test 282
   Cumulative Review 283

UNIT WRAPUP

Math Detective: The Spinning Wheel 284
Challenge: Find Mean and Median 285
Study Guide and Review 286
California Connections: San Francisco
   and San Francisco Day Trips 288

Multiply and Divide by 1-Digit Numbers

17 MULTIPLY BY 1-DIGIT NUMBERS 290

✓ **Check What You Know** 291

1 **Hands On:** Multiply 2-Digit Numbers 292

2 Record Multiplication 294
   Linkup • Reading: Choose Important Information

3 Practice Multiplication 298

4 **Problem Solving Skill:**
   Choose the Operation 300

   Chapter 17 Review/Test 302

   Cumulative Review 303

18 MULTIPLY GREATER NUMBERS 304

✓ **Check What You Know** 305

1 Mental Math: Patterns in Multiplication .. 306

2 **Problem Solving Strategy:**
   Find a Pattern 308

3 Estimate Products 310

4 Multiply 3-Digit Numbers 312
   Linkup • Math History

5 Find Products Using Money 316

6 Practice Multiplication 318

   Chapter 18 Review/Test 320

   Cumulative Review 321

Daily Review and Practice
   Quick Review
   Mixed Review and Test Prep
Intervention
   Troubleshooting,
   pp. H6, H10,
   H20–H22
Extra Practice
   pp. H48–H51

Technology Resources

Harcourt Math Newsroom Video:
Chapter 17, p. 295

E-Lab:
Chapter 17, p. 293
Chapter 19, p. 325

Mighty Math Calculating Crew:
Chapter 18, p. 317
Chapter 20, p. 347

Multimedia Glossary:
The Learning Site at
www.harcourtschool.com/
glossary/math

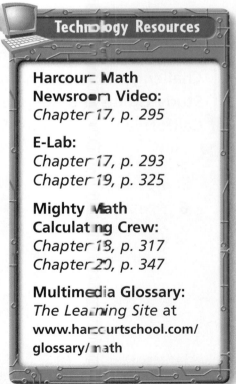

19 DIVIDE BY 1-DIGIT NUMBERS 322

✓ **Check What You Know** 323
1 **Hands On:** Divide with Remainders 324
2 Model Division of 2-Digit
Numbers • *Activity* 326
3 Record Division of 2-Digit Numbers 328
    Linkup • Social Studies
4 Practice Division 332
5 **Problem Solving Skill:**
Interpret the Remainder 334
   Chapter 19 Review/Test 336
   Cumulative Review 337

20 DIVIDE GREATER NUMBERS 338

✓ **Check What You Know** 339
1 Mental Math: Patterns in Division 340
2 Estimate Quotients 342
3 Place the First Digit in the Quotient 344
4 Practice Division of 3-Digit Numbers 346
5 **Hands On:** Divide Amounts of Money . . . 348
6 **Problem Solving Strategy:**
Solve a Simpler Problem 350
   Chapter 20 Review/Test 352
   Cumulative Review 353

UNIT WRAPUP

Math Detective: Mystery Signs 354
Challenge: Prime Numbers 355
Study Guide and Review 356
California Connections: Highways
   and The Scenic Byways 358

Geometry

21 SOLID AND PLANE FIGURES 360

✓ **Check What You Know** 361
1 Solid Figures • *Activity* 362
   Thinker's Corner • Visual Thinking
2 Combine Solid Figures 366
3 Line Segments and Angles 368
   Thinker's Corner • Clocks and Angles
4 Types of Lines 372
5 **Hands On:** Circles 374
6 **Problem Solving Strategy:**
   Break Problems into Simpler Parts 376
   Chapter 21 Review/Test 378
   Cumulative Review 379

22 POLYGONS 380

✓ **Check What You Know** 381
1 Polygons 382
2 Congruence and Symmetry • *Activity* 384
   Linkup • Science
3 Combine Plane Figures 388
4 **Problem Solving Strategy:**
   Find a Pattern 390
   Chapter 22 Review/Test 392
   Cumulative Review 393

Daily Review and Practice
   Quick Review
   Mixed Review and
      Test Prep
Intervention
   Troubleshooting,
      pp. H23–H25
Extra Practice
      pp. H52–H54

Technology Resources

Harcourt Math Newsroom Video:
Chapter 21, p. 369

E-Lab:
Chapter 21, p. 363
Chapter 22, pp. 384, 385

Mighty Math Number Heroes:
Chapter 23, p. 399

Multimedia Glossary:
The Learning Site at
www.harcourtschool.com/
glossary/math

23 TRIANGLES AND QUADRILATERALS 394

✓ **Check What You Know** 395
1 Triangles 396
2 Sort Triangles 398
   Linkup • Art
3 Quadrilaterals 402
4 Sort Quadrilaterals • *Activity* 404
   Linkup • Reading: Use Graphic Aids
5 **Problem Solving Skill:**
   Identify Relationships 408
   Chapter 23 Review/Test 410
   Cumulative Review 411

> **UNIT WRAPUP**

Math Detective: On the Edge 412
Challenge: Patterns with Plane Figures 413
Study Guide and Review 414
California Connections: Landmarks
   and Missions 416

Measurement

24 CUSTOMARY UNITS 418

✓ **Check What You Know** 419
1 Length • *Activity* 420
   Thinker's Corner • Measurement
2 **Inch, Foot, Yard, and Mile** 424
3 **Hands On:** Capacity 426
4 **Hands On:** Weight 428
5 Ways to Change Units 430
6 Algebra: Rules for Changing Units 432
7 **Problem Solving Skill:** Use a Graph 434
   Chapter 24 Review/Test 436
   Cumulative Review 437

25 METRIC UNITS 438

✓ **Check What You Know** 439
1 Length • *Activity* 440
   Thinker's Corner • Silly Circles
2 **Problem Solving Strategy:**
   Make a Table 444
3 **Hands On:** Capacity: Liters
   and Milliliters 446
4 **Hands On:** Mass: Grams and Kilograms ... 448
5 **Hands On:** Measure Temperature 450
   Chapter 25 Review/Test 452
   Cumulative Review 453

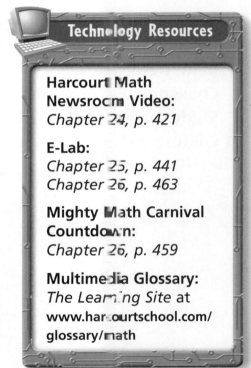

Daily Review and Practice
  Quick Review
  Mixed Review and Test Prep
Intervention
  Troubleshooting, pp. H10–H11, H26–H28
Extra Practice pp. H55–H57

Technology Resources

Harcourt Math Newsroom Video: *Chapter 24, p. 421*

E-Lab: *Chapter 25, p. 441* *Chapter 26, p. 463*

Mighty Math Carnival Countdown: *Chapter 26, p. 459*

Multimedia Glossary: *The Learning Site* at www.harcourtschool.com/glossary/math

26 PERIMETER, AREA, AND VOLUME 454

✓ **Check What You Know** 455
1 **Hands On:** Perimeter 456
2 Estimate and Find Perimeter • *Activity* ... 458
    Linkup • Social Studies
3 **Hands On:** Area of Plane Figures 462
4 Area of Solid Figures 464
5 **Problem Solving Skill:**
    Make Generalizations 466
6 Estimate and Find Volume • *Activity* 468
    Linkup • Reading: Compare
    Chapter 26 Review/Test 472
    Cumulative Review 473

UNIT WRAPUP

Math Detective: Around and Around You Go .. 474
Challenge: Find Circumference of a Circle 475
Study Guide and Review 476
California Connections: The Tournament of
    Roses® Parade and The Rose Bowl Game® .. 478

Fractions and Decimals

27 UNDERSTAND FRACTIONS 480

✓ **Check What You Know** 481
1 Count Parts of a Whole 482
  Linkup • Social Studies
2 Count Parts of a Group 486
3 Equivalent Fractions • *Activities* 488
  Linkup • Science
4 Compare and Order Fractions 492
  Thinker's Corner • Mixed Numbers
5 **Problem Solving Strategy:**
  Make a Model 496
  Chapter 27 Review/Test 498
  Cumulative Review 499

28 ADD AND SUBTRACT LIKE FRACTIONS 500

✓ **Check What You Know** 501
1 **Hands On:** Add Fractions 502
2 Add Fractions 504
  Thinker's Corner • Solve the Riddle!
3 **Hands On:** Subtract Fractions 508
4 Subtract Fractions 510
  Linkup • Geography
5 **Problem Solving Skill:**
  Reasonable Answers 514
  Chapter 28 Review/Test 516
  Cumulative Review 517

Daily Review and Practice
  Quick Review
  Mixed Review and Test Prep
Intervention
  Troubleshooting, pp. H28–H31
Extra Practice pp. H58–H61

Technology Resources

Harcourt Math Newsroom Video:
Chapter 30, p. 542

E-Lab:
Chapter 27, p. 488
Chapter 29, p. 525

Mighty Math Number Heroes:
Chapter 28, p. 506
Chapter 29, p. 520

Multimedia Glossary:
The Learning Site at
www.harcourtschool.com/glossary_math

29 DECIMALS AND FRACTIONS 518

✓ **Check What You Know** 519
1 Relate Fractions and Decimals 520
2 **Hands On:** Tenths 522
3 **Hands On:** Hundredths 524
4 Read and Write Decimals 526
5 Compare and Order Decimals 528
6 **Problem Solving Skill:**
   Reasonable Answers 530
   Chapter 29 Review/Test 532
   Cumulative Review 533

30 DECIMALS AND MONEY 534

✓ **Check What You Know** 535
1 Relate Fractions and Money 536
2 **Hands On:** Relate Decimals and Money ... 538
3 Add and Subtract Decimals and Money ... 540
   Linkup • Reading: Make Predictions
4 **Problem Solving Strategy:**
   Break Problems into Simpler Parts 544
   Chapter 30 Review/Test 546
   Cumulative Review 547

UNIT WRAPUP

Math Detective: I Found the Whole Thing 548
Challenge: Fractions and Decimals
   on a Number Line 549
Study Guide and Review 550
California Connections: Hummingbirds
   and Tucker Wildlife Sanctuary 552

Student Handbook

Table of Contents H1
Troubleshooting H2–H31
Extra Practice H32–H61
Sharpen Your Test-Taking
   Skills H62–H65
Basic Facts Tests H66–H69
Table of Measures H70
California
   Standards H71–H77
Glossary H78–H87
Index H88–H99

Welcome!

The authors of *Harcourt Math* want you to enjoy learning math and to feel confident that you can do it. We invite you to share your math book with family members. Take them on a guided tour through your book!

The Guided Tour

Choose a chapter you are interested in. Show your family some of these things in the chapter that will help you learn.

✓ Check What You Know

Do you need to review any skills before you begin the next chapter? If you do, you will find help in the Handbook in the back of your book.

🕐 The Math Lessons

✓ **Quick Review** to check the skills you need for the lesson.

✓ **Learn section** to help you study problems, models, examples, and questions that give you different ways to learn.

✓ **Check** to make sure you understood the lesson.

✓ **Practice and Problem Solving** to practice what you have just learned.

✓ **Mixed Review and Test Prep** to keep your skills sharp and to prepare you for important tests. Look back at the pages shown next to each problem to get help if you need it.

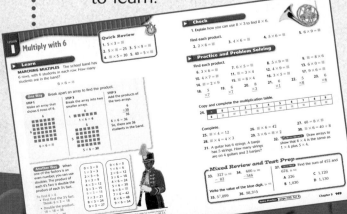

Student Handbook

Now show your family the **Student Handbook** in the back of your book. The sections will help you in many different ways.

✓ **Troubleshooting** will help you review and remember skills from last year.

✓ **Extra Practice** can be used to make sure that you are ready to move on to the next lesson.

✓ **Sharpen Your Test-Taking Skills** will help you feel confident that you can do well on a test.

✓ **Basic Facts Tests** will check whether you have memorized all of the basic facts and will show you which facts you still need to practice.

California Standards tell what you are expected to learn this year. Each lesson's standards are listed at the bottom of the lesson page.

Invite your family members to

- talk with you about what you are learning.
- help you correct errors you have made on completed work.
- help you set a time and find a quiet place to do math homework.
- help you memorize the addition, subtraction, multiplication, and division facts.
- solve problems as you play together, shop together, and do household chores.
- visit **The Learning Site** at www.harcourtschool.com
- have **Fun with Math!**

Have a great year!

The Authors

xxiii

Be a Good Problem Solver!

You need to organize your thinking. You can use problem-solving steps to stay on track.

Use these problem-solving steps. They can help you think through a problem.

Understand the problem.

What are you asked to find?	Restate the question in your own words.
What information will you use?	List all the information in the problem.
Is there information you will not use? If so, what?	You may not need all the information given.

Plan a strategy to solve.

What strategy can you use to solve the problem?	Think about some problem solving strategies you can use. Then choose one.

Solve the problem.

How can you use the strategy to solve the problem?	Follow your plan. Show your solution.

Check your answer.

Look back at the problem. Does the answer make sense? Explain.	Be sure you answered the question that is asked.
What other strategy could you use?	Solving the problem by another method is a good way to check your work.

Try It

Here's how you can use the problem-solving steps to solve a problem.

Make a Table

PROBLEM SOLVING STRATEGIES

Draw a Diagram or Picture
Make a Model or Act it Out
Make an Organized List
Find a Pattern
▶ **Make a Table or Graph**
Predict and Test
Work Backward
Solve a Simpler Problem
Write a Number Sentence
  or an Equation
Use Logical Reasoning

PROBLEM The children in Ms. Ling's class wanted to care for an animal. They could choose a fish, a rabbit, or a hamster. The children wrote their votes on slips of paper. The votes are shown here. Which animal did they choose? How can the class keep a record of their votes?

hamster rabbit rabbit fish hamster

rabbit fish rabbit hamster rabbit

hamster rabbit fish fish rabbit

hamster

 the problem.

I need to find which animal was chosen. I also need to keep a record of the votes. The votes are shown on the slips of paper.

 a strategy to solve.

I can *make a table* to show tallies of the votes. Then I can count the tallies to see which animal was chosen.

 the problem.

OUR VOTES	
Animal	**Tallies**
Fish	IIII
Rabbit	IIII II
Hamster	IIII

The rabbit has 7 votes. So, it was chosen by the class.

Check **your answer.**

I can check that the vote tallies in the table match the slips of paper shown in the problem. Since *Rabbit* has the most votes, this solves the problem.

Place Value and Number Sense

Did you know that there are more than 150 kinds of horses? Sometimes horses are used on farms to help with the work. Horses need hay, oats, and fresh water so that they will stay healthy. The pictograph shows about how much water a horse needs to drink. Skip count to find how much water a horse needs each week and each month.

HOW MUCH WATER A HORSE NEEDS

Time	Water
1 day	🪣
1 week	🪣🪣🪣🪣🪣🪣🪣
1 month	🪣🪣🪣🪣🪣🪣🪣🪣🪣🪣🪣🪣🪣🪣🪣🪣🪣🪣🪣🪣🪣🪣🪣🪣🪣🪣🪣🪣🪣🪣

Key: Each 🪣 = 10 gallons.

CHECK WHAT YOU KNOW

Use this page to help you review and remember
important skills needed for Chapter 1.

✓ ORDINAL NUMBERS (See p. H2.)

For 1–6, use the list of names.

1. Kelly is first on the list. Who is third?

2. Who is ninth on the list?

3. Who is fifth on the list?

4. In which position is Juan on the list?

5. Who is eighth on the list?

6. In which position is Julie on the list?

> **Kelly**
> **Tom**
> **Sally**
> **Susan**
> **Timothy**
> **Julie**
> **Juan**
> **Matt**
> **Maria**

✓ READ AND WRITE ONES, TENS, HUNDREDS (See p. H2.)

Write the number.

7.

8.

9.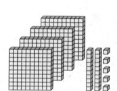

10.

11. fifty-eight

12. three hundred twenty-one

13. six hundred forty-five

14. nine hundred four

✓ PLACE VALUE WITH ONES, TENS, HUNDREDS (See p. H3.)

Write the number.

15. 6 tens 5 ones

16. 4 tens 9 ones

17. 9 tens 0 ones

18. 3 hundreds 6 tens 1 one

19. 1 hundred 7 tens 5 ones

20. 5 hundreds 2 tens 9 ones

21. 4 hundreds 2 tens 7 ones

22. 8 hundreds 0 tens 3 ones

23. 6 hundreds 8 tens 0 ones

24. 7 hundreds 3 tens 8 ones

HANDS ON
Patterns on a Hundred Chart

Quick Review

1. $8 + 2$ 2. $6 + 3$
3. $9 + 3$ 4. $10 + 5$
5. $25 + 5$

VOCABULARY
even   odd

MATERIALS
hundred chart, crayons

▶ Explore

Use the hundred chart. Start at 2.
Skip-count by twos.

STEP 1

Use a hundred chart.

STEP 2

Start at 2. Shade that box.

STEP 3

Skip-count by twos, and shade
each box you land on.

1	2	3	4	5	6	7	8	9	10
11	12	13	14	15	16	17	18	19	20
21	22	23	24	25	26	27	28	29	30
31	32	33	34	35	36	37	38	39	40
41	42	43	44	45	46	47	48	49	50
51	52	53	54	55	56	57	58	59	60
61	62	63	64	65	66	67	68	69	70
71	72	73	74	75	76	77	78	79	80
81	82	83	84	85	86	87	88	89	90
91	92	93	94	95	96	97	98	99	100

MATH IDEA **Even** numbers have a 0, 2, 4, 6,
or 8 in the ones place. **Odd** numbers have
a 1, 3, 5, 7, or 9 in the ones place.

Try It

a. Start at 2. Skip-count by twos.
   What numbers do you land on?
   Are they even or odd? How do you know?

b. Start at 3. Skip-count by threes. What numbers
   do you land on? Are they even or odd?
   How do you know?

REASONING What if you want to count by
twos and name *odd* numbers? On what
number should you start?

2, 4, 6, 8, . . . what
number do we land
on next?

CALIFORNIA STANDARDS NS 1.1 Count, read, and write whole numbers to 10,000. **MR 1.1** Analyze problems
by identifying relationships, distinguishing relevant from irrelevant information, sequencing and prioritizing information,
and observing patterns. *also* **MR 2.0, MR 3.0, MR 3.2, MR 3.3**

Connect

Look at your shaded chart. What pattern do you see?

All the even numbers are shaded. Even numbers have a 0, 2, 4, 6, or 8 in the ones place.

The odd numbers are not shaded. Odd numbers have a 1, 3, 5, 7, or 9 in the ones place.

1	2	3	4	5	6	7	8	9	10
11	12	13	14	15	16	17	18	19	20
21	22	23	24	25	26	27	28	29	30
31	32	33	34	35	36	37	38	39	40
41	42	43	44	45	46	47	48	49	50
51	52	53	54	55	56	57	58	59	60
61	62	63	64	65	66	67	68	69	70
71	72	73	74	75	76	77	78	79	80
81	82	83	84	85	86	87	88	89	90
91	92	93	94	95	96	97	98	99	100

Practice

TECHNOLOGY LINK

More Practice: Use E-lab, *Number Patterns.*

www.harcourtschool.com/elab2002

Use the hundred chart. Tell whether the number is *odd* or *even*.

1. 7   **2.** 8   **3.** 12   **4.** 30

5. 27   **6.** 98   **7.** 19   **8.** 45

9. 44   **10.** 81   **11.** 76   **12.** 100

Use the hundred chart.

13. Start at 5. Skip-count by fives. Move 4 skips. What number do you land on? Is it odd or even?

14. Start at 10. Skip-count by tens. Move 6 skips. What number do you land on? Is it odd or even?

15. The houses on Quinn's street are numbered 4, 8, 12, 16, and 20. What are the next three house numbers? Explain.

16. **REASONING** Marcos skip-counted. He started at 2. He landed on 15. Could he be skip-counting by twos? Why?

Mixed Review and Test Prep

17. $\begin{array}{r} 64 \\ +21 \\ \hline \end{array}$

18. $\begin{array}{r} 78 \\ -54 \\ \hline \end{array}$

19. $25 + 11 = $ ■

20. $37 - 10 = $ ■

21. **TEST PREP** Jon had 12 red marbles and 17 blue marbles. How many marbles did he have in all?

A 30   **B** 29   **C** 19   **D** 5

Understand Place Value

▶ **Learn**

FARM FACTS The symbols 0, 1, 2, 3, 4, 5, 6, 7, 8, and 9 are **digits**. Numbers are made up of digits.

On Mr. Sam's farm there are 248 chickens. What does the number 248 mean?

VOCABULARY
digits
standard form
expanded form
word form

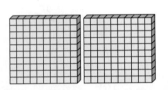

HUNDREDS	TENS	ONES
2	4	8

So, 248 means 2 hundreds + 4 tens + 8 ones or 200 + 40 + 8.

MATH IDEA You can write a number in different ways: standard form, expanded form, and word form.

Standard form: 248

Expanded form: 200 + 40 + 8

Word form: two hundred forty-eight

• What does the number 527 mean?

▶ **Check**

USE DATA For 1–2, use the table.

1. **Tell** the value of the digit 3 in the number of cows on the farm.

2. What is the expanded form for the number of pigs on the farm?

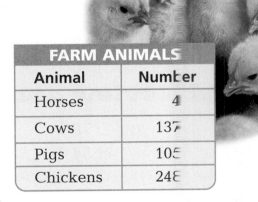

FARM ANIMALS	
Animal	**Number**
Horses	4
Cows	137
Pigs	105
Chickens	248

CALIFORNIA STANDARDS NS 1.0 Students understand the place value of whole numbers. **O—n NS 1.5** Use expanded notation to represent numbers. *also* **NS 1.1, O—n NS 1.3, MR 2.0, MR 2.3, MR 3.0, MR 3.2, MR 3.3**

Write each number in standard form.

3.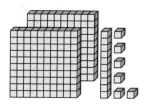

4.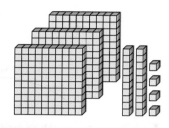

5.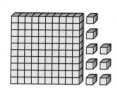

▶ Practice and Problem Solving

Write each number in standard form.

6. $100 + 50 + 3$

7. $400 + 70 + 6$

8. $600 + 30 + 9$

9. $900 + 2$

10. 4 hundreds 2 tens 1 one

11. 6 hundreds 8 tens 3 ones

12. 7 hundreds 2 tens 3 ones

13. 4 hundreds 5 ones

14. one hundred three

15. three hundred forty-five

16. six hundred eleven

17. nine hundred seventy-one

Write the value of the blue digit.

18. 846

19. 267

20. 493

21. 923

22. Mr. Sam put 297 bales of hay in one barn. There are still 86 bales of hay in the field. How many more bales of hay are in the barn than in the field?

23. There are 100 cows in one field and 30 cows in another field. There are 7 cows in the barn. How many cows are there in all?

24. REASONING What is the greatest 3-digit number you can write using the digits 8, 9, and 4?

25. REASONING I am a digit in each of the numbers 312, 213, and 132. My value is different in all three numbers. What digit am I? What is my value in each number?

Mixed Review and Test Prep

26. $24 + 24 = $ ■   **27.** $24 - 24 = $ ■

28. Pablo came in second, Jacob came in third, and Lauren came in ahead of Pablo. Who won the race?

29. What number continues the pattern? (p. 2)

3,　6,　9,　12,　15,　■

30. **TEST PREP** Which number is greater than 54?

A 34　**B** 45　**C** 53　**D** 55

Extra Practice page H32, Set A

Understand Numbers to 10,000

Quick Review

Write the value of the blue digit.

1. 246   2. 394   3. 714

4. 520   5. 302

▶ **Learn**

BUILDING BLOCKS You can show 100 with a base-ten hundreds block or with a 10-by-10 paper grid. How can you show 1,000?

Activity

Materials: base-ten blocks, 10-by-10 grid paper, paste, stapler

Model 1,000. Stack hundreds blocks until you have built a cube of 1,000. Then make a book of 1,000 squares.

STEP 1

Paste 10-by-10 paper grids onto pieces of paper. Use one grid for each piece of paper. Number each page at the bottom. Staple the pages together.

STEP 2

Number the squares of the grids from 1 to 1,000.

A thousands cube and a book of 1,000 squares are models for 1,000.

• How many 10-by-10 grids did it take to make the 1,000 book? Explain.

CALIFORNIA STANDARDS NS 1.0 Students understand the place value of whole numbers.
NS 1.1 Count, read, and write whole numbers to 10,000. *also* O━n NS 1.3, O━n NS 1.5,
MR 1.1, MR 2.3, MR 3.2, MR 3.3

Building with Thousands

What if you put 3 books of one thousand squares together? How many squares will you have in all?

You can add the squares.

$1,000 + 1,000 + 1,000 = 3,000$

So, the 3 books equal 3,000 squares.

You can skip-count by thousands.

1,000   2,000   3,000

• How many books would you need to have 5,000 squares?

Here are some ways to show 2,346.

Examples

A With base-ten blocks

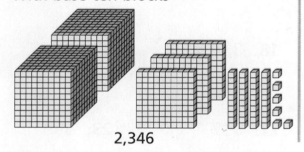

2,346

B On a place-value chart

Thousands	Hundreds	Tens	Ones
2,	3	4	6

C In standard form:

2,346
↑
A comma separates the thousands and hundreds.

D In expanded form:

$2,000 + 300 + 40 + 6$

E In word form:

two thousand, three hundred forty-six

► Check

1. Tell how to write the expanded form for 5,403.

Write in standard form.

2.

3.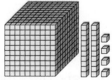

Write in expanded form.

4. 5,632   **5.** 7,401   **6.** 8,011   **7.** 3,462

LESSON CONTINUES ▶

Write in standard form.

8.

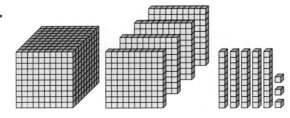

9.

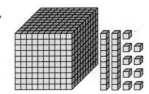

10. 5,000 + 400 + 50

11. 2,000 + 300 + 90 − 7

12. 3,000 + 700 + 20 + 3

13. 1,000 + 10 + 8

14. two thousand, four hundred eighty-three

15. six thousand, one hundred ninety-four

Write in expanded form.

16. 1,234   17. 4,321

18. 3,016   19. 8,367

Write in words.

20. 87   21. 148

22. 317   23. 599

Complete.

24. 1,000 + 500 + ■ + 8 = 1,548

25. 3,000 + ■ + 90 + 7 = 3,897

26. **USE DATA** The pictograph shows what Molly saw at the farm. How many animals did Molly see in all?

27. **REASONING** I am a 3-digit number. My hundreds digit equals the sum of my tens digit and my ones digit. My tens digit is 4 more than my ones digit. My ones digit is 1. What number am I?

28. What is the least possible number you can write with the digits 2, 9, 4, and 7? Use each digit only once.

29. ▬ **Write About It** Why do you have to use a zero when you write one thousand, six hundred four in standard form?

Horses	🏠🏠🏠
Lambs	🏠🏠🏠🏠
Cows	🏠🏠🏠🏠🏠

FARM ANIMALS

Key: Each 🏠 = 2 animals.

30. The number 124 is an even number. Write 5 more even numbers including one with 3 digits.

31. Show that each of the even numbers 6, 8, 10, and 12 can be written as the sum of a group of 2's.

Mixed Review and Test Prep

Tell whether the number is *even* or *odd*. (p. 2)

32. 61   **33.** 18   **34.** 58

35. 97   **36.** 102   **37.** 183

Find the sum.

38. 6 + 3   **39.** 4 + 7   **40.** 12 + 4

41. 9 + 8   **42.** 10 + 5   **43.** 11 + 10

Find the difference.

44. 14 − 5   **45.** 16 − 8   **46.** 13 − 7

47. 17 − 7   **48.** 18 − 6   **49.** 11 − 6

Find the sum.

50.	**51.**	**52.**
6	9	8
5	7	3
+3	+1	+7

53. **TEST PREP** Which number is eight hundred four in standard form? (p. 4)

A 940   **C** 804

B 840   **D** 84

54. **TEST PREP** Which number is the expanded form for 432? (p. 4)

F 40 + 2   **H** 400 + 30

G 400 + 2   **J** 400 + 30 + 2

Thinker's Corner

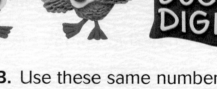

REASONING

Materials: number cards 0–9, paper and pencil

A. Play the game with a partner. Choose 3 number cards. Make a 3-digit number. Record it.

B. Use these same number cards to make and record as many *different* 3-digit numbers as you can.

C. Players get one point for each different 3-digit number.

D. Players take turns choosing 3 cards and making all the possible numbers.

The player who scores 25 points first wins the game.

CHALLENGE: Play the game again, choosing 4 number cards and making 4-digit numbers.

Extra Practice page H32, Set B

Understand 10,000

Quick Review

Write in expanded form.

1. 384   2. 51

3. 677   4. 9,240

5. 3,818

▶ **Learn**

BUILDING BOOKS   How many 1,000 books do you need to show 10,000 squares?

Activity

MATERIALS: books of 1,000 squares

STEP 1

Put 2 books of 1,000 together.

STEP 2

Continue putting books of 1,000 together until they show 10,000.

Remember
Put a comma between the thousands place and the hundreds place.
23,680
↑
comma

• How many books of 1,000 squares would you need to have 15,000 squares? to have 23,000 squares?

What does each digit in 23,680 mean?

Use a place-value chart.

TEN THOUSANDS	THOUSANDS	HUNDREDS	TENS	ONES
2	3,	6	8	0

Standard form: 23,680

Expanded form: 20,000 + 3,000 + 600 + 80

Word form: twenty-three thousand, six hundred eighty

• What is the value of the 3 in 34,152?

CALIFORNIA STANDARDS NS 1.0 Students understand the place value of whole numbers. **NS 1.1** Count, read, and write whole numbers to 10,000. *also* ○ⁿ **NS 1.3,** ○ⁿ **NS 1.5, MR 2.0, MR 3.0, MR 3.2, MR 3.3**

▶ Check

1. Tell the expanded form for 21,694.

Write in standard form.

2. 20,000 + 6,000 + 700 + 30 + 4

3. thirty-five thousand, nine hundred forty-seven

Write in expanded form.

4. 16,723   **5.** 52,019

▶ Practice and Problem Solving

Write in standard form.

6. 20,000 + 6,000 + 700 + 30 + 4   **7.** 10,000 + 400 + 8

8. forty-two thousand, three hundred fifteen

9. eighteen thousand, nine hundred

Write in expanded form.

10. 16,723   **11.** 55,119   **12.** 11,012   **13.** 49,207

Write the value of the blue digit.

14. 81,465   **15.** 26,817   **16.** 43,912   **17.** 19,273

Complete.

18. 10,000 + 2,000 + ■ + 50 + 1 = 12,651

19. 60,000 + ■ + 300 + 10 + 9 = 62,319

20. **? What's the Question?** Mrs. Chung wrote the number 37,895. The answer is 7,000.

21. **? What's the Error?** Karla wrote eleven thousand, forty-five as 1,145. Explain her error. Write the number correctly in standard form.

Mixed Review and Test Prep

22.   77
       −52

23.   72
       +19

24. 25 + ■ = 29

25. 43 + ■ = 49

26. **TEST PREP** What is the value of the blue digit in 5,789? (p. 6)

   A 7   C 700

   B 70   D 7,000

Extra Practice page H32, Set C

Problem Solving Strategy
Use Logical Reasoning

Understand → Plan → Solve → Check

Quick Review

1. ■ + 700 + 40 + 5 = 8,745
2. 400 + ■ − 9 = 499
3. ■ + 60 + 3 = 763
4. 900 + 30 + ■ = 936
5. ■ + 400 − 7 = 10,407

PROBLEM I am a 2-digit number. The sum of my digits is 17. The tens digit is odd. The ones digit is even. What number am I?

Understand

- What are you asked to find?

- Is there information you will not use? If so, what?

Plan

- What strategy can you use to solve the problem?

  You can *use logical reasoning.*

Solve

- How can you use the strategy to solve the problem?

  Use a hundred chart. Find the numbers with digits whose sum equals 17. The numbers are 89 and 98.
  Then decide which number has a tens digit that is odd and a ones digit that is even.
  That number is 98.

So, the number is 98.

1	2	3	4	5	6	7	8	9	10
11	12	13	14	15	16	17	18	19	20
21	22	23	24	25	26	27	28	29	30
31	32	33	34	35	36	37	38	39	40
41	42	43	44	45	46	47	48	49	50
51	52	53	54	55	56	57	58	59	60
61	62	63	64	65	66	67	68	69	70
71	72	73	74	75	76	77	78	79	80
81	82	83	84	85	86	87	88	89	90
91	92	93	94	95	96	97	98	99	100

Check

- Look at the problem. Does your answer make sense? Explain.

- Explain how you know that 89 and 98 are the only 2-digit numbers whose digits have a sum that equals 17.

CALIFORNIA STANDARDS MR 1.0 Students make decisions about how to approach problems. **MR 1.1** Analyze problems by identifying relationships, distinguishing relevant from irrelevant information, sequencing and prioritizing information, and observing patterns. *also* **NS 1.0, MR 2.0, MR 2.3, MR 2.6, MR 3.1, MR 3.2**

Use logical reasoning and solve.

1. **What if** the riddle said that the tens digit is even, and the ones digit is odd and the sum of the digits is 5? What would be the answer to this riddle?

2. I am a number on the hundred chart. Both of my digits are odd. Both of my digits are the same. What numbers can I be?

3. I am a number on the hundred chart. The sum of my digits is 12. My tens digit is 7. What number am I?

   A 72   C 77
   B 75   D 79

4. I am a number in the third row on the hundred chart. My ones digit is 5 more than my tens digit. What number am I?

   F 22   H 27
   G 24   J 30

PROBLEM SOLVING STRATEGIES

Draw a Diagram or Picture
Make a Model or Act It Out
Make an Organized List
Find a Pattern
Make a Table or Graph
Predict and Test
Work Backward
Solve a Simpler Problem
Write an Equation
▶ **Use Logical Reasoning**

Mixed Strategy Practice

5. Write the greatest possible four-digit number using the digits 3, 4, 5, and 6. Write the least possible four-digit number.

6. **REASONING** A number has the same number of ones, tens, and hundreds. If the sum of the digits is 9, what is the number?

7. **USE DATA** The pictograph shows Tim's pets. How many more lambs than ponies does Tim have? How many animals does Tim have in all?

8. ✎ **Write a problem** about a number riddle. Use the hundred chart to help you. Tell how to solve the problem.

TIM'S PETS	
Ponies	∩∩
Calves	∩
Lambs	∩∩∩∩∩∩∩∩∩∩∩∩∩∩

Key: Each ∩ = 1 animal.

Review/Test

✅ CHECK VOCABULARY AND CONCEPTS

Choose the best term from the box.

1. A number that ends in 0, 2, 4, 6, or 8 is an __?__ number. (p. 2)

2. The symbols 0, 1, 2, 3, 4, 5, 6, 7, 8, and 9 are called __?__ . (p. 4)

> odd
> even
> digits
> expanded form

Write each number in standard form. (pp. 4–9)

3.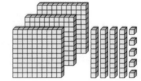

4.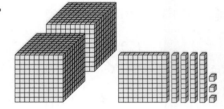

✅ CHECK SKILLS

Write whether the number is *odd* or *even*. (pp. 2–3)

5. 317
6. 200
7. 1,348
8. 12,999

Write in standard form. (pp. 4–11)

9. $800 + 60 + 9$

10. $3,000 + 700 + 10 + 1$

11. $8,000 + 500 + 20 + 2$

12. $30,000 + 4,000 + 700 + 5$

13. two thousand, thirty-nine

14. fifteen thousand, sixty-five

Write the value of the blue digit. (pp. 4–5, 10–11)

15. 863
16. 9,845
17. 12,053
18. 32,859

✅ CHECK PROBLEM SOLVING

Solve. Use a hundred chart. (pp. 12–13)

19. I am a 2-digit number. My tens digit is two more than my ones digit. My ones digit is between 4 and 6. What number am I?

20. I am a 2-digit number. I am greater than 40 but less than 60. My tens and ones digits are the same. I am an odd number. What number am I?

Cumulative Review

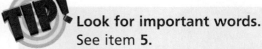

Look for important words.
See item **5**.

Not is an important word. **Not** odd means a number that is even.

Also see problem **2**, p. H63.

For 1–11, choose the best answer.

1. What does the 6 mean in 461?

- **A** 6
- **B** 60
- **C** 600
- **D** 6,000

2. How is 6 hundreds 2 tens 3 ones written in standard form?

- **F** 263
- **G** 326
- **H** 623
- **J** 6,023

3. How is 2,047 written in expanded form?

- **A** 2,000 + 40 + 7
- **B** 2,000 + 400 + 7
- **C** 20 + 40 + 7
- **D** 2,000 + 47

4. Which number makes the number sentence true?

$$2,000 + 400 + \blacksquare + 5 = 2,475$$

- **F** 7
- **G** 70
- **H** 700
- **J** 7,000

5. Which number is **not** odd?

- **A** 35
- **B** 41
- **C** 50
- **D** 77

6.
$$\begin{array}{r} 43 \\ 21 \\ +19 \\ \hline \end{array}$$

- **F** 73
- **G** 83
- **H** 713
- **J** NOT HERE

7. The sum of the digits of a number is 10. Both of the digits are even. The ones digit is 2 more than the tens digit. What is the number?

- **A** 19
- **B** 28
- **C** 37
- **D** 46

8. Which of these has a 7 in the thousands place?

- **F** 87,532
- **G** 35,872
- **H** 23,587
- **J** 83,752

9. Start with 5 and skip-count by fives to 100. Which number is **not** said?

- **A** 44
- **B** 55
- **C** 65
- **D** 80

10. What is the value of the 2 in 23,790?

- **F** 20
- **G** 200
- **H** 2,000
- **J** NOT HERE

11. A number has the same number of ones, tens, and hundreds. The sum of the digits is 24. What is the number?

- **A** 699
- **B** 777
- **C** 888
- **D** NOT HERE

Compare, Order, and Round Numbers

Trees have different shapes, colors, and sizes. You can identify many trees by the shape of their leaves and the roughness of their bark. You can also identify trees by their height. Look at the chart to compare the heights of some kinds of trees.

HEIGHTS OF TREES	
Tree	Height (in feet)
Apple	20 to 30
Beech	60 to 80
Maple	75 to 100
Pine	175 to 200
Redwood	200 to 275

Redwood trees are among the tallest trees in the world.

CHECK WHAT YOU KNOW ✓

Use this page to help you review and remember
important skills needed for Chapter 2.

✓ **COMPARE 2-DIGIT NUMBERS** (See p. H3.)

Write the words *greater than* or *less than*.

1. 78 is _?_ 41.　　**2.** 35 is _?_ 55.　　**3.** 41 is _?_ 45.

4. 63 is _?_ 68.　　**5.** 37 is _?_ 31.　　**6.** 56 is _?_ 58.

7. Which number is greater, 98 or 89?

✓ **ORDER NUMBERS** (See p. H4.)

Write the number that is just after, just before,
or between.

8. 7, ■　　　　**9.** ■, 45　　　　**10.** 21, ■, 23

11. 307, ■　　**12.** ■, 768　　**13.** 871, ■, 873

14. 454, ■　　**15.** ■, 133　　**16.** 189, ■, 191

Write the number each model shows.

17. 　　**18.** 　　**19.**

20. Which model above shows the greatest number?

Write the number each model shows.

21. 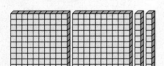　　**22.** 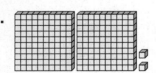　　**23.**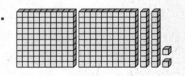

24. Which model shows the greatest number?

Write the numbers in order from least to greatest.

25. 74, 79, 73　　　**26.** 19, 16, 18　　　**27.** 49, 57, 51

28. 117, 127, 111　　**29.** 181, 178, 176　　**30.** 342, 349, 344

31. 248, 255, 250　　**32.** 75, 69, 84　　　**33.** 428, 423, 425

Size of Numbers

▶ Learn

HOW MANY? Numbers that help you estimate the number of objects without counting them are called **benchmark numbers**. Any useful number can be a benchmark.

About how many beans are in Jar B?
You can use 25 as a benchmark to estimate.

There are 25 beans in Jar A.

A

There are ■ beans in Jar B.

B

There are about two times as many beans in Jar B.

So, there are about 50 beans in Jar B.

Think about the number of students in your class, your grade, and your school. Which has about 20 students? Which has about 100 students? Which has about 1,000 students?

BENCHMARK	NUMBER TO BE ESTIMATED
20	students in your class
100	students in your grade
1,000	students in your school

• There are 5 third-grade classes in Nora's school. About how many students are in the third grade? What benchmark can you use to help you?

CALIFORNIA STANDARDS NS 1.1 Count, read, and write whole numbers to 10,000. **MR 2.1** Use estimation to verify the reasonableness of calculated results. *also* **MR 1.0, MR 2.0, MR 2.3**

1. **Explain** the benchmark you would use to estimate the number of girls in your class.

Estimate the number of beans in each jar. Use Jars A and B as benchmarks.

Jar A has 10 beans.

Jar B has about 50 beans.

A

B

2.

10 or 50?

3.

25 or 50?

4.

100 or 200?

► **Practice and Problem Solving**

5. Estimate the number of beans in the jar at the right. Use Jars A and B as benchmarks.

Choose a benchmark of 10, 100, or 1,000 to estimate each.

6. the number of players on a soccer team

7. the number of pretzels in a large bag

8. the number of sheets in a package of notebook paper

9. the number of leaves on a large tree in summer

10. Juan wants about 100 jelly beans. How could he use a benchmark instead of counting them?

11. **Write a problem** in which a benchmark is used to estimate. Solve.

Mixed Review and Test Prep

Write each number in expanded form. (p. 4)

12. 268 **13.** 354 **14.** 420 **15.** 679

16. **TEST PREP** Choose the value of the blue digit in 15,688. (p. 10)

A 6   **B** 60   **C** 600   **D** 700

Compare Numbers

▶ **Learn**

HOW NEAR? HOW FAR? Beth lives 262 miles from Homer and 245 miles from Lakewood. Which city does she live closer to?

Compare numbers to decide which of two numbers is greater. Use these symbols.

VOCABULARY
compare
is less than <
is greater than >
is equal to =

is less than	**is greater than**	**is equal to**
<	>	=

You can use base-ten blocks to compare 262 and 245.

STEP 1

Show 262 and 245 with base-ten blocks.

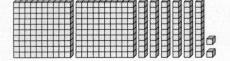 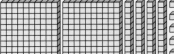

STEP 2

Compare from left to right. First compare the hundreds. Since they are the same, compare the tens.

6 tens is greater than 4 tens. So, 262 is greater than 245.

Beth lives closer to Lakewood since 262 > 245.

You can use a number line to compare numbers.

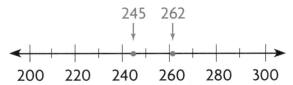

245   262

200   220   240   260   280   300

From left to right, the numbers on a number line are in order from *least* to *greatest*.

So, 245 < 262 or 262 > 245.

CALIFORNIA STANDARDS AF 1.0 Students select appropriate symbols, operations, and properties to represent, describe, simplify, and solve simple number relationships. **NS 1.2** Compare and order whole numbers to 10,000. *also* **NS 1.0, NS 1.1, O━┓NS 1.3, O━┓AF 1.1, AF 1.2, AF 1.3, MR 1.0, MR 1.━, MR 2.0, MR 2.3, MR 2.4**

Comparing Thousands

You can use a place-value chart to compare numbers.

Compare 3,165 and 3,271 starting from the left.

THOUSANDS	HUNDREDS	TENS	ONES
3,	1	6	5
3,	2	7	1

↑ Thousands are the same. ↑ 200 > 100

TECHNOLOGY LINK

To learn more about *Comparing Numbers*, watch the **Harcourt Math Newsroom Video**, *Chicago Skyscraper.*

So, 3,271 > 3,165 or 3,165 < 3,271.

MATH IDEA Compare numbers by using base-ten blocks, a number line, or a place-value chart.

▶ Check

1. **Explain** how to use base-ten blocks to compare 341 and 300 + 40 + 1. What do you notice?

2. Use the number line on page 20 to compare 268 and 279. Which number is greater? Explain.

Compare the numbers. Write <, >, or = for each ●.

3.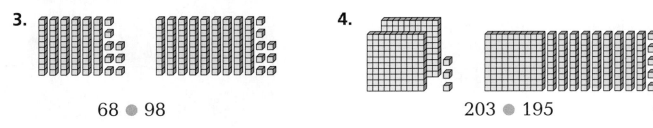

68 ● 98

4.

203 ● 195

5.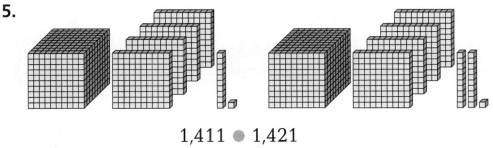

1,411 ● 1,421

LESSON CONTINUES ▶

▶ Practice and Problem Solving

Compare the numbers. Write <, >, or = for each ●.

6.

T	O
9	2
8	3

92 ● 83

7.

H	T	O
1	0	1
1	1	0

101 ● 110

8.

H	T	O
4	2	8
4	3	8

428 ● 438

9. 629 ● 631

10. 758 ● 750

11. 439 ● 438

12. 3,425 ● 3,799

13. 5,712 ● 5,412

14. 2,412 ● 2,412

15. 894 ● 2,139

16. 348 ● 348

17. 7,393 ● 7,396

18. 151 + 200 ● 350

19. 274 + 128 ● 402

20. 475 + 52 ● 537

21. 344 − 103 ● 244

22. 786 − 231 ● 555

23. 696 − 418 ● 296

24. What is the greatest place-value position in which the digits of 831 and 819 are different? Compare the numbers.

25. Compare the numbers 5,361 and 3,974. How can you tell which number is greater?

For 26–28, use the numbers on the box.

26. List all the numbers that are less than 575.

27. List all the numbers that are greater than 830.

28. List all the numbers that are greater than 326 and less than 748.

29. **? What's the Question?** Louis read 125 pages. Tom read 137 pages. The answer is 12.

30. **Write About It** You have 3 four-digit numbers. The digits in the thousands, hundreds, and ones places are the same. Which digit would you use to compare the numbers? Explain.

31. The numbers 456 and 564 have the same digits in a different order. Do they both have the same value? Explain.

32. Kim modeled 6 hundreds, 2 tens, and 4 ones, and Jon modeled 6 hundreds, 8 tens, and 5 ones. Which shows the greater number?

Mixed Review and Test Prep

Tell whether the number is odd or even. (p. 2)

33. 13 **34.** 46 **35.** 187

36. 35 **37.** 2,721 **38.** 544

39. 736 **40.** 4,922 **41.** 6,571

Write the number in standard form. (p. 10)

42. 50,000 + 8,000 + 300

43. 30,000 + 700 + 5

TEST PREP Choose the letter for the number in standard form. (p. 10)

44. fifty-three thousand, six hundred seventy-two

    A 53,072   **C** 53,627

    B 53,602   **D** 53,672

45. thirty-four thousand, five hundred twenty

    F 3,452   **H** 34,502

    G 34,052   **J** 34,520

LINKUP to Reading

STRATEGY • COMPARE Early settlements were often built near rivers. Rivers provided people with food, water, and easy trade routes. Even today, many goods are shipped on riverboats rather than by truck or train.

When you *compare* things, you decide how they are alike. When you *contrast* things, you decide how they are different. To solve some problems, you need to compare and contrast the information.

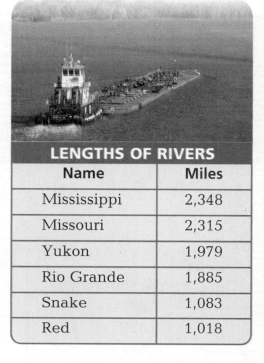

LENGTHS OF RIVERS	
Name	**Miles**
Mississippi	2,348
Missouri	2,315
Yukon	1,979
Rio Grande	1,885
Snake	1,083
Red	1,018

USE DATA For 1–4, use the table.

1. The Mississippi River is the longest river in the United States. In which place-value positions are the lengths of the Mississippi and Missouri Rivers different?

2. Compare the lengths of the Rio Grande and the Yukon River. Which river is longer?

3. How much longer is the Mississippi River than the Missouri River?

4. Explain how to compare the lengths of the Snake River and the Red River.

Order Numbers

▶ **Learn**

HOW TALL? HOW SHORT? The table lists the heights of three mountains in the United States.

Which mountain is the tallest?

Use a number line to order the numbers.

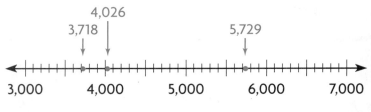

$$3{,}718 < 4{,}026 < 5{,}729$$

So, Mount Rogers is the tallest.

You can order numbers by comparing the digits in the same place-value positions.

Order 7,613; 7,435; and 7,551.

Quick Review

Tell which number is greater.

1. 37 or 29   2. 21 or 32

3. 58 or 65   4. 120 or 99

5. 235 or 253

MOUNTAIN HEIGHTS

State	Mountain	Height
Maine	Coburn Mountain	3,718 ft
Kansas	Mount Sunflower	4,026 ft
Virginia	Mount Rogers	5,729 ft

STEP 1	**STEP 2**	**STEP 3**
Compare the thousands.	Compare the hundreds.	Write the numbers in order from greatest to least.
7,613	7,613	
7,435	7,435	
7,551	7,551	
↑	↑	
The digits are the same.	They are not the same.	$7{,}613 > 7{,}551 > 7{,}435$
	$6 > 5 > 4$	

MATH IDEA Order numbers by comparing the digits in the same place-value position from left to right. List the numbers from least to greatest or greatest to least.

CALIFORNIA STANDARDS NS 1.0 Students understand the place value of whole numbers. **NS 1.2** Compare and order whole numbers to 10,000. *also* **NS 1.1, O¬n NS 1.3, AF 1.0, O¬n AF 1.1, AF 1.2, MR 1.0, MR 2.0, MR 2.2, MR 2.3**

1. Explain how you can order the numbers
1,432; 1,428; and 1,463 from greatest to least.

Write the numbers in order from least to greatest.

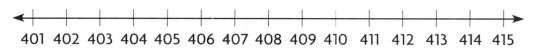

2. 408, 413, 411   **3.** 403, 410, 407   **4.** 415, 405, 409

▶ **Practice and Problem Solving**

Write the numbers in order from least to greatest.

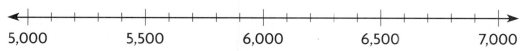

5. 5,200; 6,500; 5,900   **6.** 6,750; 6,125; 6,500

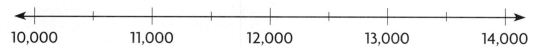

7. 10,500; 13,000; 12,500   **8.** 12,240; 11,845; 13,156

Write the numbers in order from greatest to least.

9. 384, 526, 431   **10.** 793, 728, 756

11. 37, 18, 24   **12.** 583, 791, 476

13. 7,837; 5,126; 3,541   **14.** 17,655; 22,600; 9,860

15. REASONING Write in order from least to greatest the six numbers whose digits are 2, 8, and 9.

16. ? What's the Error? Jason put the following numbers in order from least to greatest: 3,545; 3,556; 3,554. What was his error?

Mixed Review and Test Prep

17. What number continues the pattern? 27, 23, 19, 15, ▨ (p. 2)

18. Write the standard form for 60,000 + 500 + 9. (p. 10)

19. 45 + 10   **20.** 56 + 20

21. TEST PREP Which number is greater than 567? (p. 20)

    A 560   **C** 562

    B 549   **D** 657

Problem Solving Skill
Identify Relationships

Understand ➤ Plan ➤ Solve ➤ Check

Quick Review

Compare. Write <, >, or = for each ●.

1. 124 ● 118
2. 229 ● 232
3. 244 ● 184
4. 156 ● 156
5. 371 ● 372

FOLLOW THE TRAIL   At Yosemite National Park, Nancy and Emilio saw Mount Gibbs, Mount Dana, Matterhorn Peak, and Sentinel Rock. They want to hike to the mountain that is higher than Matterhorn Peak but not as high as Mount Dana. Which one should they choose?

Knowing how the numbers are related can help you solve the problem.

STEP 1

Identify how the numbers are related.

13,053 > 12,764 > 12,264 > 7,038

The mountain heights in the table are listed in order from greatest to least.

STEP 2

Find all the mountains that are higher than Matterhorn Peak.

Mount Gibbs and Mount Dana

MOUNTAINS IN YOSEMITE NATIONAL PARK	
Name	Height
Mount Dana	13,053 ft
Mount Gibbs	12,764 ft
Matterhorn Peak	12,264 ft
Sentinel Rock	7,038 ft

STEP 3

Find all the mountains that are not as high as Mount Dana.

Mount Gibbs, Matterhorn Peak, and Sentinel Rock

STEP 4

Find the mountain that is listed in both Step 2 and Step 3.

Mount Gibbs is the only mountain listed in both steps.

So, Nancy and Emilio chose Mount Gibbs.

Talk About It

• How does the height of Sentinel Rock compare to that of Matterhorn Peak?

CALIFORNIA STANDARDS NS 1.2 Compare and order whole numbers to 10,000. **MR 1.1** Analyze problems by identifying relationships, distinguishing relevant from irrelevant information, sequencing and prioritizing information, and observing patterns. *also* **NS 1.0, O—n NS 1.3, AF 1.0, AF 1.2, MR 1.0, MR 2.0, MR 2.3, MR 3.2**

Problem Solving Practice

1. **What if** Emilio and Nancy hiked to the mountain that is higher than Sentinel Rock but not as high as Mount Gibbs? To which mountain did they hike?

2. What is the highest mountain under 12,000 feet that Emilio and Nancy saw?

3. **What if** the mountains were listed in order of heights from least to greatest? What would be the order of the mountains?

USE DATA For 4–5, use the table at the right.

4. Which state is larger than Connecticut but smaller than New Jersey?

   A Hawaii   **C** New Jersey
   B Rhode Island   **D** Delaware

5. Name the three smallest states.

   F Delaware, Rhode Island, Hawaii

   G New Jersey, Hawaii, Connecticut

   H Hawaii, Rhode Island, New Jersey

   J Rhode Island, Delaware, Connecticut

SMALLEST STATES IN THE U.S.	
Name	**Size in Square Miles**
Connecticut	5,006
New Jersey	7,790
Rhode Island	1,213
Hawaii	6,459
Delaware	2,026

Mixed Applications

6. Louis had base-ten blocks that showed 6 hundreds, 7 tens, 3 ones. Tom gave him 2 hundreds, 1 ten, 5 ones. Using standard form, write the number that shows the value of Louis's blocks now.

7. Celia lives in a town with a population of 12,346. Last year there were 1,000 fewer people living in the town. How many people lived in the town last year?

8. Tony had 24 postcards. Arlo gave him some more postcards. Now Tony has 46 postcards. How many postcards did Arlo give to Tony?

9. There were 53 students that went to the Science Museum. Of them, 27 were girls. How many were boys?

10. **Write About It** Explain how you would compare 4,291; 4,921; and 4,129 to put them in order from greatest to least.

Round to Nearest 10 and 100

▶ **Learn**

HOW CLOSE? There are 43 third graders and 47 fourth graders going on a field trip to the San Diego Zoo. About how many third graders and how many fourth graders are going on the field trip?

You can round numbers when you want to know *about how many*. **Rounding** is one way to estimate.

A number line can help you.

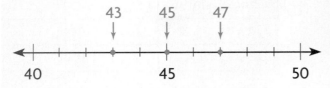

43 is closer to 40 than to 50.
43 rounds to 40.

47 is closer to 50 than to 40.
47 rounds to 50.

45 is halfway between 40 and 50. If a number is halfway between two tens, round to the greater ten. 45 rounds to 50.

So, about 40 third graders and about 50 fourth graders are going on the field trip.

You can round 3-digit numbers to the nearest ten or hundred.

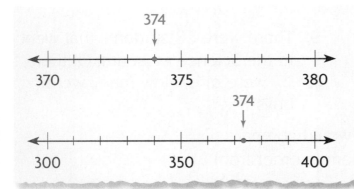

Round to the nearest 10.
374 is closer to 370 than to 380.
374 rounds to 370.

Round to the nearest 100.
374 is closer to 400 than to 300.
374 rounds to 400.

CALIFORNIA STANDARDS NS 1.4 Round off numbers to 10,000 to the nearest ten, hundred, and thousand. *also* **NS 1.0, O—ㅠ NS 1.3, MR 2.0, MR 2.3, MR 2.5, MR 3.2**

Quick Review

Write the numbers in order from least to greatest.

1. 17, 87, 57

2. 23, 19, 16   3. 37, 31, 23

4. 29, 33, 32   5. 59, 57, 58

VOCABULARY
rounding

1. **Explain** how you can round 350 to the nearest hundred using the number line.

Round to the nearest hundred and the nearest ten.

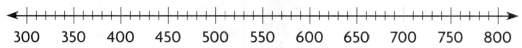

| 300 | 350 | 400 | 450 | 500 | 550 | 600 | 650 | 700 | 750 | 800 |

2. 643 **3.** 377 **4.** 845 **5.** 518 **6.** 750

► **Practice and Problem Solving**

Round to the nearest ten.

7. 16 **8.** 74 **9.** 53 **10.** 5 **11.** 78

12. 37 **13.** 44 **14.** 78 **15.** 94 **16.** 98

Round to the nearest hundred and the nearest ten.

17. 363 **18.** 405 **19.** 115 **20.** 165 **21.** 952

22. 237 **23.** 917 **24.** 385 **25.** 456 **26.** 883

USE DATA For 27–28, use the table.

27. To the nearest hundred, about how many kinds of birds does the zoo have?

28. REASONING The number of _?_ + the number of _?_ < the number of _?_ .

29. Kim rounded 348 to the nearest ten and said it was 350. She rounded 348 to the nearest hundred and said it was 400. Was this correct? Explain.

ZOO ANIMALS	
Type	**Number**
Mammals	214
Birds	428
Reptiles	174

30. Write a problem about animals. Use rounding to the nearest ten or to the nearest hundred in your problem.

Mixed Review and Test Prep

Find 100 more. (p. 6)

31. 877 **32.** 352

33. 461 **34.** 208

35. TEST PREP What is the value of the blue digit in 16,230? (p. 10)

A 10 **C** 1,000

B 100 **D** 10,000

Round to Nearest 1,000

Quick Review

Round to the nearest ten.

1. 52   2. 15   3. 22

4. 68   5. 54

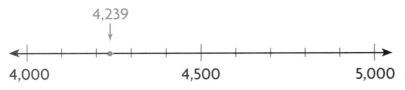

▶ **Learn**

ABOUT HOW MANY? The San Diego Zoo has the largest collection of mammals, birds, and reptiles in North America. It has 4,239 animals.

To the nearest thousand, how many animals are in the zoo?

4,239

↓

```
4,000          4,500          5,000
```

4,239 is closer to 4,000 than to 5,000.
4,239 rounds to 4,000.

So, there are about 4,000 animals in the zoo.

You can use rounding rules to round numbers.

Examples

A Round 2,641 to the nearest *thousand*.

2,641

↑

Look at the hundreds digit. Since 6 > 5, the 2 thousands rounds to 3 thousands. So, 2,641 rounds to 3,000.

B Round 2,641 to the nearest *hundred*.

2,641

↑

Look at the tens digit. Since 4 < 5, the 6 hundreds stays the same. So, 2,641 rounds to 2,600.

Rounding Rules

- Find the place to which you want to round.
- Look at the digit to its right.
- If the digit is less than 5, the digit in the rounding place stays the same.
- If the digit is 5 or more, the digit in the rounding place increases by 1.

CALIFORNIA STANDARDS NS 1.4 Round off numbers to 10,000 to the nearest ten, hundred, and thousand.
also **NS 1.0, O─ n NS 1.3, MR 2.0, MR 2.3, MR 2.5, MR 3.2**

1. **Tell** how you would use the rounding rules to round 2,641 to the nearest ten.

Round to the nearest thousand.

2. 6,427   3. 2,500   4. 4,526   5. 1,670

► **Practice and Problem Solving**

Round to the nearest thousand.

6. 8,312   7. 4,500   8. 674   9. 9,478

10. 1,611   11. 5,920   12. 2,543   13. 4,444

Round to the nearest thousand, the nearest hundred, and the nearest ten.

14. 3,581   15. 6,318   16. 2,350   17. 8,914

18. 4,624   19. 5,337   20. 1,273   21. 2,845

USE DATA For 22–24, use the table.

22. To the nearest thousand pounds, about how much does the African elephant weigh?

23. Round the weights of the giraffe and rhinoceros to the nearest thousand pounds. About how many giraffes would it take to equal the weight of the rhinoceros?

24. **Write About It** Tell how to round the weight of the hippopotamus to the nearest thousand, hundred, and ten.

HEAVIEST LAND MAMMALS

Animal	Weight in Pounds
African elephant	11,023
Indian rhinoceros	8,818
Hippopotamus	4,409
Giraffe	2,646

Mixed Review and Test Prep

Write the value of the blue digit. (p. 10)

25. 8,251   26. 87,668

Write in expanded form. (p. 10)

27. 337   28. 12,982

29. **TEST PREP** 68 + 42 + 36 = ■

A 136   C 142

B 138   D 146

Extra Practice page H33, Set E

Review/Test

✓ CHECK VOCABULARY AND CONCEPTS

Choose the best term from the box.

> greatest to least
>
> compare
>
> round

1. You can use <, >, or = to _?_ numbers. (p. 20)

2. One way to estimate is to _?_ numbers. (p. 28)

Suppose you want to round 371 to the nearest hundred. (pp. 28–29)

3. Which hundreds is 371 between?

4. Which hundred is 371 closer to?

✓ CHECK SKILLS

Compare the numbers. Write <, >, or = for each ●. (pp. 20–23)

5. 532 ● 523   6. 3,246 ● 325   7. 7,583 ● 7,583

Write the numbers in order from least to greatest. (pp. 24–25)

8. 143, 438, 92   9. 7,304; 7,890; 7,141   10. 23,256; 23,161; 23,470

11. Round 85 to the nearest ten. (pp. 28–29)

12. Round 824 to the nearest hundred. (pp. 28–29)

13. Round 3,721 to the nearest thousand and hundred. (pp. 30–31)

✓ CHECK PROBLEM SOLVING

USE DATA For 14–15, use the table. (pp. 26–27)

14. On which night was the number of tickets sold greater than the number sold on Monday but less than the number sold on Wednesday?

15. On which night was the number of tickets sold less than the number sold on Friday but greater than the number sold on Wednesday?

TICKET SALES	
Day	Number
Monday	1,079
Tuesday	1,580
Wednesday	1,493
Thursday	1,208
Friday	2,112

Cumulative Review

Understand the problem.
See item **3**.

There is more than one number greater than 240 and less than 250. Find digits that add up to 7 from these numbers.

Also see problem **1**, p. H62.

For 1–11, choose the best answer.

1. 38
    −13

    A 24   **C** 51
    B 25   **D** NOT HERE

2. Which number is greater than 672 and less than 683?

    F 671   **H** 685
    G 682   **J** 687

3. A number is greater than 240 and less than 250. The sum of its digits is 7. What is the number?

    A 142   **C** 241
    B 214   **D** 242

4. Which number is **not** even?

    F 753   **H** 3,558
    G 1,350   **J** 6,904

5. Which shows the numbers in order from least to greatest?

    A 587, 590, 571
    B 590, 587, 571
    C 587, 571, 590
    D 571, 587, 590

6. What is 451 rounded to the nearest ten?

    F 400   **H** 460
    G 450   **J** 500

7. What is 781 rounded to the nearest hundred?

    A 700   **C** 790
    B 780   **D** 800

8. On Friday, 8,656 people attended the football game. What is that number rounded to the nearest thousand?

    F 9,000   **H** 8,600
    G 8,700   **J** 8,000

9. What is the value of the 9 in 79,524?

    A 90,000   **C** 900
    B 9,000   **D** NOT HERE

10. Which shows the numbers in order from greatest to least?

    F 13,456; 43,165; 14,653
    G 13,456; 14,653; 43,165
    H 43,165; 14,653; 13,456
    J 43,165; 13,653; 14,653

11. 49
    +23

    A 612   **C** 72
    B 73   **D** NOT HERE

Addition

Dogs were first used as watchdogs, herding dogs, and hunting dogs. Now, more dogs are pets than workers. There are still dogs that work, though. Some dogs are trained to help disabled people. Look at the chart below. How many dogs in all graduated in 1997 and 1998?

NUMBER OF CANINE COMPANION GRADUATES

138	139	132	105
1996	1997	1998	1999

Year

CHECK WHAT YOU KNOW ✓

Use this page to help you review and remember
important skills needed for Chapter 3.

✓ VOCABULARY

Choose the best term from the box.

1. The answer to an addition problem is called the ? .

2. In $8 + 4 = 12$, the 8 and 4 are ? .

| addends |
| difference |
| sum |

✓ ADDITION FACTS (See p. H4.)

Add.

3.   2
   +7

4.   9
   +4

5.   6
   +5

6.   3
   +8

7.   8
   +7

8.   5
   +7

9.   8
   +2

10.   7
    +9

11.   5
    +3

12.   4
    +3

13.   7
    +7

14.   3
    +9

15.   6
    +6

16.   7
    +3

17.   8
    +8

✓ 2-DIGIT ADDITION (See p. H5.)

Add.

18.
tens	ones
1	2
+ 1	5

19.
tens	ones
1	6
+ 1	8

20.
tens	ones
4	3
+ 3	6

21.   11
    +65

22.   53
    +18

23.   27
    +19

24.   26
    +35

25.   15
    +45

✓ MENTAL MATH: ADD 2-DIGIT NUMBERS (See p. H5.)

Use mental math to find the sum.

26. $23 + 10 =$ ■

27. $24 + 11 =$ ■

28. $32 + 14 =$ ■

29. $15 + 25 =$ ■

30. $22 + 35 =$ ■

31. $45 + 21 =$ ■

Column Addition

▶ **Learn**

PET PORTIONS Maria bought 5 pounds of cat food, 2 pounds of birdseed, and 8 pounds of dog food. How many pounds of pet food did she buy?

$$5 + 2 + 8 = \blacksquare$$

Quick Review

1. $3 + 2 + 3$

2. $4 + 5 + 6$

3. $8 + 2 + 3$

4. $6 + 1 + 9$

5. $8 + 7 + 3$

VOCABULARY

Grouping Property of Addition

MATH IDEA The Grouping Property of Addition states that you can group addends in different ways. The sum is always the same.

$$5 + (2 + 8) = \blacksquare \qquad (5 + 2) + 8 = \blacksquare$$
$$\downarrow \qquad\qquad\qquad \downarrow$$
$$5 + \ 10 \ = 15 \qquad 7 \ + 8 = 15$$

Add the numbers in ( ) first.

So, Maria bought 15 pounds of pet food.

Example Find $13 + 18 + 27$.

STEP 1

Add ones.
$10 + 8 = 18$ ones

```
 1
13
18  >10
+27
 8
```
Group 3 and 7 to make a ten.

STEP 2

Add tens.
$1 + 4 = 5$ tens

```
 1
13
18
+27
58
```

CALIFORNIA STANDARDS O—πNS 2.1 Find the sum or difference of two whole numbers between 0 and 10,000.
AF 1.0 Students select appropriate symbols, operations, properties to represent, describe, simplify, and solve simple number relationships. *also* NS 1.0, O—πNS 1.3, NS 2.0, MR 1.0, MR 2.0, MR 2.3, MR 2.4

▶ Check

1. Explain what happens to the sum when you group addends in different ways.

Find the sum.

2. $(5 + 5) + 6 = \blacksquare$ **3.** $8 + (2 + 8) = \blacksquare$ **4.** $(7 + 13) + 2 = \blacksquare$

▶ Practice and Problem Solving

Find the sum.

5. $(7 + 3) + 6 = \blacksquare$ **6.** $7 + (5 + 5) = \blacksquare$ **7.** $(8 + 2) + 4 = \blacksquare$

8. $8 + (5 + 3) = \blacksquare$ **9.** $(1 + 4) + 11 = \blacksquare$ **10.** $(9 + 8) + 20 = \blacksquare$

Use the Grouping Property to find the sum.

11.	12.	13.	14.	15.	16.
8	5	4	3	21	34
9	5	9	9	45	27
+2	+8	+8	+7	+32	+11

17. $8 + 2 + 6 = \blacksquare$ **18.** $5 + 18 + 5 = \blacksquare$ **19.** $12 + 8 + 9 = \blacksquare$

USE DATA For 20-21, use the table.

20. How many pounds did the puppy gain in three months?

21. How many more pounds did the puppy gain in September and October than in November?

PUPPY POUNDS GAINED		
September	October	November
6	4	5

22. **Write About It** Does $18 + 4 + 3 = 5 + 2 + 10 + 8$? Explain.

23. **Write a problem** with three addends. Use the Grouping Property to help you solve.

Mixed Review and Test Prep

24. Add 20 and 15.

25. Write 16,072 in expanded form. (p. 10)

26. What is the least number you can write with the digits 4, 8, 7, and 2? (p. 6)

27. What is the place-value position of the digit 6 in $16,083$? (p. 10)

28. **TEST PREP** Which is greater than $7,855$? (p. 20)

A 6,999 **C** 7,847

B 7,755 **D** 7,901

Extra Practice page H34, Set A

Estimate Sums

Quick Review

1. 5 + 25
2. 30 + 20
3. 70 + 20
4. $4 + $6
5. 20 + 55

▶ **Learn**

MORE FOR THE MANATEES Manny and Marsha are manatees that live at a sea park. In one day, Manny ate 91 pounds of food and Marsha ate 88 pounds. About how much did they eat in all?

To find *about* how much, you can **estimate**.

VOCABULARY

estimate

STEP 1

Round each number to the nearest ten.

$$88 \rightarrow 90$$
$$+91 \rightarrow +90$$

STEP 2

Find the estimated sum.

$$90$$
$$+90$$
$$\overline{180}$$

So, they ate about 180 pounds of food.

MATH IDEA When you do not need an exact answer, you can estimate.

Examples

A Round to the nearest hundred.

$$174 \rightarrow 200$$
$$+123 \rightarrow +100$$
$$\overline{300}$$

B Round to the nearest dollar.

$$\$7.80 \rightarrow \$8$$
$$+\$4.35 \rightarrow +\$4$$
$$\overline{\$12}$$

C Round to the nearest thousand.

$$3,260 \rightarrow 3,000$$
$$+2,755 \rightarrow +3,000$$
$$\overline{6,000}$$

▲ A manatee will eat from 32 to 108 pounds of plants every day, depending on its size.

▶ **Check**

1. **Explain** why it makes sense for the sea park to estimate how much food the manatees need.

CALIFORNIA STANDARDS O━n NS 2.1 Find the sum or difference of two whole numbers between 0 and 10,000. NS 1.4 Round off numbers to 10,000 to the nearest ten, hundred, and thousand. *also* NS 1.0, C━n NS 1.3, NS 2.8, MR 2.3, MR 2.4, MR 2.5, MR 3.0

Estimate the sum.

2. 67
+19

3. 410
+890

4. $5.30
+$3.80

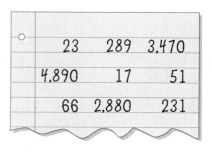

▶ Practice and Problem Solving

Estimate the sum.

5. 12
+23

6. 14
+28

7. 24
+78

8. 518
+305

9. 206
+668

10. $6.38
+$1.04

11. 480
+240

12. 379
+325

13. 109
+515

14. 922
+111

15. $5.90
+$2.99

16. 2,610
+3,397

17. 7,800
+1,620

18. 1,340
+5,600

19. $4.70
+$2.90

20. REASONING Josh put 21 heads of lettuce in each basket to feed to the manatees. About how many heads of lettuce would there be in 3 baskets?

21. Erica earned $2.90 on Monday. If she earns about the same amount Tuesday and Wednesday, can she buy a $13.00 CD? Explain.

For 22–25 use the numbers at the right.

Choose two numbers whose sum is about:

22. 70.

23. 500.

24. 8,000.

23	289	3,470
4,890	17	51
66	2,880	231

25. ✏️ Write a problem in which you estimate the sum of two or more numbers from the list.

Mixed Review and Test Prep

Write <, >, or = for each ⬤. (p. 20)

26. 27 ⬤ 38

27. 723 ⬤ 726

28. Write in order from least to greatest: 3,291; 3,245; 3,311. (p. 24)

29. Write 40,000 + 3,000 + 700 + 9 in standard form. (p. 10)

30. **TEST PREP** Which is sixty thousand, two hundred forty written in standard form? (p. 10)

A 60,024

C 60,240

B 60,204

D 62,040

HANDS ON
Add 3-Digit Numbers

▶ **Explore**

Make a model to add 134 and 279.

$$\begin{array}{r} 134 \\ +279 \\ \hline \end{array}$$

MATERIALS
base-ten blocks

STEP 1

Add ones.
$4 + 9 = 13$ ones
Regroup 13 ones
as 1 ten 3 ones.

STEP 2

Add tens.
$1 + 3 + 7 = 11$ tens
Regroup 11 tens
as 1 hundred 1 ten.

We are adding
116 and 144. We now
have 10 ones. What
should we do next?

STEP 3

Add hundreds.
$1 + 1 + 2 = 4$ hundreds

So, $134 + 279 = 413$.

Try It

Use base-ten blocks to find each sum.

a. $116 + 144 = $ ▮

b. $269 + 358 = $ ▮

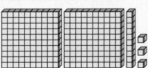

▶ Connect

Here is a way to record addition. To add 137 and 264, first line up hundreds, tens, and ones.

STEP 1

Add the ones.
Regroup.
11 ones = 1 ten 1 one.

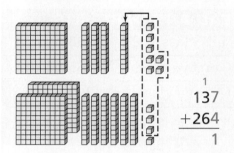

```
  1
  137
+ 264
    1
```

STEP 2

Add the tens.
Regroup.
10 tens = 1 hundred 0 tens

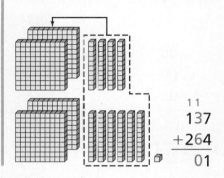

```
 11
  137
+ 264
   01
```

STEP 3

Add the hundreds.

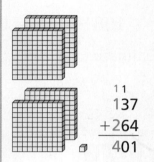

```
 11
  137
+ 264
  401
```

▶ Practice

Use base-ten blocks to find each sum.

1. 134 + 217 = ■

2. 265 + 423 = ■

3. 368 + 416 = ■

4. 333 + 128 = ■

5. 295 + 382 = ■

6. 192 + 439 = ■

7. 493 + 256 = ■

8. 563 + 139 = ■

9. 612 + 308 = ■

USE DATA For 10–11, use the table.

10. A trucker drove from San Diego to San Francisco. How many miles did she drive?

11. ✏️ **Write a problem** in which you add distances. Exchange with a partner. Solve.

DISTANCES BETWEEN CALIFORNIA CITIES	
San Diego to Los Angeles	133 miles
Los Angeles to San Francisco	441 miles
San Francisco to Crescent City	367 miles

Mixed Review and Test Prep

Compare. Use <, >, or = for each ●. (p. 20)

12. 89 ● 78

13. 324 ● 342

14. 142 ● 412

15. 8 + 3 ● 20 − 7

16. **TEST PREP** Which group of numbers is in order from least to greatest? (p. 24)

A 394, 379, 380

B 521, 539, 540

C 427, 450, 431

D 201, 263, 229

Add 3-Digit Numbers

▶ **Learn**

BUNCHES OF BOOKS How many books did Mr. Chi's and Mrs. Garcia's classes read in all?

$$198 + 165 = \blacksquare$$

Estimate.	198 →	200
	+165 →	+200
		400

READ-A-THON RESULTS	
Mr. Chi's class	198 books
Mrs. Garcia's class	165 books
Mrs. Miller's class	203 books

STEP 1

Add the ones. Regroup.
13 ones = 1 ten 3 ones.

```
  1
 198
+165
-----
   3
```

STEP 2

Add the tens. Regroup.
16 tens = 1 hundred 6 tens.

```
 1 1
 198
+165
-----
  63
```

STEP 3

Add the hundreds.

```
 1 1
 198
+165
-----
 363
```

So, the two classes read 363 books in all. Since 363 is close to 400, the answer is reasonable.

MATH IDEA Estimate to see if your answer is reasonable.

Examples

A

```
  1
 325
+ 67
-----
 392
```

B

```
   2
 591
 173
+290
-----
1,054
```

C

```
   1
 $4.83
+$2.74
------
 $7.57
   ↑
```
decimal point

• Add money like whole numbers.
• Then use a decimal point to separate dollars and cents.

CALIFORNIA STANDARDS O⎯ⁿNS 2.1 Find the sum or difference of two whole numbers between 0 and 10,000. **MR 2.1** Use estimation to verify the reasonableness of calculated results. *also* **NS 1.0**, O⎯ⁿNS 1.3, **NS 2.0**, O⎯ⁿNS 3.3, **MR 2.4**, **MR 2.6**, **MR 3.1**

1. **Explain** whether you would regroup to find how many books Mr. Chi's and Mrs. Miller's classes read.

Find the sum. Estimate to check.

2. 224	3. 298	4. $9.07	5. 468
+511	+172	+$1.25	+ 89

► **Practice and Problem Solving**

Find the sum. Estimate to check.

6. 321	7. 505	8. 173	9. 561	10. 299
+268	+228	+368	+246	+ 66

11. 284	12. $7.44	13. 629	14. $1.42	15. 152
+325	+$5.02	+ 67	+$5.61	+339

16. 297	17. 579	18. $7.39	19. 896	20. 177
+580	+486	+$2.51	+424	+695

21. 219 + 316 + 222 = ■   **22.** 267 + 741 + 109 = ■

23. $\frac{a+b}{c}$ **Algebra** Write the missing addend. 230 + ■ + 50 = 282

24. Manuel spent $3.65 for lunch on Monday and $2.78 on Tuesday. How much did he spend in all?

25. **? What's the Question?** Eva read to page 112 in her book. There are 67 more pages in the book. The answer is 179 pages.

26. **? What's the Error?** Sharon added 458 and 83 like this.

   458   Describe her error and
  +83   solve.
 ─────
 1,288

Mixed Review and Test Prep

27. 19 + 24 = ■   **28.** 30 + 62 = ■

29. Explain the pattern. (p. 2)
   86, 81, 76, 71, 66

30. What is the value of the blue digit in 34,241? (p. 10)

31. **TEST PREP** 50 − 27 = ■
   A 21   C 32
   B 23   D 33

Problem Solving Strategy
Predict and Test

Understand → Plan → Solve → Check

Quick Review

1. $21 + 6$
2. $45 + 15$ 3. $9 + 36$
4. $72 + 8$ 5. $12 + 13$

PROBLEM The third-grade classes bought 75 containers of food for the animal shelter. They had 15 more cans than bags of food. How many bags and cans did the classes buy?

- What are you asked to find?
- What information will you use?
- Is there any information you will not use?

- What strategy can you use to solve the problem?

  You can *predict and test* to find the number of bags and cans the classes bought.

- How can you use the strategy to solve the problem?

  Predict the number of bags the classes bought. Add 15 to that number for the number of cans. Then test to see if the sum is 75.

BAGS	CANS	TOTAL	NOTES
20	20+15=35	20+35=55	too low
50	50+15=65	50+65=115	too high
30	30+15=45	30+45=75	just right

So, the classes bought 30 bags and 45 cans of food.

- How can you use the first two predictions to make a better prediction?

CALIFORNIA STANDARDS MR 2.0 Students use strategies, skills, and concepts in finding solutions. **MR 3.2** Note the method of deriving the solution and demonstrate a conceptual understanding of the derivation by solving similar problems. *also* **NS 1.0, NS 2.0, NS 2.1, MR 1.1, MR 2.3, MR 2.4, MR 3.1**

PROBLEM SOLVING STRATEGIES

Draw a Diagram or Picture
Make a Model or Act It Out
Make an Organized List
Find a Pattern
Make a Table or Graph
► **Predict and Test**
Work Backward
Solve a Simpler Problem
Write an Equation
Use Logical Reasoning

Use *predict and test* to solve.

1. **What if** the classes bought 120 containers and had 30 more cans than bags? How many bags and how many cans did they buy?

2. Pilar has 170 stamps in her collection. Her first book of stamps has 30 more stamps in it than her second book. How many stamps are in each book?

Two numbers have a sum of 27. Their difference is 3. What are the two numbers?

3. Which is a reasonable prediction for one of the numbers?

    A 3   **C** 27

    B 10   **D** 30

4. What solution answers the question?

    F 3 and 27   **H** 10 and 17

    G 10 and 13   **J** 12 and 15

Mixed Strategy Practice

USE DATA For 5–6, use the table.

5. The number of pounds used in Week 2 was greater than in Week 1, but less than in Week 3. The number of pounds used in Week 2 is an odd number that does not end in 5. How many pounds were used in Week 2?

6. ✎ **Write a problem** about the dog food used at the shelter in which the difference is greater than 5.

DOG FOOD USED AT SHELTER	
February	**Pounds**
Week 1	73
Week 2	▪
Week 3	79
Week 4	81

7. The sum of two numbers is 55. Their difference is 7. What are the numbers?

8. There are 4 students in line. Max is before Keiko but after Liz. Adam is fourth. Who is first?

Add Greater Numbers

▶ Learn

PADDLE POWER Tomas and Eli took a kayak trip. They paddled from White Rock to Bear Corner. Then they paddled from Bear Corner to Raccoon Falls. How many yards did they paddle in all?

$$4,365 + 3,852 = \blacksquare$$

Estimate. $\begin{array}{r} 4,365 \rightarrow 4,000 \\ +3,852 \rightarrow +4,000 \\ \hline 8,000 \end{array}$

STEP 1		**STEP 2**	
Add the ones.	$\begin{array}{r} 4,365 \\ +3,852 \\ \hline 7 \end{array}$	Add the tens. Regroup. 11 tens = 1 hundred 1 ten	$\begin{array}{r} {}^{1} \\ 4,365 \\ +3,852 \\ \hline 17 \end{array}$
STEP 3		**STEP 4**	
Add the hundreds. Regroup. 12 hundreds = 1 thousand 2 hundreds	$\begin{array}{r} {}^{1\ 1} \\ 4,365 \\ +3,852 \\ \hline 217 \end{array}$	Add the thousands.	$\begin{array}{r} {}^{1\ 1} \\ 4,365 \\ +3,852 \\ \hline 8,217 \end{array}$

So, Tomas and Eli paddled 8,217 yards. Since 8,217 is close to 8,000, the answer is reasonable.

Examples

A $\begin{array}{r} {}^{1\ 1\ 1} \\ 5,673 \\ +\ \ 997 \\ \hline 6,670 \end{array}$

B $\begin{array}{r} {}^{2\ 1} \\ 2,094 \\ 8,167 \\ +5,041 \\ \hline 15,302 \end{array}$

C $\begin{array}{r} {}^{1\ 1\ 1} \\ \$37.56 \\ +\$52.54 \\ \hline \$90.10 \end{array}$

• How does the estimated sum help you see if the answer is reasonable?

CALIFORNIA STANDARDS ○┳NS 2.1 Find the sum or difference of two whole numbers between 0 and 10,000. **MR 2.1** Use estimation to verify the reasonableness of calculated results. *also* **NS 2.0,** ○┳**NS 3.3, MR 2.3, MR 2.4, MR 3.1**

Adding More Thousands

Suppose you paddled 7,438 feet in one day and 5,623 feet on the next day. How many feet did you paddle in all?

Hillary and Terrance each solved the problem.

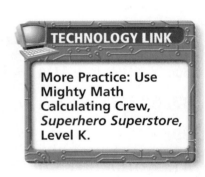

Hillary

$$\begin{array}{r} \overset{1}{7},\overset{1}{4}38 \\ +5{,}623 \\ \hline 13{,}061 \end{array}$$

Terrance

$$\begin{array}{r} \overset{1}{7},\overset{1}{4}38 \\ +\ 5{,}623 \\ \hline 1{,}361 \end{array}$$

- Which student solved the problem correctly?

- What error did the other student make?

- How could Terrance have used estimation to tell that his answer was not reasonable?

MATH IDEA Use the same regrouping rules you learned with 2- and 3-digit numbers to add greater numbers. Use estimation and number sense to check your work.

TECHNOLOGY LINK

More Practice: Use Mighty Math Calculating Crew, *Superhero Superstore,* Level K.

▶ Check

1. **Explain** how you could tell that $2{,}294 + 8{,}167$ is a 5-digit number without finding the sum.

Find the sum. Estimate to check.

2. $\begin{array}{r} 5{,}347 \\ +2{,}091 \\ \hline \end{array}$

3. $\begin{array}{r} 1{,}348 \\ +\ 721 \\ \hline \end{array}$

4. $\begin{array}{r} 4{,}919 \\ +1{,}592 \\ \hline \end{array}$

5. $\begin{array}{r} 7{,}625 \\ +4{,}032 \\ \hline \end{array}$

6. $\begin{array}{r} \$73.95 \\ +\$86.89 \\ \hline \end{array}$

7. $1{,}032 + 5{,}198 = \blacksquare$

8. $\$69.81 + \$23.11 = \blacksquare$

9. $3{,}035 + 989 + 4{,}918 = \blacksquare$

10. $2{,}354 + 4{,}526 + 831 = \blacksquare$

11. Add 6,144 and 7,902.

12. Add 5,832 and 6,288.

LESSON CONTINUES ▶

Find the sum. Estimate to check.

13.   6,709
      +1,226

14. $22.78
      +$ 5.01

15.   1,821
      +6,744

16. $33.58
      +$42.65

17.   1,259
      +6,074

18.   2,458
      +1,939

19. $27.35
      +$ 7.96

20.   3,624
      + 947

21.   3,769
      +2,347

22. $64.11
      +$88.49

23.   4,641
      +7,989

24.   4,329
      +8,117

25.   5,492
      +4,508

26.   1,895
      +1,799

27.   9,294
      +2,136

28.   2,164
      +8,235

29.   4,921
      +8,700

30.   7,493
      +2,020

31.   4,733
      +3,377

32.   8,284
      +1,223

33. 6,592 + 2,135 = ■

34. 5,209 + 3,810 = ■

35. 2,699 + 1,104 + 1,009 = ■

36. 4,462 + 3,574 = ■

37. $23.65 + $57.77 = ■

38. 2,594 + 6,563 + 1 241 = ■

39. 1,090 + 9,311 = ■

40. 900 + 1,788 + 35 = ■

41. Add 3,901 and 7,189.

42. Add 3,052 and 609.

43. **NUMBER SENSE** Write a number less than 3,425 + 8,630 but greater than 7,614 + 4,429.

44. **a+b/c Algebra** Write the missing addend. 4,020 + ■ = 4,222

45. **ESTIMATION** Allie estimates that 5,109 + 4,995 is about 1,000. Do you agree or disagree? Explain.

46. **? What's the Error?** Sergio found 8,235 + 986 like this. Describe his error. Find the sum.

    ```
      1 11
      8,235
    +   986
      9,211
    ```

47. **USE DATA** Use the price list. If Craig mows and rakes 2 lawns, how much has he earned?

CRAIG'S PRICE LIST
Weed Garden $5.00
Mow Lawn $7.50
Rake Lawn $4.50

48. Can you add two 4-digit numbers and get a sum greater than 20,000? Explain.

49. $6 + 8 + 4 = $ ▪ (p. 36)

50. $24 + 36 + 19 = $ ▪ (p. 36)

51. $\$15.76 + \$22.19 + \$3.00 = $ ▪
(p. 36)

Choose $<$, $>$, **or** $=$ **for each** ●. (p. 42)

52. $305 + 281$ ● 550

53. $416 + 966$ ● $1,600$

54. $223 + 100$ ● $50 + 50 + 223$

55. The sum of two numbers is 20.
The difference of the numbers is
10. What are the numbers? (p. 44)

   A 10, 10   **C** 12, 8
   B 20, 10   **D** 15, 5

For 56, use the graph.

FAVORITE SEASONS	
Summer	☀ ☀ ☀ ☀ ☀
Winter	☀ ☀ ☀
Spring	☀ ☀ ☀ ☀
Fall	☀ ☀

Key: Each ☀ = 2 votes.

56. **TEST PREP** How many students did
NOT vote for summer?

  F 12   **G** 18   **H** 20   **I** 24

Thinker's Corner

Try to make a greater sum than your partner.

MATERIALS: index cards numbered 0–9

Sum It Up

A. Player 1 chooses 4 cards and uses
the digits to write two 4-digit addends. Each digit
should be used twice. Player 1 replaces the cards.

B. Player 2 repeats Step A.

C. Both players find the sum. The player with the
greatest sum wins. Play this game several times.
See if you can find a winning strategy.

D. Repeat this game. Try to make the least sum.

1. When making the greatest sum, where is the
best position to put a 9?

2. When making the least sum, where is the
best place to put your highest digit?

Review/Test

CHECK VOCABULARY AND CONCEPTS

Choose the best term from the box.

> regroup
> Grouping Property of
>   Addition
> estimate

1. To find *about* how much, you can ? . (p. 38)

2. The ? states that the sum is the same no matter how you group the addends. (p. 36)

For 3, think of how to model 120 + 138. (pp. 40-41)

3. Do you need to regroup to find the sum 120 + 138? Explain.

CHECK SKILLS

Use the Grouping Property to find the sum. (pp. 36–37)

4.	5.	6.	7.
47	26	65	87
39	34	32	91
+21	+18	+55	+63

Estimate the sum. (pp. 38–39)

8.	9.	10.	11.
76	267	$5.92	3,200
+17	+193	+$3.25	+1,900

Find the sum. (pp. 42–43, 46–49)

12. $419 + 451 =$ ■   13. $321 + 683 =$ ■   14. $127 + 315 + 299 =$ ■

15.	16.	17.	18.
4,782	5,436	$64.33	3,764
+3,917	+7,695	+$27.98	+8,109

CHECK PROBLEM SOLVING

Solve. (pp. 44–45)

19. Two numbers have a sum of 47. Their difference is 5. What are the two numbers?

20. Mr. Samuel has 150 pennies in two jars. There are 40 more pennies in one jar than in the other. How many pennies are in each jar?

Cumulative Review

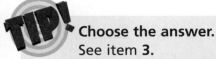

Choose the answer.
See item **3**.

If your answer doesn't match one of the choices, check your computation. If your computation is correct, mark NOT HERE.

Also see problem **6,** p. H65.

For 1–10, choose the best answer.

1. Jenna got 8 cat stickers from her aunt. She got 7 dog stickers from her friend. Then she bought 9 monkey stickers. How many stickers did she have in all?

   A 20 **C** 24
   B 22 **D** 25

2. Which number is less than 4,562?

   F 4,565 **H** 4,572
   G 4,560 **J** 4,625

3.   3,562
    +3,779

   A 6,241 **C** 7,241
   B 6,331 **D** NOT HERE

4. What is the sum of 637 and 295?

   F 822 **H** 931
   G 832 **J** 932

5. There were 596 people at the fair on Saturday and 246 people at the fair on Sunday. How many people in all were at the fair on these days?

   A 742 **C** 842
   B 832 **D** NOT HERE

6. The sum of two numbers is 54. Their difference is 8. What are the two numbers?

   F 20 and 34
   G 23 and 31
   H 24 and 32
   J 50 and 4

7. Which two numbers have a sum of about 300?

   A 144 + 66
   B 175 + 123
   C 127 + 88
   D 134 + 116

8. Which number has the least value?

   F 1,198 **H** 1,527
   G 2,571 **J** 1,275

9. Barry put 23 red marbles, 46 green marbles, and 34 blue marbles in a bag. How many marbles did he put in the bag in all?

   A 103 **C** 107
   B 104 **D** 113

10. A number is between 10 and 20 and it is even. The difference between the digits is 5. What is the number?

   F 14 **H** 16
   G 15 **J** 27

Subtraction

Scientists know many things about sea lions and seals. They know what these animals eat, where they live, and how much they weigh. Of course, not every sea lion or seal weighs the same amount! The table shows the maximum, or greatest, weight for each animal. What is the difference between the weight of the male and female California sea lions?

California sea lions

MAXIMUM WEIGHT (IN POUNDS)		
Animal	Male	Female
Northern Fur Seal	600	110
California Sea Lion	850	250
Steller Sea Lion	2,000	600
Northern Elephant Seal	5,000	2,000

CHECK WHAT YOU KNOW ✓

Use this page to help you review and remember
important skills needed for Chapter 4.

✓ VOCABULARY

Choose the best term from the box.

> difference
> estimate
> regroup

1. When you want to know *about* how many,
you can find an ? .

2. In $15 - 8 = 7$, the 7 is the ? .

✓ SUBTRACTION FACTS (See p. H6.)

3. $\begin{array}{r} 12 \\ -\ 9 \\ \hline \end{array}$	**4.** $\begin{array}{r} 15 \\ -\ 7 \\ \hline \end{array}$	**5.** $\begin{array}{r} 11 \\ -\ 5 \\ \hline \end{array}$	**6.** $\begin{array}{r} 14 \\ -\ 6 \\ \hline \end{array}$	**7.** $\begin{array}{r} 13 \\ -\ 5 \\ \hline \end{array}$
8. $\begin{array}{r} 10 \\ -\ 3 \\ \hline \end{array}$	**9.** $\begin{array}{r} 10 \\ -\ 6 \\ \hline \end{array}$	**10.** $\begin{array}{r} 12 \\ -\ 7 \\ \hline \end{array}$	**11.** $\begin{array}{r} 13 \\ -\ 4 \\ \hline \end{array}$	**12.** $\begin{array}{r} 16 \\ -\ 7 \\ \hline \end{array}$
13. $\begin{array}{r} 11 \\ -\ 4 \\ \hline \end{array}$	**14.** $\begin{array}{r} 18 \\ -\ 9 \\ \hline \end{array}$	**15.** $\begin{array}{r} 17 \\ -\ 8 \\ \hline \end{array}$	**16.** $\begin{array}{r} 16 \\ -\ 9 \\ \hline \end{array}$	**17.** $\begin{array}{r} 11 \\ -\ 3 \\ \hline \end{array}$

✓ 2-DIGIT SUBTRACTION (See p. H6.)

Subtract.

18.
tens	ones
4	6
−1	4

19.
tens	ones
8	2
−2	8

20.
tens	ones
7	0
−5	6

21. $\begin{array}{r} 33 \\ -18 \\ \hline \end{array}$	**22.** $\begin{array}{r} 90 \\ -48 \\ \hline \end{array}$	**23.** $\begin{array}{r} 98 \\ -17 \\ \hline \end{array}$	**24.** $\begin{array}{r} 57 \\ -27 \\ \hline \end{array}$	**25.** $\begin{array}{r} 25 \\ -19 \\ \hline \end{array}$

✓ MENTAL MATH: SUBTRACT 2-DIGIT NUMBERS (See p. H7.)

Use mental math to find the difference.

26. $38 - 10 = \blacksquare$ **27.** $43 - 11 = \blacksquare$ **28.** $52 - 12 = \blacksquare$ **29.** $28 - 15 = \blacksquare$

30. $36 - 15 = \blacksquare$ **31.** $65 - 14 = \blacksquare$ **32.** $48 - 24 = \blacksquare$ **33.** $75 - 25 = \blacksquare$

Estimate Differences

Quick Review

1. $50 - 30$ 2. $40 - 10$

3. $35 - 10$ 4. $60 - 20$

5. $50 - 40$

▶ Learn

SEEN IN THE SEA Scientists keep track of how many baby sea otters, or pups, are seen in California each spring. About how many more pups were seen in 1995 than in 1998?

To find *about* how many more, you can estimate.

STEP 1	STEP 2
Round each number to the nearest hundred.	Find the estimated difference.
$282 \rightarrow 300$ $-159 \rightarrow -200$	$\begin{array}{r} 300 \\ -200 \\ \hline 100 \end{array}$

SEA OTTER PUP SIGHTINGS	
Year	Pups Seen
1995	282
1996	315
1997	310
1998	159

So, about 100 more pups were seen in 1995.

Examples

A Round to the nearest ten.

$$\begin{array}{r} 61 \rightarrow 60 \\ -48 \rightarrow -50 \\ \hline 10 \end{array}$$

B Round to the nearest dollar.

$$\begin{array}{r} \$8.95 \rightarrow \$9 \\ -\$3.35 \rightarrow -\$3 \\ \hline \$6 \end{array}$$

C Round to the nearest thousand.

$$\begin{array}{r} 6,860 \rightarrow 7,000 \\ -4,655 \rightarrow -5,000 \\ \hline 2,000 \end{array}$$

Sometimes it makes sense to round to a different place. Try estimating $341 - 265$.

$$\begin{array}{r} 341 \rightarrow 300 \\ -265 \rightarrow -300 \\ \hline 0 \end{array}$$

Round to the nearest ten to get a closer estimate.

$$\begin{array}{r} 341 \rightarrow 340 \\ -265 \rightarrow -270 \\ \hline 70 \end{array}$$

MATH IDEA To estimate a difference, round to the place value that makes sense.

CALIFORNIA STANDARDS O--n NS 2.1 Find the sum or difference of two whole numbers between 0 and 10,000. **NS 1.4** Round off numbers to the nearest ten, hundred, and thousand. *also* **NS 1.0,** O--n NS 1.3, MR 1.0, MR 2.3, MR 2.4, MR 3.2

▶ Check

1. Tell how you would estimate 203 − 181.

Estimate the difference.

2.	3.	4.	5.	6.
87 −32	478 −115	813 −491	$9.01 −$2.60	5,020 −1,750

▶ Practice and Problem Solving

Estimate the difference.

7.	8.	9.	10.	11.
42 −19	51 −29	84 −28	69 −43	91 −23

12.	13.	14.	15.	16.
470 −110	613 −371	$9.08 −$3.80	880 −114	625 −489

17.	18.	19.	20.	21.
$5.17 −$1.01	322 −199	3,288 −1,255	1,970 −1,050	4,819 −1,766

Estimate the difference. Round to the place value that makes sense.

22.	23.	24.	25.	26.
422 −394	201 −178	4,411 −3,509	1,001 − 599	2,312 −1,845

27. The table shows how many pounds of food two sea otters ate in one week. About how much more did the 55-pound otter eat than the 45-pound otter?

28. 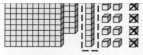 Write a problem about estimating. Use the table at the right.

POUNDS OF FOOD EATEN

Sea Otter Weight	Amount of Food Eaten
45 pounds	79 pounds
55 pounds	96 pounds

Mixed Review and Test Prep

Write the value of the blue digit.
(pp. 6 and 10)

29. 2,051　　**30.** 1,321

31. 37,820　　**32.** 52,902

33. **TEST PREP** Amy had 34 shells. On Friday she found more and had a total of 45 shells. How many shells did she find on Friday?

    A 9　　**B** 11　　**C** 55　　**D** 79

Extra Practice page H35, Set A

Chapter 4　**55**

STEP 3

Subtract the hundreds.

$$\begin{array}{r} \overset{\overset{10}{\scriptstyle 1\ 0\ 15}}{2\ \cancel{1}\ 5} \\ -1\ 2\ 6 \\ \hline 8\ 9 \end{array}$$

TECHNOLOGY LINK

More Practice: Use E-Lab, *Subtraction of Three-Digit Numbers*.

www.harcourtschool.com/elab2002

▶ Practice

Use base-ten blocks to find each difference.

1. 94 − 28 = ■　　**2.** 223 − 122 = ■　　**3.** 437 − 243 = ■

4. 183 − 159 = ■　　**5.** 329 − 87 = ■　　**6.** 223 − 135 = ■

7. REASONING There are 32 more apples than oranges at a fruit stand. How many apples and oranges could there be?

Mixed Review and Test Prep

8. 9 − 5　　**9.** 17 − 8

10. 6 + 7　　**11.** 8 + 4

12. **TEST PREP** 3 + (8 + 12) (p. 36)

  A 11　　　　**C** 23
  B 20　　　　**D** 26

HANDS ON
Subtract 3-Digit Numbers

▶ Explore

Use models to subtract
195 from 324.

$$\begin{array}{r} 324 \\ -195 \end{array}$$

Quick Review
1. 30 − 20
2. 120 − 10
3. 62 − 32
4. 92 − 31
5. 35 − 15

MATERIALS
base-ten blocks

STEP 1	STEP 2
Show 324. 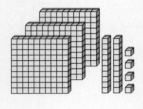	Try to subtract 5 ones. Since there are not enough ones, regroup 1 ten as 10 ones. Subtract 5 ones.

STEP 3

Try to subtract 9 tens.
Since there are not enough tens, regroup 1 hundred as 10 tens. Subtract 9 tens.

STEP 4

Subtract 1 hundred.

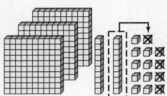

We are subtracting 93 from 181. We regrouped 8 tens 1 one as 7 tens 11 ones. What should we do next?

Subtract 3-Digit Numbers

Quick Review
1. 12 − 5 2. 16 − 9
3. 11 − 7 4. 15 − 6
5. 15 − 8

▶ Learn

TURTLE TALK A sea park has two green sea turtles, Tara and Tom. Tom weighs 332 pounds and Tara weighs 198 pounds. How much more does Tom weigh than Tara?

$$332 - 198 = \blacksquare$$

Estimate.
$$\begin{array}{r} 332 \to \ \ 300 \\ -198 \to -200 \\ \hline 100 \end{array}$$

▲ A green sea turtle weighs between 150 and 410 pounds.

STEP 1	STEP 2	STEP 3
Subtract the ones. 8 > 2 Regroup. 3 tens 2 ones = 2 tens 12 ones	Subtract the tens. 9 > 2 Regroup. 3 hundreds 2 tens = 2 hundreds 12 tens	Subtract the hundreds.
$$\begin{array}{r} 2\ 12 \\ 3\ 3\ 2 \\ -1\ 9\ 8 \\ \hline 4 \end{array}$$	$$\begin{array}{r} 12 \\ 2\ 2\ 12 \\ 3\ 3\ 2 \\ -1\ 9\ 8 \\ \hline 3\ 4 \end{array}$$	$$\begin{array}{r} 12 \\ 2\ 2\ 12 \\ 3\ 3\ 2 \\ -1\ 9\ 8 \\ \hline 1\ 3\ 4 \end{array}$$

So, Tom weighs 134 pounds more than Tara. Since 134 is close to 100, the answer is reasonable.

Examples

A
$$\begin{array}{r} 2\ 15 \\ 3\ 5\ 9 \\ -\ \ 8\ 4 \\ \hline 2\ 7\ 5 \end{array}$$

B
$$\begin{array}{r} 13 \\ 7\ 3\ 16 \\ 8\ 4\ 6 \\ -6\ 9\ 8 \\ \hline 1\ 4\ 8 \end{array}$$

C
$$\begin{array}{r} 6\ 12 \\ \$4.\ 7\ 2 \\ -\$1.\ 1\ 5 \\ \hline \$3.\ 5\ 7 \end{array}$$

Use a decimal point to separate dollars and cents.

Subtract Across Zeros

What if Tom weighed 300 pounds and Tara weighed 178 pounds? How much more does Tom weigh?

$300 - 178 = \blacksquare$

Estimate. $300 - 180 = 120$

STEP 1

$8 > 0$. Since there are 0 tens, regroup hundreds.
3 hundreds 0 tens =
2 hundreds 10 tens

```
  2 10
  3 0 0
- 1 7 8
```

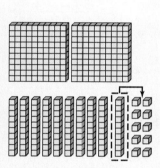

STEP 2

Regroup tens.
10 tens 0 ones =
9 tens 10 ones

```
      9
  2 10 10
  3 0 0
- 1 7 8
```

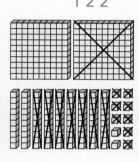

STEP 3

Subtract the ones.
Subtract the tens.
Subtract the hundreds.

```
      9
  2 10 10
  3 0 0
- 1 7 8
  1 2 2
```

So, Tom weighs 122 pounds more than Tara. Since 122 is close to 120, the answer is reasonable.

MATH IDEA When you subtract across zeros, you may need to regroup the hundreds first.

Examples

A
```
     9
  1 10 14
  2 0 4
-   8 7
  1 1 7
```

B
```
      9
  4 10 13
  $5. 0 3
- $1. 2 4
  $3. 7 9
```

C
```
      9
  6 10 10
  7 0 0
- 4 8 1
  2 1 9
```

CHECK ✓
```
  219
+481
  700
```
You can add to check your answer.

▶ Check

1. Explain why 9 is written above the crossed-out 10 in Example A.

LESSON CONTINUES

Find the difference. Estimate to check.

2. 595 −242	**3.** 336 −191	**4.** 300 − 84	**5.** $4.00 −$2.83	**6.** 607 −349

▶ Practice and Problem Solving

Find the difference. Estimate to check.

7. 574 −412	**8.** 584 −250	**9.** 438 −119	**10.** 672 −468	**11.** $9.63 −$4.05
12. 294 −137	**13.** 891 − 86	**14.** $6.57 −$4.98	**15.** 372 −196	**16.** 730 −217
17. 800 −585	**18.** 506 −439	**19.** $7.00 −$2.11	**20.** 805 − 99	**21.** $9.06 −$4.08
22. 354 −148	**23.** 942 −817	**24.** 647 −435	**25.** 461 −178	**26.** 724 −536

Subtract. Use addition to check.

27. 308 − 149 = ■   **28.** 900 − 312 = ■   **29.** 604 − 485 = ■

30. 401 − 173 = ■   **31.** 304 − 255 = ■   **32.** 300 − 92 = ■

33. $\frac{a+b}{c}$ **Algebra** Write the missing addend. 255 + ■ = 305

35. Sherrie is thinking of a number. It is 128 less than 509. What number is she thinking of?

36. REASONING Would you have to regroup to find 255 − 125? Explain.

USE DATA For 37–38, use the graph.

37. How much less does Tim Turtle weigh than Talia Turtle?

38. How much do Tim, Ted, and Talia weigh altogether?

34. **? What's the Error?** Michael wrote a subtraction problem like this. Describe his error. Find the difference.

$$\begin{array}{r} {\scriptstyle 16} \\ 2\overset{}{6}1 \\ -170 \\ \hline 191 \end{array}$$

WEIGHTS OF TURTLES

Mixed Review and Test Prep

Choose <, >, or = for each ⬤. (p. 20)

39. 72 ⬤ 27   **40.** 321 ⬤ 213

41. 66   **42.** 148   **43.** 399
    −32 (p. 42) +134 (p. 42) +722

44. **TEST PREP** Find 95 + 37.

   A 58   **C** 217
   B 132   **D** 300

45. Carlos found 112 green bottles, 115 tomato cans, and 129 plastic cups. How many cups and cans did he find? (p. 42)

46. 182   **47.** 325   **48.** 456
(p. 42) + 28 (p. 42) +149 (p. 42) +344

49. **TEST PREP** Find the sum of 342 and 160. (p. 42)

   F 182   **H** 500
   G 402   **J** 502

LINKUP to Science

Turtles are the only reptiles with shells. Most turtles can pull their head, legs, and tail into their shells. The female turtle digs a hole on land, lays her eggs, and covers them. The heat from the sun hatches the eggs.

1. There are about 250 kinds of turtles. About 50 kinds live in the United States and in Canada. About how many kinds don't live in the United States and Canada?

2. One of the largest green sea turtles ever measured weighed 871 pounds. If a green sea turtle weighs 395 pounds, how much heavier is the largest green sea turtle?

3. If a flatback sea turtle swam 752 miles in the spring and 374 miles in the summer, how much farther did the turtle swim in the spring?

4. If a leatherback sea turtle traveled 989 miles one year and 873 miles the next year, how far did it travel in the two years?

Extra Practice page H35, Set B

Chapter 4 **61**

Problem Solving Skill
Estimate or Exact Answer

Understand → Plan → Solve → Check

Quick Review

1. $(3 + 2) + 1$
2. $4 + (5 + 5)$
3. $7 + (3 + 4)$
4. $2 + 3 + 4$
5. $32 + 6 + 14$

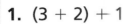

PLANNING AHEAD Alicia invited 9 boys, 19 girls, and 8 adults to a meeting with the park ranger. She must decide whether to hold the meeting in the classroom, the library, or the cafeteria.

Since she doesn't need to know exactly how many people are coming, she estimates.

$$
\begin{array}{rcr}
9 & \to & 10 \\
19 & \to & 20 \\
+\ 8 & \to & +10 \\
\hline
 & & 40
\end{array}
$$

MEETING ROOMS	
Place	**Number of Seats**
Classroom	25
Library	50
Cafeteria	100

So, she needs a room with about 40 seats. She can hold the meeting in the library.

What if Alicia must make a name tag for each person at the meeting? Alicia must decide how many name tags she needs to make.

Since she doesn't want to make too many or too few name tags, she finds an exact answer.

$$9 + 19 + 8 = 36$$

So, she must make 36 name tags.

Talk About It

• Why was an estimate good enough to decide which room to use?

• Why did Alicia need an exact answer to decide how many name tags to make?

CALIFORNIA STANDARDS MR 2.0 Students use strategies, skills, and concepts in finding solutions.
MR 2.5 Indicate the relative advantages of exact and approximate solutions to problems and give answers to a specified degree of accuracy. *also* NS 1.0, NS 2.0, O—π NS 2.1, MR 1.0, MR 1.1, MR 2.3, MR 2.4, MR 3.1, MR 3.2

Problem Solving Practice

USE DATA Use the table for 1–2. Write whether you need an exact answer or an estimate. Then solve.

SUPPLIES	
Item	**Price**
Glitter	$3.50
Glue	$1.79
Hole punch	$3.99

1. Roberta has $15. Can she buy glitter and a hole punch? Explain.

2. Lorenzo pays for glue with a $5 bill. How much change will he get?

Clarissa is planning a picnic for two scout troops. There are 29 scouts in one troop and 17 scouts in the other troop.

3. Clarissa must make a name tag for each scout. Which sentence shows how many name tags she must make?

   A $30 + 20 = 50$
   B $29 + 17 = 46$
   C $29 + 17 = 36$
   D $29 - 17 = 12$

4. A package of cupcakes holds 10 cupcakes. If Clarissa wants each scout to have at least 1 cupcake, how many packages should she buy?

   F 1   **H** 4
   G 2   **J** 5

Mixed Applications

5. Wesley has 4 more hockey cards than baseball cards. If he has 28 cards in all, how many hockey cards does he have?

6. Each person at a picnic needs about 2 cups of punch. There will be 19 girls and 29 boys at the picnic. About how many cups should Jeff make?

7. I am a number greater than 110 and less than 120. The sum of my digits is 9. What number am I?

8. Mitch has 43 bottle caps and 12 buttons. If he finds 12 more bottle caps, how many caps will he have in all?

9. **? What's the Question?** Last week Luann ran 50 miles and Patrice ran 15 miles. The answer is 35 miles.

10. **REASONING** Joel had 32 inches of ribbon. He cut 10 inches off each end of the ribbon. What is the length of the ribbon he has left?

Algebra: Expressions and Number Sentences

▶ **Learn**

LUNCH LINE In the morning, visitors bought 34 packets of food for the animals in the petting zoo. In the afternoon, visitors bought 58 packets. How many food packets were bought in all?

You can write an expression for this problem.

$$34 \text{ packets plus } 58 \text{ packets}$$
$$\downarrow \qquad\qquad \downarrow \quad \downarrow$$
$$34 \qquad\qquad + \quad 58$$

An **expression** is part of a number sentence. It combines numbers and operation signs. It does not have an equal sign.

$34 + 58 = 92$ is a number sentence.

92 is the number of food packets bought in all.

MATH IDEA A number sentence can be true or false.

$$4 + 3 = 7 \text{ is true.}$$
$$4 - 3 = 7 \text{ is false.}$$

Mike spent \$12 for a basketball and \$18 for a soccer ball. How much more did the soccer ball cost?

$$\$18 \ ● \ \$12 = \$6$$

Which symbol will make the sentence *true*?

Try + $\$18 + \$12 = \$6$ *Not* true.
Try − $\$18 - \$12 = \$6$ **True**.

So, the correct operation symbol is −.

Write an expression for each.

1. Takeo had 273 cards. He gave away 35. How many cards does he have left?

2. Mia had 13 apples. She bought 7 more. How many does she have in all?

Write + or − to make the number sentence true.

3. 12 ● 2 = 10   4. 37 ● 11 = 48   5. 126 ● 79 = 47   6. 367 ● 43 = 410

► **Practice and Problem Solving**

Write an expression for each.

7. Gwen bought 12 red pencils, 2 blue pencils, and 22 yellow pencils. How many blue and red pencils did she buy?

8. Ned has 17 crayons. He has 15 pens. How many more crayons than pens does he have?

Write + or − to make the number sentence true.

9. 4 ● 3 = 1   10. 28 ● 9 = 37   11. 329 ● 87 = 242

12. 559 ● 50 = 609   13. 74 ● 47 = 17 + 10   14. 444 ● 6 = 460 − 10

Write the missing number that makes the number sentence true.

15. ■ + 3 = 14   16. 140 + 5 = ■   17. 45 − ■ = 25

18. 309 − ■ = 209   19. 215 − ■ = 120   20. ■ − 125 = 318

21. **REASONING** Blair says, "12 + 3 + 1 = 5 + 11 is a true number sentence." Do you agree or disagree? Explain.

22. Selma wrote 18 > 27. Is this true? If not, rewrite her sentence to make it true.

Mixed Review and Test Prep

Find each sum. (p. 36)

23.
```
  64
  91
+26
```
24.
```
  15
  73
+22
```
25.
```
  41
  66
+19
```

26. Find 1,035 − 887. (p. 62)

27. (**TEST PREP**) Benjy had 62 marbles. He gave 15 to Lisa, 14 to Lenny, and 9 to Jake. He kept the rest. Who has the most marbles?

A Lisa   C Benjy

B Lenny   D Jake

Review/Test

✔ CHECK VOCABULARY AND CONCEPTS

Choose the correct term from the box.

1. A part of a number sentence that combines numbers and operation signs is an ? . (p. 68)

> estimate
> expression

Tell whether you need to regroup to find each difference. Explain. (pp. 56–57)

2.	342	3.	312	4.	162
	−214		−181		− 51

✔ CHECK SKILLS

Estimate the difference. (pp. 54–55)

5.	67	6.	967	7.	748	8.	4,175	9.	8,596
	−29		−283		−599		−1,832		−3,714

Find the difference. (pp. 58–65)

10. $341 - 133 = $ ■

11. $837 - 247 = $ ■

12. $\$3.73 - \$2.08 = $ ■

13. $645 - 347 = $ ■

14.	$4.02	15.	800	16.	602	17.	200	18.	703
	−$2.56		−364		−531		− 95		−155

19.	4,250	20.	9,308	21.	8,029	22.	7,000	23.	5,789
	−2,872		−5,970		−6,047		−1,074		− 898

✔ CHECK PROBLEM SOLVING

24. Abe wants each person at his party to have about 1 cup of punch. If he invites 18 children and 9 adults, about how many cups of punch should he make? (pp. 66–67)

25. Barry collected 29 cans on Monday, 12 cans on Tuesday, and 17 cans on Friday. Write an expression to show how to find how many cans Barry collected. (pp. 68–69)

Cumulative Review

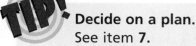
Decide on a plan.
See item **7.**

The question asks how many in all. Look for an expression that shows the correct operation for putting groups together.

Also see problem **4,** p. H64.

For 1–10, choose the best answer.

1. Which two numbers have a difference of about 500?

   A 647 and 374
   B 521 and 110
   C 782 and 367
   D 659 and 188

2. 528 − 265 = ▇

   F 263   **H** 363
   G 343   **J** 793

3. Tara is thinking of a number. It is 294 less than 638. What is the number?

   A 344   **C** 454
   B 364   **D** 932

4. What is 4,612 rounded to the nearest thousand?

   F 3,000   **H** 5,000
   G 4,000   **J** 6,000

5. Joan's family traveled 4,502 miles on their vacation this year and 2,810 miles last year. How many more miles did they travel this year than last year?

   A 1,392 miles   **C** 2,612 miles
   B 1,692 miles   **D** 7,312 miles

6. What is the sum of 539 and 649?

   F 1,177   **H** 1,187
   G 1,178   **J** 1,188

7. Jay's family collects picture postcards. They have 147 cards from the United States and 64 cards from other countries. Which expression tells how many postcards they have in all?

   A 147 − 64
   B 147 + 147
   C 147 + 64
   D 64 − 64

8. Natalie bought a pen for 37 cents, a notebook for 49 cents, and an eraser for 10 cents. How much did she spend in all?

   F 86¢   **H** 96¢
   G 88¢   **J** 98¢

9.   6,003
   −2,336
   ─────

   A 3,667   **C** 4,773
   B 3,763   **D** NOT HERE

10. A number is odd and the sum of its digits is 15. The number is greater than 80 and less than 96. What is the number?

   F 86   **H** 89
   G 87   **J** NOT HERE

MATH DETECTIVE

Putting It Together

For each case, copy the squares on grid paper. Then fill in the missing numbers.

Remember
A hundred chart has numbers from 1 to 100 arranged in ten rows and ten columns.

Case 1

1. If the hundred chart were cut apart into the pieces shown, what numbers would go in the blank squares?

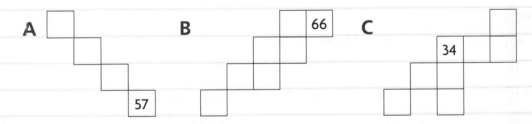

A ▢

B 66

C 34

57

Case 2

2. What if you made a two hundred chart by showing twenty rows and ten columns. What numbers would go in the blank squares?

A ▢ ▢
127

B 131

C

146

STRETCH YOUR THINKING If this square were cut out of a hundred chart, how would you use the value of the number in box E to figure out the other numbers?

A	B	C
D	E	F
G	H	I

CASE CLOSED

Challenge

Understand 100,000

The distance from Earth to the moon is about 238,857 miles. Use a place-value chart to show the value of each digit in this number.

238,857 miles

Hundred Thousands	Ten Thousands	Thousands	Hundreds	Tens	Ones
2	3	8,	8	5	7

Standard form: 238,857

Expanded form:
200,000 + 30,000 + 8,000 + 800 + 50 + 7

Word form:
two hundred thirty-eight thousand, eight hundred fifty-seven

Talk About It

• What is the value of the 2 in the number 238,857?

Try It

Write in standard form.

1. 300,000 + 20,000 + 6,000 + 700 + 40 + 4

2. 800,000 + 5,000 + 400 + 90 + 2

3. seven hundred forty-six thousand, one hundred eighteen

4. five hundred twelve thousand, twenty-two

Write in expanded form.

5. 312,456

6. 605,127

Write the value of the blue digit.

7. 312,876   8. 897,546   9. 185,254   10. 302,170

Study Guide and Review

STUDY AND SOLVE

Chapter 1

Identify numbers.

Even numbers end in 0, 2, 4, 6, or 8.
Odd numbers end in 1, 3, 5, 7, or 9.

Understand place value.

Ten Thousands	Thousands	Hundreds	Tens	Ones
4	2,	1	0	5

Standard form: 42,105
Expanded form:
40,000 + 2,000 + 100 + 5
Word form:
forty-two thousand, one hundred five

Tell whether the number is *odd* or *even*. (pp. 2-3)

1. 9   **2.** 23   **3.** 40

Write in standard form. (pp. 4-11)

4. seven hundred twenty-eight

5. two hundred eighty

6. 3,000 + 200 + 20 + 5

7. 10,000 + 2,000 + 500 + 20 + 2

Write in expanded form. (pp. 4-11)

8. 4,542   **9.** 71,061

Chapter 2

Compare and order numbers.

Write in order from greatest to least.
2,761; 1,793; 5,219

Compare thousands.
5 > 2 > 1

5,219; 2,761; 1,793

Write <, >, or = for each ●. (pp. 20-23)

10. 739 ● 728   **11.** 461 ● 461

12. 3,125 ● 452   **13.** 1,203 ● 1,209

Write in order from greatest to least. (pp. 24-25)

14. 423, 432, 417

15. 9,005; 5,009; 5,010

Round numbers.

Round 4,483 to the nearest thousand.
4,483 is between 4,000 and 5,000.
It is closer to 4,000.
So, 4,483 rounded to the nearest
thousand is 4,000.

Round to the nearest hundred. (pp. 28-29)

16. 144   **17.** 653   **18.** 247

Round to the nearest thousand. (pp. 30-31)

19. 3,609   **20.** 1,289

Chapter 3

Add two or more numbers.

$6 + (3 + 1) = 10$ and $(6 + 3) + 1 = 10$

Find the sum. (pp. 36-37)

21. $(5 + 3) + 8 = \blacksquare$

22. $(4 + 6) + 3 = \blacksquare$

Add 3- and 4-digit numbers.

$$\begin{array}{r} 1 \\ 437 \\ +155 \\ \hline 592 \end{array}$$

$$\begin{array}{r} 1\ 1 \\ 3{,}987 \\ +2{,}532 \\ \hline 6{,}519 \end{array}$$

Find the sum. Estimate to check.

(pp. 42-43, 46-49)

23. $\begin{array}{r} 192 \\ +432 \end{array}$ **24.** $\begin{array}{r} 643 \\ +289 \end{array}$ **25.** $\begin{array}{r} 534 \\ +846 \end{array}$

26. $\begin{array}{r} 4{,}276 \\ +1{,}071 \end{array}$ **27.** $\begin{array}{r} 2{,}008 \\ +6{,}439 \end{array}$ **28.** $\begin{array}{r} 5{,}976 \\ +8{,}668 \end{array}$

Chapter 4

Estimate differences.

$$\begin{array}{r} 689 \rightarrow 700 \\ -408 \rightarrow -400 \\ \hline 300 \end{array}$$

• Round to the nearest hundred.

Estimate the difference. (pp. 54-55)

29. $\begin{array}{r} 58 \\ -19 \end{array}$ **30.** $\begin{array}{r} 311 \\ -196 \end{array}$ **31.** $\begin{array}{r} 3{,}032 \\ -1{,}114 \end{array}$

Subtract 3- and 4-digit numbers.

$$\begin{array}{r} 8\,13 \\ 8\,\cancel{9}\,\cancel{3} \\ -5\,0\,8 \\ \hline 3\,8\,5 \end{array}$$

$$\begin{array}{r} 9 \\ 2\ \cancel{10}10 \\ \cancel{3}{,}0\,\cancel{0}\,\cancel{0} \\ -1{,}6\,5\,0 \\ \hline 1{,}3\,5\,0 \end{array}$$

Subtract. (pp. 58-65)

32. $\begin{array}{r} 562 \\ -313 \end{array}$ **33.** $\begin{array}{r} 430 \\ -287 \end{array}$ **34.** $\begin{array}{r} 406 \\ -\ 89 \end{array}$

35. $\begin{array}{r} 6{,}314 \\ -2{,}509 \end{array}$ **36.** $\begin{array}{r} 2{,}006 \\ -1{,}734 \end{array}$ **37.** $\begin{array}{r} 3{,}508 \\ -2{,}779 \end{array}$

PROBLEM SOLVING PRACTICE

Solve. (pp. 12-13, 44-45)

38. I am a number less than 100. My ones digit is 6. The sum of my digits is 11. What number am I?

39. Susan has 10 more goldfish than Gary. Together, they own 50 goldfish. How many goldfish does each have?

California Connections

SQUAW VALLEY USA

The 1960 Olympic Winter Games were held at Squaw Valley USA, in California's Sierra Nevada.

▲ Squaw Valley USA, in the Sierra Nevada is one of the largest ski areas in the United States.

USE DATA For 1–5, use the table.

PEAKS AT SQUAW VALLEY	
Peak	**Height (in feet)**
Broken Arrow	8,020
Emigrant	8,700
Granite Chief	9,050
KT-22	8,200
Snow King	7,550
Squaw Peak	8,900

1. To the nearest hundred feet, about how high is Snow King?

2. To the nearest thousand feet, about how high is Emigrant?

3. Which mountain peak is higher, Broken Arrow or KT-22? How many feet higher is it?

4. Order these peaks from lowest to highest: Emigrant, Granite Chief, Snow King, Squaw Peak.

5. **REASONING** Mt. Whitney is the highest point in the United States outside of Alaska. The 14,494-foot peak is also in the Sierra Nevada. How many feet higher is Mt. Whitney than the highest peak in Squaw Valley USA?

76

SKIING IN THE SIERRA NEVADA

Some of the slopes at Squaw Valley USA overlook Lake Tahoe and the Sierra Nevada.

◄ At Squaw Valley USA, there are 30 lifts that can carry 49,000 skiers per hour.

1. During the first half hour that the lifts were open, one ski lift carried 359 skiers and another lift carried 297 skiers. How many skiers did the two lifts carry in all?

2. During one hour, the main lift on KT-22 carried 730 skiers and the main lift on Granite Chief carried 492 skiers. How many more skiers rode the KT-22 lift?

3. One afternoon all the ski lifts together carried 45,391 skiers. To the nearest thousand, about how many skiers did the lifts carry?

4. In December, 43 inches of snow fell at one ski area. Machines made 28 more inches of snow. How much snow was there altogether in December?

5. **REASONING** Joel teaches two morning ski classes with 12 skiers in each class. He teaches three afternoon ski classes, one with 9 skiers, one with 14 skiers, and one with 15 skiers. How many skiers does Joel teach during the day? Explain.

Use Money

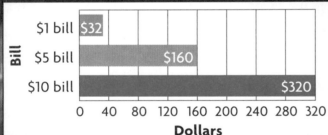

VALUE OF 1 SHEET OF PRINTED CURRENCY

Bill	Dollars
$1 bill	$32
$5 bill	$160
$10 bill	$320

0 40 80 120 160 200 240 280 320

Dollars

In one hour, presses at the Bureau of Engraving and Printing can print 8,000 sheets of currency. Each sheet contains 32 bills. That means that 256,000 bills are printed every hour. The graph shows how much one sheet of each type of bill is worth. How much are two sheets of each bill worth?

CHECK WHAT YOU KNOW ✓

Use this page to help you review and remember
important skills needed for Chapter 5.

✓ VOCABULARY

Choose the best term from the box.

| quarter |
| dime |
| nickel |

1. The value of a ? is $0.10.

2. The value of a ? is $0.25.

✓ MONEY: COUNT BILLS AND COINS (See p. H7.)

Count and write the amount.

3.

4.

5.

6.

7. 2 quarters, 1 dime,
   3 nickels, and 2 pennies

8. 1 quarter, 3 dimes,
   2 nickels, and 4 pennies

List the coins you would use to make each amount.

9. 39¢   10. 72¢   11. 68¢   12. 46¢

Count and write the amount. List the coins you would
use to make the same amount with the fewest coins.

13.

14.

HANDS ON
Make Equivalent Sets

Quick Review

List the coins to make each amount.

1. 11¢   **2.** 35¢   **3.** 55¢
4. 75¢   **5.** 80¢

VOCABULARY
equivalent

MATERIALS
play bills and coins

▶ Explore

Sets of money that have the same value are **equivalent**.

Examples A, B, C, and D are equivalent sets. Each set of coins has a value of 10¢.

Examples

A   B   C   D

Use play money to make equivalent sets.

This set of coins has a value of $1.25.

$0.25 $0.50 $0.75 $1.00 $1.25

• Make at least three equivalent sets with a value of $1.25. One set should use the fewest bills and coins. Draw pictures to record each set.

> What are two other ways we could show $2.40?

Try It

Find two equivalent sets for:
a. $2.40   **b.** $1.85

CALIFORNIA STANDARDS MR 1.1 Analyze problems by identifying relationships, distinguishing relevant from irrelevant information, sequencing and prioritizing information, and observing patterns. **MR 2.3** Use a variety of methods, such as words, numbers, symbols, charts, graphs, tables, diagrams, and models, to explain mathematical reasoning. *also* **O—n NS 3.3, AF 2.0, MR 1.0, MR 1.2, MR 2.4**

Make two equivalent sets with a value of $6.13.
The first set uses the fewest bills and coins.

THINK:		NOW I HAVE:
one $5 bill	→	$5.00
plus one $1 bill	→	$6.00
plus 1 dime	→	$6.10
plus 3 pennies	→	$6.13

THINK:		NOW I HAVE:
six $1 bills	→	$6.00
plus 2 nickels	→	$6.10
plus 3 pennies	→	$6.13

MATH IDEA You can make equivalent sets of money by using different combinations of bills and coins.

TECHNOLOGY LINK

More Practice:
Use E-Lab, *Equivalent Sets of Coins.*
www.harcourtschool.com/elab2000

▶ **Practice**

Make two equivalent sets for each amount.
List the bills and coins you used.

1. $1.35   **2.** $5.50   **3.** $2.46   **4.** $6.92

5. $3.75   **6.** $5.03   **7.** $2.25   **8.** $8.04

9. If Rosa has 5 nickels, how many pennies could she trade them for? Copy and complete the table.

nickels	1	2	3	4	5
pennies	5	■	■	■	■

10. **REASONING** Fiona has 2 quarters and 1 nickel. Jake has an equivalent amount in dimes and nickels. Jake has 8 coins. How many dimes and nickels does he have?

11. List the fewest bills and coins you can use to make $2.37.

Mixed Review and Test Prep

Round to the nearest hundred. (p. 28)

12. 87   **13.** 267   **14.** 142

15. Which digit is in the thousands place of 5,723? (p. 6)

16. **TEST PREP** Which digit is in the hundreds place of 6,295? (p. 6)

A 2   **C** 6
B 5   **D** 9

Problem Solving Strategy
Make a Table

Understand ➡ Plan ➡ Solve ➡ Check

Quick Review

1. 2 quarters = ■ pennies

2. 10 dimes = ■ quarters

3. 4 nickels = ■ cimes

4. 1 quarter = ■ nickels

5. 10 pennies = ■ nickels

PROBLEM Aretha has four $1 bills, 3 quarters, 5 dimes, 1 nickel, and 5 pennies. How many different equivalent sets of bills and coins can she use to pay for a magazine that costs $4.75?

Understand

• What are you asked to find?

• What information will you use?

Plan

• What strategy can you use?

   You can *make a table* to find sets of bills and coins with a value of $4.75.

Solve

• How can you use the strategy to solve the problem?

   Make a table to show equivalent sets of money.

$1 BILLS	QUARTERS	DIMES	NICKELS	PENNIES	VALUE

4	3				$4.75
4	2	2	1		$4.75
4	2	2		5	$4.75
4	1	5			$4.75
4	1	4	1	5	$4.75

So, there are 5 equivalent sets.

Check

• How can you decide if your answer is correct?

CALIFORNIA STANDARDS MR 2.3 Use a variety of methods, such as words, numbers, symbols, charts, graphs, tables, diagrams, and models, to explain mathematical reasoning. **MR 1.0** Students make decisions about how to approach problems. *also* **NS 2.0, ⚬┐ NS 3.3, ⚬┐ AF 1.1, MR 1.1, MR 2.0, MR 2.4, MR 2.6**

Problem Solving Practice

PROBLEM SOLVING STRATEGIES

Draw a Diagram or Picture
Make a Model or Act It Out
Make an Organized List
Find a Pattern
▶ **Make a Table or Graph**
Predict and Test
Work Backward
Solve a Simpler Problem
Write an Equation
Use Logical Reasoning

Make a table to solve.

1. **What if** Aretha's magazine costs $5.25? How many different equivalent sets of bills and coins can she use?

2. Tyler has one $1 bill, 5 quarters, 1 dime, and 2 nickels. How many ways can he pay for a goldfish that costs $1.35?

Kevin has 7 quarters, 4 dimes, and 1 nickel. He wants to buy a bookmark that costs $1.80.

3. Kevin wants to keep 1 quarter. Which set of coins should he use?

   A 6 quarters, 2 dimes, 1 nickel
   B 6 quarters, 3 dimes
   C 7 quarters, 1 nickel
   D 6 quarters, 3 dimes, 1 nickel

4. If Kevin uses the fewest coins, which type of coin will he NOT use?

   F quarters
   G dimes
   H nickels
   J none of the above

Mixed Strategy Practice

USE DATA For 5–7, use the table.

5. Laura has one $5 bill, four $1 bills, 7 quarters, 2 dimes, 2 nickels, and 4 pennies. How many different ways can she pay for the flashlight?

6. Paco has only quarters and nickels in his pocket. If he uses 9 coins to buy the can opener, what coins does he use?

Camping Equipment	
Flashlight	$5.99
Canteen	$4.65
Can Opener	$1.05
Bug Spray	$2.49

7. Fran, Geri, Harold, and Ivan each buy a different item. Use the clues to decide what each person buys.

   Fran pays with one $1 bill and 1 nickel. Ivan pays without using pennies. Geri pays with three $1 bills.

8. **Write About It** Betty has three $1 bills, 5 quarters, 7 dimes, and 2 nickels. Explain how Betty can trade some of her bills and coins for a $5 bill.

Compare Amounts of Money

▶ **Learn**

MONEY MATTERS Ming and Ben have these sets of bills and coins. Who has more money?

Count each amount and compare.

Ming has $5.75.
Ben has $5.50.

$5.75 > $5.50.
So, Ming has more money.

Ming's money

 Ben's money

Examples Compare. Which amount is greater?

A

Since $2.73 = $2.73, the amounts are equal.

B

Since $3.54 < $4.12, then $4.12 is the greater amount.

MATH IDEA To compare amounts of money, count each set and decide if one is greater than, less than, or equal to the other.

• **REASONING** Is a set of bills and coins always worth more than a set that has fewer bills and coins? Explain.

CALIFORNIA STANDARDS ⊶ **AF 1.1** Represent relationships of quantities in the form of mathematical expressions, equations, or inequalities. **AF 1.3** Select appropriate operational and relational symbols to make an expression true. *also* **AF 1.0, MR 1.0, MR 1.1, MR 2.3, MR 2.4**

► Check

1. **Explain** how you can use what you know about comparing whole numbers to compare amounts of money.

Use > or < to compare the amounts of money.

2. a.

   b.

► Practice and Problem Solving

Use > or < to compare the amounts of money.

3. a.

   b.

4. a.

   b.

5. a.

   b.

6. Setsuo sells lemonade for 25¢ a glass. He has 9 quarters, 6 dimes, and 3 nickels. How many glasses of lemonade did he sell? Draw a picture to explain.

7. **? What's the Error?** Janice says that $4.87 is greater than $6.21 because 87 cents is greater than 21 cents. Describe her error. Explain which is greater.

Mixed Review and Test Prep

8. (p. 36)
$$\begin{array}{r} 21 \\ 98 \\ +45 \\ \hline \end{array}$$

9. (p. 36)
$$\begin{array}{r} 18 \\ 24 \\ +42 \\ \hline \end{array}$$

10. (p. 42)
$$\begin{array}{r} 256 \\ +148 \\ \hline \end{array}$$

11. Describe and continue the pattern.
13, 23, 33, 43, ■, ■, ■ (p. 12)

12. **TEST PREP** What is the value of the blue digit? 15,271 (p. 10)

A 5   **B** 50   **C** 500   **D** 5,000

HANDS ON
Make Change

▶ Explore

Jessica buys a kitty toy at Pal's Pet Store. She pays with a $1 bill. How much change will she get?

To find Jessica's change, **count on** from the cost of the kitty toy to the amount paid.

| $0.77 | $0.78 | $0.79 | $0.80 | $0.90 | $1.00 |

4 pennies and 2 dimes equal $0.24.
So, Jessica will get $0.24 in change.

Try It

Each person pays with a $1 bill. Use play money to make change. Draw a picture to show the change each person will get.

a. Tony buys a bird seed bell.

b. Marian buys fish food. Show her change, using the fewest coins.

c. Emma buys a chew bone. Show at least two different ways to make change.

• Why does it help to count on with pennies first when making change for $0.63 from a $1 bill?

MATERIALS
play coins

PET SUPPLIES

Dog Leash	$5.99
Dog Shampoo	$3.68
Kitty Toy	$0.76
Fish Food	$0.89
Chew Bone	$0.59
Bird Seed Bell	$0.63

$0.64, $0.65, $0.75 . . .
What should I count next
to make change?

86

CALIFORNIA STANDARDS O—n NS 3.3 Solve problems involving addition, subtraction, multiplication, and division of money amounts in decimal notation and multiply and divide money amounts in decimal notation by using whole-number multipliers and divisors. *also* **MR 1.0, MR 2.2, MR 2.4, MR 3.0, MR 3.2**

▶ Connect

Dog shampoo costs $3.68. Anton pays with a $5 bill. How much change will he get?

You can count on from the cost of the dog shampoo to the amount paid.

| $3.69 | $3.70 | $3.75 | $4.00 | $5.00 |

2 pennies, 1 nickel, 1 quarter, and one $1 bill equal $1.32. So, Anton will get $1.32 in change.

TECHNOLOGY LINK

Mighty Math
More Practice: Use Mighty Math Calculating Crew, *Superhero Superstore* Levels B and C.

▶ Practice

Copy and complete the table. Use play money.

	COST OF ITEM	AMOUNT PAID	CHANGE IN COINS AND BILLS	TOTAL AMOUNT OF CHANGE
1.	$0.54	$1.00	■	■
2.	$3.23	$5.00	■	■
3.	$2.69	$3.00	■	■

4. REASONING Dana buys a rock for her fish tank for $0.65. She pays with a $1 bill. The clerk doesn't have any quarters for change. What is the least number of coins Dana can get? List the coins.

5. ? What's the Question? Evan bought a dog bowl for $2.85. He paid with a $5 bill. The answer is $2.15 in change.

Mixed Review and Test Prep

6. 375 (p. 42)
+499

7. 4,359 (p. 62)
−2,276

8. Order 2,743; 987; and 8,143 from least to greatest. (p. 24)

9. What is one hundred more than 3,420? (p. 6)

10. TEST PREP What is the standard form of seven thousand, nine hundred twenty-three? (p. 6)

A 7,092   C 7,923
B 7,903   D 70,923

Add and Subtract Money

Quick Review

1. 863
+219

2. 673
−482

3. 457
+361

4. 1,073
− 845

5. 920
+549

▶ **Learn**

CHECK YOUR CHANGE Ryan bought a dog collar for $3.95 and a leash for $4.64. How much money did Ryan spend?

Add. $3.95 + $4.64 = ■

STEP 1	STEP 2	STEP 3
Estimate the sum. Round to the nearest dollar.	Add money amounts as you add whole numbers.	Write the sum in dollars and cents.
$3.95 → $4.00 +$4.64 → +$5.00 $9.00	$3.95 → ¹395 +$4.64 → +464 859	$3.95 +$4.64 $8.59

So, Ryan spent $8.59. Since $8.59 is close to $9.00, the answer is reasonable.

Ryan paid for the collar and leash with a $10 bill. How much change should he get?

Subtract. $10.00 − $8.59 = ■

STEP 1	STEP 2	STEP 3
Estimate the difference. Round to the nearest dollar.	Subtract money amounts as you subtract whole numbers.	Write the difference in dollars and cents.
$10.00 → $10.00 −$ 8.59 → −$ 9.00 $1.00	9 9 0 10 10 10 $10.00 → 1,000 −$ 8.59 → − 859 1 4 1	$10.00 −$ 8.59 $ 1.41

So, Ryan should get $1.41 in change. $1.41 is close to $1.00, so the answer is reasonable.

CALIFORNIA STANDARDS O➞NS 3.3 Solve problems involving addition, subtraction, multiplication, and division of money amounts in decimal notation and multiply and divide money amounts in decimal notation by using whole-number multipliers and divisors. *also* NS 2.0, NS 2.8, AF 1.0, AF 1.3, MR 2.1, MR 2.4, MR 2.6, MR 3.1

▶ Check

1. **Explain** how you can check the subtraction to be sure Ryan got the correct change.

Find the sum or difference. Estimate to check.

2. $1.45
    +$2.32

3. $3.67
    +$1.19

4. $4.83
    +$2.51

5. $1.76
    −$0.45

▶ Practice and Problem Solving

Find the sum or difference. Estimate to check.

6. $2.63
    +$1.24

7. $4.55
    +$0.38

8. $4.64
    −$1.80

9. $1.85
    −$0.73

10. $3.52
     −$2.34

11. $8.26
     +$4.81

12. $7.69
     +$5.95

13. $10.00
     −$ 5.25

14. $4.28 + $2.59 = ▇

15. $6.72 − $3.94 = ▇

16. $3.26 − $1.09 = ▇

$\frac{a+b}{c}$ **Algebra** Write <, >, or = for each ●.

17. $5.00 ● $3.94 + $1.06

18. $4.57 − $1.14 ● $5.71

USE DATA For 19–21, use the table.

19. How much more does the kitty perch cost than the cat bed?

20. Do a cat bed and a bowl cost more or less than a kitty perch and a ball? Explain.

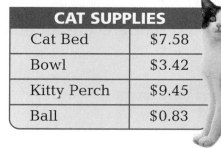

CAT SUPPLIES	
Cat Bed	$7.58
Bowl	$3.42
Kitty Perch	$9.45
Ball	$0.83

21. **Estimation** Tara has $5.00. Does she have enough money for a bowl and a ball? Explain.

22. **? What's the Error?** Roberto wrote $7.40 + $5.60 = $1.80. Describe his error. What should the sum be?

Mixed Review and Test Prep

Write <, >, or = for each ●. (p. 20)

23. 127 ● 171   24. 682 ● 659

25. Round 463 to the nearest hundred. (p. 28)

26. 723 − 581 = ▇ (p. 58)

27. **TEST PREP** Which shows the numbers in order from least to greatest? (p. 24)

A 481, 419, 527   C 521, 518, 509
B 519, 531, 656   D 704, 698, 549

Extra Practice page H36, Set B

Review/Test

✓ CHECK VOCABULARY AND CONCEPTS

Choose the best term from the box.

1. Sets of money that have the same value are ___?___ .
   (p. 80)

> equivalent
>
> change

Find two equivalent sets for each amount. List the bills and coins you can use. (pp. 80–81)

2. $0.78   **3.** $3.65   **4.** $5.17   **5.** $8.42

Copy and complete the table. (pp. 86–87)

	COST OF ITEM	AMOUNT PAID	CHANGE IN COINS AND BILLS	AMOUNT OF CHANGE
6.	$0.62	$1.00	■	■
7.	$3.49	$5.00	■	■

✓ CHECK SKILLS

Use > or < to compare the amounts of money. (pp. 84–85)

8. a.     **b.**

Find the sum or difference. (pp. 88–89)

9. $2.46   **10.** $6.39   **11.** $6.74   **12.** $8.00   **13.** $9.05
      +$3.37   −$1.81   +$4.46   −$4.72   −$2.88

✓ CHECK PROBLEM SOLVING

Solve. (pp. 82–83)

14. Michelle has two $1 bills, 5 quarters, 3 dimes, and 4 nickels. How many ways can she make $3.50?

15. Brian has one $5 bill, one $1 bill, 5 quarters, 3 dimes, and 3 nickels. How many ways can he make $7.35?

Cumulative Review

TIP! Get the information you need.
See item **4**.

Before you can find which is greatest, you need to find the amounts to compare. After you find the amount in cents for each answer choice, you can compare and order them.

Also see problem **3**, p. H63.

For 1–10, choose the best answer.

1. Which set of coins is worth 76 cents?

A 3 quarters, 1 nickel, 1 penny
B 2 quarters, 3 dimes, 6 pennies
C 3 quarters, 1 penny
D 7 dimes, 5 pennies

2. How many equivalent sets with a value of 21 cents can be made?

F 3　　　　H 5
G 4　　　　J More than 5

3. Jonathan wants to buy a book that costs $1.95. Which set of bills and coins makes exactly $1.95?

A $1 bill, 3 quarters, 2 dimes
B $1 bill, 9 quarters, 5 nickels
C $1 bill, 5 dimes, 6 nickels
D $1 bill, 5 dimes, 2 nickels

4. Which is the greatest amount of money?

F 3 quarters, 5 dimes, 1 nickel
G 3 quarters, 3 dimes, 3 nickels
H 4 quarters, 3 nickels
J 5 quarters

5. What is the sum of 762 and 498?

A 1,150　　　C 1,250
B 1,160　　　D 1,260

6. Barry bought some candy for $3.47. He paid with a $5 bill. How much change did he get?

F $1.53　　　H $1.67
G $1.63　　　J $2.47

7. Find the sum.

$8.45
+$3.87

A $11.22　　　C $12.32
B $11.32　　　D NOT HERE

8. Emma has $9.32. Jenelle has $6.65. How much more money does Emma have?

F $2.63　　　H $3.33
G $2.67　　　J NOT HERE

9. Mindy's school has collected 872 cans for the food drive. What is that number rounded to the nearest hundred?

A 700　　　C 900
B 800　　　D 1,000

10. Tara practiced dancing for 45 minutes. She talked on the phone for 15 minutes and read for 22 minutes. How much time did she spend on these activities?

F 80 minutes　　H 85 minutes
G 82 minutes　　J 87 minutes

Understand Time

Living things grow at different rates. The time it takes flowers to grow depends on many things such as temperature, soil, and light. Use the flower chart to compare the time it takes some flowers to sprout and to bloom. Which flower sprouts the fastest? Which flower takes the longest to bloom?

GROWTH TIMES OF FLOWERS		
Flower	Days to Sprout	Days to Bloom
Black-Eyed Susan	10–15	60–90
Snapdragon	8–14	55–100
Zinnia	5–10	35–60
Sunflower	10–18	120
Petunia	4–10	45–80
Marigold	7–18	70
Shasta Daisy	14–21	180–300

Flower clock

CHECK WHAT YOU KNOW

Use this page to help you review and remember
important skills needed for Chapter 6.

✔ VOCABULARY

Choose the best term from the box.

1. The short hand on the clock is the _?_ .

2. The long hand on the clock is the _?_ .

> time
> minute hand
> hour hand

✔ TELL TIME (See p. H8.)

Read and write the time.

3.

4.

5.

6.

7.

8.

9.

10.

11.

✔ CALENDAR (See p. H8.)

For 12–14, use the calendar.

12. There are _?_ Fridays in November.

13. The second Monday in November is _?_ .

14. The third Wednesday in November is _?_ .

November						
Sun	Mon	Tue	Wed	Thu	Fri	Sat
					1	2
3	4	5	6	7	8	9
10	11	12	13	14	15	16
17	18	19	20	21	22	23
24	25	26	27	28	29	30

HANDS ON
Time to the Minute

▶ **Explore**

The hands, numbers, and marks on a clock help you tell what time it is.

VOCABULARY

minute

MATERIALS

clocks with movable hands

In five minutes, the minute hand moves from one number to the next.

In one **minute**, the minute hand moves from one mark to the next.

To find the number of minutes after the hour, count by fives and ones to where the minute hand is pointing.

5 minutes
10 minutes
15 minutes
20 minutes
25 minutes
26 minutes

Remember

minute hand   hour hand

In one *hour*, the hour hand moves from one number to the next. There are 60 minutes in 1 hour.

Read: nine twenty-six or 26 minutes after nine

Write: 9:26

• Show 7:42 and 9:07 on your clock.

Try It

Read and write each time.

a.

b.

Show each time on your clock. Then read each time in two ways.

c.  `5:32`

d.  `10:46`

CALIFORNIA STANDARDS MG 1.4 Carry out simple unit conversions within a system of measurement.
MR 2.3 Use a variety of methods, such as words, numbers, symbols, charts, graphs, tables, diagrams, and models, to explain mathematical reasoning. *also* **MR 2.0, MR 2.4**

▶ Connect

When a clock shows 31 or more minutes *after* the hour, you can read the time as a number of minutes *before* the next hour.

Count back by fives and ones to where the minute hand is pointing.

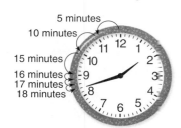

5 minutes
10 minutes
15 minutes
16 minutes
17 minutes
18 minutes

Read: 18 minutes before two

Write: 1:42

TECHNOLOGY LINK

More Practice:
Use E-Lab, *Time to the Minute.*

www.harcourtschool.com/
elab2002

▶ Practice

Read and write each time.

1.

2.

3.

4.

Show each time on your clock. Then write two ways you can read each time.

5.

1:20

6.

11:13

7.

7:52

8.

4:37

9. **REASONING** Does it take about 1 minute or about 5 minutes to tie your shoe? to make your lunch?

10. **REASONING** How many minutes are in 2 hours? Explain.

Mixed Review and Test Prep

11.   600 (p. 58)
    −343

12.   3,191 (p. 46)
    + 980

Round to the nearest thousand. (p. 30)

13. 5,439

14. 3,572

15. **TEST PREP** What is the value of the 2 in 56,297? (p. 10)

A 2   C 200

B 20   D 2,000

A.M. and P.M.

▶ **Learn**

IT'S ABOUT TIME Using A.M. and P.M. helps you know what time of the day or night it is. Times from midnight to noon are in the A.M. Times from noon to midnight are in the P.M.

12:00 in the day is **noon**.
12:00 at night is **midnight**.

Here are some ways to read and write times.

quarter to midnight
eleven forty-five P.M.
11:45 P.M.

quarter past seven
seven fifteen A.M.
7:15 A.M.

half past three
three thirty P.M.
3:30 P.M.

MATH IDEA The hours between midnight and noon are A.M. hours. The hours between noon and midnight are P.M. hours.

▶ **Check**

1. **Name** something you do at 9:00 A.M. and something you do at 9:00 P.M. Explain how you know these times are not the same.

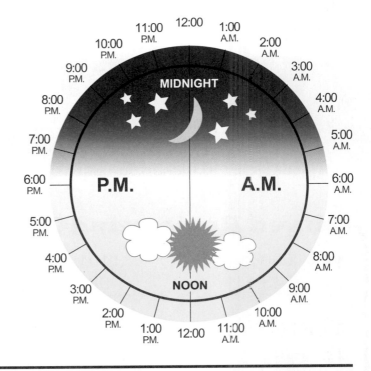

CALIFORNIA STANDARDS MR 2.3 Use a variety of methods, such as words, numbers, symbols, charts, graphs, tables, diagrams, and models, to explain mathematical reasoning. **MR 2.0** Students use strategies, skills, and concepts in finding solutions. *also* **MR 1.1, MR 2.4, MR 3.3**

Write the time, using A.M. or P.M.

2.

school starts

3.

eat lunch

4.

do homework

5.

6:30

sun sets

▶ Practice and Problem Solving

Write the time, using A.M. or P.M.

6.

get ready for school

7.

go to the store

8.

recess

9.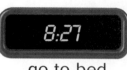

8:27

go to bed

Write two ways you can read each time.
Then write the time, using A.M. or P.M.

10.

play softball

11.

moon shines

12.

sun rises

13.

6:15

eat dinner

14. **? What's the Error?** Ty says that 11:45 A.M. is close to midnight. Explain his error. Then give a time that is close to midnight.

15. **Write About It** List three things you do at both A.M. and P.M. times. Write both times for each thing.

Mixed Review and Test Prep

Write + or − to make the sentence true. (p. 68)

16. 5 ● 3 = 2   **17.** 15 ● 9 = 24

18. 129 ● 5 = 134

19. 637 ● 42 = 595

20. **TEST PREP** Which set of coins is equivalent to $0.31? (p. 80)

A 1 dime, 3 pennies
B 1 quarter, 1 dime, 1 penny
C 1 quarter, 1 nickel, 1 penny
D 3 dimes, 1 nickel

HANDS ON
Elapsed Time

Quick Review

Write the time 2 hours later.

1. 1:00   2. 6:30

3. 3:15   4. 8:55

5. 11:25

▶ **Explore**

Cameron read a book from 8:15 P.M. to 8:45 P.M. How long did Cameron read?

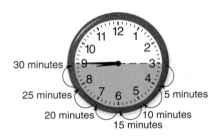

Start:
8:15

Count the minutes:
30 minutes

So, Cameron read for 30 minutes.

VOCABULARY
elapsed time

MATERIALS
clocks with movable hands

TECHNOLOGY LINK

More Practice:
Use E-Lab, *Elapsed Time: Minutes and Hours.*

www.harcourtschool.com/elab2002

Elapsed time is the amount of time that passes from the start of an activity to the end of that activity.

- Find the elapsed time from 1:15 P.M. to 1:45 P.M.

The start time was 9:15. The end time is 10:00. How much time has elapsed?

Try It

Use a clock to find the elapsed time.

a. start: 9:15 A.M.
   end: 10:00 A.M.

b. start: 3:45 P.M.
   end: 4:30 P.M.

c. start: 10:00 A.M.
   end: 11:15 A.M.

d. start: 11:45 A.M.
   end: 1:15 P.M.

CALIFORNIA STANDARDS MR 2.3 Use a variety of methods, such as words, numbers, symbols, charts, graphs, tables, diagrams, and models, to explain mathematical reasoning. **MR 2.4** Express the solution clearly and logically by using the appropriate mathematical notation and terms and clear language; support solutions with evidence in both verbal and symbolic work. *also* **MR 2.0, MR 3.2**

Soccer practice starts at 11:00 A.M. It lasts
2 hours 15 minutes. What time does practice end?

Start: 11:00   Count the hours.   Count the minutes.

So, practice ends at 1:15 P.M.

MATH IDEA If you know when an activity starts and
how long it takes, you can find the time it ends.

▶ **Practice**

Use a clock to find the elapsed time.

1. start: 7:00 A.M.
   end: 7:45 A.M.

2. start: 5:15 P.M.
   end: 6:20 P.M.

3. start: 11:30 A.M.
   end: 12:15 P.M.

Use a clock to find the end time.

4. start: 4:15 P.M.
   elapsed time: 1 hour
   15 minutes

5. start: 11:45 A.M.
   elapsed time: 20 minutes

6. **What's the Question?** The
   basketball game started at
   11:30 A.M. It ended at 1:15 P.M.
   The answer is 1 hour 45 minutes.

7. **Write About It** Explain how you
   can use a clock to find the
   elapsed time from 11:30 A.M. to
   1:15 P.M.

Mixed Review and Test Prep

Describe the pattern. Then continue
the pattern. (p. 2)

8. 3, 6, 9, 12, ▓

9. 7, 14, 21, 28, ▓

10. 4, 8, 12, 16, ▓

11. Order 297, 179, and 253 from
    greatest to least. (p. 24)

12. **TEST PREP** Which is 100 less than
    461? (p. 4)

    A 361   B 460   C 451   D 561

Quick Review

1. $15 + 5 = \blacksquare$

2. $15 + \blacksquare = 25$

3. $\blacksquare + 15 = 30$

4. $15 + 30 = \blacksquare$

5. 30 minutes after 1:15 is $\blacksquare$.

VOCABULARY
schedule

▶ Learn

RIGHT ON TIME A **schedule** is a table that lists activities or events and the times they happen.

You can use what you know about elapsed time to finish Stacy's schedule.

☆ STACY'S SCHEDULE ☆

Activity	Time	Elapsed Time
🍎 Eat snack	3:45 P.M.–4:05 P.M.	20 minutes
✏️ Do homework	4:05 P.M.–5:10 P.M.	⬜
🐾 Walk dog	5:10 P.M.–⬜	25 minutes

How long will Stacy do homework?

> **Think:** Find the elapsed time.
> 4:05 P.M. to 5:05 P.M.   1 hour
> 5:05 P.M. to 5:10 P.M.   5 minutes

So, Stacy will do homework for 1 hour 5 minutes.

When will Stacy walk her dog?

> **Think:** Find the end time.
> Start: 5:10 P.M.
> Count on 25 minutes to 5:35 P.M.

So, Stacy will walk her dog from 5:10 P.M. to 5:35 P.M.

MATH IDEA You can use a schedule to find elapsed times of events. If you know the elapsed times, you can find start or end times on a schedule.

TECHNOLOGY LINK

To learn more about schedules, watch the Harcourt Math Newsroom Video *Kangaroo Foster Mom.*

▶ Check

1. **Explain** why Stacy won't start walking her dog at 5:00 P.M.

USE DATA For 2–3, use Stacy's schedule above.

2. What time does Stacy finish her homework?

3. How long does it take Stacy to do her homework and walk the dog?

CALIFORNIA STANDARDS MR 2.3 Use a variety of methods, such as words, numbers, symbols, charts, graphs, tables, diagrams, and models, to explain mathematical reasoning. **MR 2.4** Express the solution clearly and logically by using the appropriate mathematical notation and terms and clear language; support solutions with evidence in both verbal and symbolic work. *also* **MR 1.0, MR 1.1, MR 2.0, MR 3.2**

▶ Practice and Problem Solving

USE DATA For 4–6, use the class schedule.

4. Which activities last 45 minutes each?

5. Which activity is the longest?

6. **ESTIMATION** About how long are the reading and math activities altogether?

MORNING CLASS SCHEDULE	
Activity	Time
Reading	8:30 A.M. – 9:15 A.M.
Math	9:15 A.M. – 10:15 A.M.
Recess	10:15 A.M. – 10:35 A.M.
Music	10:35 A.M. – 11:20 A.M.
Art	11:20 A.M. – 12:05 P.M.

Copy and complete the schedule.

	THE SCIENCE CHANNEL SCHEDULE		
	Program	Time	Elapsed Time
7.	Animals Around Us	6:00 P.M. – 7:00 P.M.	■
8.	Wonderful Space	7:00 P.M. – ■	25 minutes
9.	Weather in Your Town	7:25 P.M. – 7:30 P.M.	■
10.	Earthly Treasures	7:30 P.M. – ■	30 minutes

For 11–12, use the schedule you completed.

11. *Weather in Your Town* begins ■ minutes after 7:00 P.M.

12. *Earthly Treasures* begins ■ hour and ■ minutes after *Animals Around Us* begins.

13. **REASONING** Sean needs at least 30 minutes to get ready for school. If he leaves for school at 8:05 A.M., what is the latest he can start getting ready?

14. **Write About It** Think about activities you do on a school day and how much time each takes. Make a schedule. Be sure to include start and end times.

Mixed Review and Test Prep

15. Write $3,000 + 400 + 8$ in standard form. (p. 6)

16. (p. 42)
$$\begin{array}{r} 345 \\ 102 \\ +593 \\ \hline \end{array}$$

17. (p. 42)
$$\begin{array}{r} 199 \\ 255 \\ +175 \\ \hline \end{array}$$

18. Kate has 5 quarters, 1 dime, and 2 nickels. Jim has one $1 bill and 1 half dollar. Who has more money? (p. 84)

19. **TEST PREP** The difference between $4.35 and $1.67 is ■. (p. 88)

A $6.02 C $3.32

B $3.78 D $2.68

Extra Practice page H37, Set B

Use a Calendar

Quick Review

1. 7 pennies = ▮¢

2. 3 nickels = ▮¢

3. 4 dimes = ▮¢

4. 1 quarter = ▮¢

5. __?__ is the month after September.

► Learn

DAYS AND DATES   A **calendar** shows the days, weeks, and months of a year.

MATH IDEA You can use a calendar to find elapsed time in days, weeks, and months.

The Youngs are taking a trip. They left on July 21 and will return on August 6. How long will they be gone?

VOCABULARY
calendar

July						
Sun	Mon	Tue	Wed	Thu	Fri	Sat
	1	2	3	4	5	6
7	8	9	10	11	12	13
14	15	16	17	18	19	20
21	22	23	24	25	26	27
28	29	30	31			

August						
Sun	Mon	Tue	Wed	Thu	Fri	Sat
				1	2	3
4	5	6	7	8	9	10
11	12	13	14	15	16	17
18	19	20	21	22	23	24
25	26	27	28	29	30	31

Think: Start: Sunday, July 21.
Move down 2 weeks to August 4.
Then count on 2 days to August 6.

So, the Youngs will be gone for 2 weeks and 2 days, or 16 days.

Remember
Units of Time
7 days = 1 week
12 months = 1 year

• **What if** the Youngs are gone for 2 months? In what month will they return?

► Check

1. **Explain** how you know 1 week and 2 days is the same amount of time as 9 days.

For 2, use the calendars above.

2. How many weeks are there from July 30 to August 20? How many days?

CALIFORNIA STANDARDS MR 2.3 Use a variety of methods, such as words, numbers, symbols, charts, graphs, tables, diagrams, and models, to explain mathematical reasoning. **MG 1.4** Carry out simple unit conversions within a system of measurement. *also* **MR 1.0, MR 1.1, MR 2.0, MR 2.4, MR 3.2**

For 3–6, use the calendars.

February						
Sun	Mon	Tue	Wed	Thu	Fri	Sat
					1	2
3	4	5	6	7	8	9
10	11	12	13	14	15	16
17	18	19	20	21	22	23
24	25	26	27	28		

March						
Sun	Mon	Tue	Wed	Thu	Fri	Sat
					1	2
3	4	5	6	7	8	9
10	11	12	13	14	15	16
17	18	19	20	21	22	23
24/31	25	26	27	28	29	30

April						
Sun	Mon	Tue	Wed	Thu	Fri	Sat
	1	2	3	4	5	6
7	8	9	10	11	12	13
14	15	16	17	18	19	20
21	22	23	24	25	26	27
28	29	30				

3. Doug went on vacation from February 16 to March 9. How many weeks was his vacation?

4. Chantel started her art project on March 1. She worked for 3 weeks and 4 days. When did she finish?

5. **ESTIMATION** Ms. Horner's class rehearsed for a play from February 11 to March 9. About how long did the class rehearse?

6. Mr. Todd left for a 4-week trip on February 4. He came home for a week and then left again for 10 days. Did he return by March 19? Explain.

USE DATA For 7–9, use the calendars above and the pictograph.

7. "Raisins in the Sun" first hit number 1 on March 13. When did it lose its first-place spot?

8. Which song was number 1 for the longest time?

9. **REASONING** How many more days was "Dance and Sing" number 1 than "Rabbit in the Moon?"

TOP-OF-THE-CHART SONGS	
Dance and Sing	♪ ♪ ♪ ♪
Raisins in the Sun	♪ ♪
Rabbit in the Moon	♪ ♪ ♪

Key: Each ♪ = 1 week.

Mixed Review and Test Prep

10. Round 4,812 to the nearest thousand. (p. 30)

11. (p. 58)
$$457 - 186$$

12. (p. 88)
$$\$7.83 + \$5.09$$

13. (p. 88)
$$\$1.62 - \$0.85$$

14. **TEST PREP** Jay buys a book for $4.65. He pays with a $5 bill. Which coins can he get for change? (p. 86)

A 1 quarter, 1 nickel

B 3 dimes, 1 nickel

C 3 nickels, 5 pennies

D 6 nickels

Problem Solving Skill
Sequence Events

Understand ➤ Plan ➤ Solve ➤ Check

GOING CAMPING Mr. Cob is planning a camping trip. Use the list of things to do. Help him decide what to do first.

To Do List

Today's Date: October 22
Trip Date: November 16

Things to do:
• 3 days before the trip, check the weather.

• Tell mail carrier about the trip 1 week from today.

• Rent camping gear 2 weeks before the trip.

October						
Sun	Mon	Tue	Wed	Thu	Fri	Sat
		1	2	3	4	5
6	7	8	9	10	11	12
13	14	15	16	17	18	19
20	21	22	23	24	25	26
27	28	29	30	31		

November						
Sun	Mon	Tue	Wed	Thu	Fri	Sat
					1	2
3	4	5	6	7	8	9
10	11	12	13	14	15	16
17	18	19	20	21	22	23
24	25	26	27	28	29	30

Use the calendars to find the date for each.

Find the date 3 days before the trip.

Start: November 16. Count back 3 days.

On November 13, check the weather.

Find the date 1 week from today.

Start: October 22. Count on 1 week.

Tell mail carrier on October 29.

Find the date 2 weeks before the trip.

Start: November 16. Count back 2 weeks.

Rent gear on November 2.

Write the things to do in order.

Things to Do	Date
Tell mail carrier.	October 29
Rent gear.	November 2
Check weather.	November 13

So, Mr. Cob should start by telling the mail carrier about the trip.

Talk About It

• How can you find the date in 3 weeks?

For 1–4, use the calendars and the list.

April						
Sun	Mon	Tue	Wed	Thu	Fri	Sat
	1	2	3	4	5	6
7	8	9	10	11	12	13
14	15	16	17	18	19	20
21	22	23	24	25	26	27
28	29	30				

May						
Sun	Mon	Tue	Wed	Thu	Fri	Sat
			1	2	3	4
5	6	7	8	9	10	11
12	13	14	15	16	17	18
19	20	21	22	23	24	25
26	27	28	29	30	31	

To Do List

Today's Date: April 25
Date of Party: May 18

Things to do:
• Rent tables and chairs
  2 weeks before the party.

• Send invitations in
  5 days.

• Order cake 10 days after
  sending the invitations.

1. Use the list of things to do to help plan the party. Write what needs to be done, in order, and tell the date for each.

2. What if the date of the party changes to May 11? List what needs to be done, in order, and tell the date for each.

Mr. Hulin's class will go to the science museum on April 26, to the history museum on April 5, and to the art museum on May 9.

3. Which shows the museum field trips in order?
   A science, history, art
   B history, science, art
   C science, art, history
   D history, art, science

4. How many days are there between the history trip and the science trip?
   F 14 days   **H** 31 days
   G 21 days   **J** 35 days

Mixed Applications

5. Terry will need chairs for 8 adults and 14 children at her party. Can Terry estimate to decide how many chairs to rent? Explain.

6. Ms. Tate leaves on May 15. She must reserve a room 10 days before she leaves and a flight 2 weeks before she leaves. Which should she do first?

7. The sum of two numbers is 72. Their difference is 24. What are the numbers?

8. **Write a problem** you can solve by using a calendar. Trade problems with a partner and solve.

Review/Test

✅ CHECK VOCABULARY

Choose the best term from the box.

1. 12:00 at night is ? . (p. 96)

2. The amount of time that passes from the start of an activity to the end of that activity is ? . (p. 98)

> midnight
> noon
> elapsed time

✅ CHECK SKILLS

Write two ways you can read each time.
Then write the time, using A.M. or P.M. (pp. 94–97)

3.

dance class

4.

bedtime

5.

eat lunch

6.

eat breakfast

USE DATA For 7–9, use the schedule. (pp. 98–101)

7. At what time does Ramon start cleaning his room?

8. How long does Ramon read a book?

9. Which activity is the shortest?

RAMON'S SATURDAY SCHEDULE	
Activity	**Time**
Breakfast	7:30 A.M. – 8:00 A.M.
Clean room	8:00 A.M. – 8:45 A.M.
Read book	8:45 A.M. – 9:15 A.M.
Play softball	9:15 A.M.–11:15 A.M.
Lunch	11:15 A.M.–11:35 A.M.

✅ CHECK PROBLEM SOLVING

For 10, use the calendar. (pp. 102–105)

10. The soap-box derby will be on January 30. It will take Ken 2 weeks to build his car and 1 week to paint it. What is the latest date Ken should begin work on his car?

January						
Sun	Mon	Tue	Wed	Thu	Fri	Sat
		1	2	3	4	5
6	7	8	9	10	11	12
13	14	15	16	17	18	19
20	21	22	23	24	25	26
27	28	29	30	31		

Cumulative Review

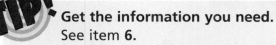

Get the information you need.
See item **6.**
Use the schedule to find the number of minutes each group is in class. Then compare and order them.

Also see problem **3**, p. H63.

For 1–9, choose the best answer.

1. A number has 2 digits. The tens digit is 4 more than the ones digit. The sum of the digits is 8. What is the number?

   A 53   **C** 73
   B 62   **D** 84

2. What time does the clock show?

   F 6:40   **H** 7:41
   G 7:36   **J** 8:41

3. Which of these activities would **not** likely happen at 7:30 P.M.?

   A get ready for bed
   B eat dinner
   C study for a test
   D eat breakfast

4. Josh watched a baseball game from 7:15 P.M. to 9:00 P.M. How long did Josh watch the game?

   F 1 hour 15 minutes
   G 1 hour 30 minutes
   H 1 hour 45 minutes
   J 2 hours

5. $8.69 − $4.78 = ▇

   A $3.91   **C** $4.91
   B $4.11   **D** NOT HERE

For 6–7, use the schedule.

SWIMMING LESSONS	
Group	Schedule
Ages 4–6	9:40 A.M.–10:05 A.M.
Ages 7–9	10:05 A.M.–10:35 A.M.
Ages 10–12	10:35 A.M.–11:10 A.M.
Ages 13–15	11:10 A.M.–11:50 A.M.

6. Which age group has the longest lesson?

   F Ages 4–6   **H** Ages 10–12
   G Ages 7–9   **J** Ages 13–15

7. How long is the lesson for Ages 10–12?

   A 40 minutes   **C** 30 minutes
   B 35 minutes   **D** 25 minutes

June						
Sun	Mon	Tue	Wed	Thu	Fri	Sat
		1	2	3	4	5
6	7	8	9	10	11	12
13	14	15	16	17	18	19
20	21	22	23	24	25	26
27	28	29	30			

8. Camp begins on June 4th and lasts for 3 weeks. When does camp end?

   F June 11   **H** June 25
   G June 18   **J** June 29

9. Julie saved $5.86 in April and $8.77 in May. How much did she save in all?

   A $14.63   **C** $14.73
   B $14.65   **D** $15.63

Study Guide and Review

VOCABULARY

Choose the best term from the box.

1. A table that lists activities or events and the times they happen is called a ? . (p. 100)

2. The hours between midnight and noon are ? hours. (p. 96)

> calendar
> schedule
> A.M.
> P.M.

STUDY AND SOLVE

Chapter 5

Make equivalent sets of money.

Examples A and B are equivalent sets with a value of $1.40.

A

B

Find two equivalent sets for each amount. List the bills and coins you can use. (pp. 80–81, 84–85)

3. $1.65   4. $3.70

5. $5.37   6. $5.82

7. $2.13   8. $0.49

Count on to make change.

A comb costs $0.69. Nikki pays with a $1 bill. How much change should she get?

$0.69 $0.70 $0.75 $1.00

1 penny, 1 nickel, and 1 quarter equal $0.31. So, Nikki will get $0.31 in change.

Copy and complete the table. (pp. 86–87)

	COST OF ITEM	AMOUNT PAID	AMOUNT OF CHANGE
9.	$1.25	$2.00	▪
10.	$3.72	$5.00	▪
11.	$2.34	$3.00	▪
12.	$0.17	$1.00	▪

Add or subtract money amounts.

Add or subtract as with whole numbers. Then write the sum or difference in dollars and cents.

$$\begin{array}{r} \overset{1}{}\$2.59 \\ +\$3.17 \\ \hline \$5.76 \end{array} \qquad \begin{array}{r} \overset{3\;12}{\$7.42} \\ -\$2.23 \\ \hline \$5.19 \end{array}$$

Find the sum or difference. Estimate to check. (pp. 88–89)

13. $\begin{array}{r} \$3.58 \\ +\$2.21 \end{array}$ **14.** $\begin{array}{r} \$8.64 \\ -\$7.75 \end{array}$ **15.** $\begin{array}{r} \$2.46 \\ +\$6.54 \end{array}$

16. $\begin{array}{r} \$5.75 \\ -\$2.30 \end{array}$ **17.** $\begin{array}{r} \$5.16 \\ +\$3.95 \end{array}$ **18.** $\begin{array}{r} \$8.07 \\ -\$6.49 \end{array}$

Chapter 6

Tell time to the minute.

Read:
three fifteen
15 minutes after three
quarter past three
Write: 3:15

Write two ways you can read each time. (pp. 94–97)

19.    **20.**

Find elapsed time.

Jerry played basketball from 11:30 A.M. to 1:45 P.M. How long did Jerry play?

Think:
From 11:30 A.M. to 1:30 P.M. is 2 hours.
From 1:30 P.M. to 1:45 P.M. is 15 minutes.

So, Jerry played for 2 hours 15 minutes.

Copy and complete the schedule. (pp. 98–101)

	Activity	Time	Elapsed Time
21.	Reading	11:45 A.M.–12:30 P.M.	■
22.	Lunch	12:30 P.M.–1:00 P.M.	■
23.	Soccer	1:00 P.M.–■	1 hour 30 minutes

PROBLEM SOLVING PRACTICE

Solve. (pp. 82–83, 104–105)

24. Ann has one $1 bill, 8 dimes, and 7 nickels. She wants to buy a notebook for $1.35. List the fewest bills and coins she can use.

25. Jim's recital is on November 12. His soccer game is on October 19. His class project is due October 6. List Jim's activities in order.

California Connections

MONTEREY BAY AQUARIUM

At the Monterey Bay Aquarium, you can learn about the marine plants and animals that live in the bay.

1. The sea otters are fed at 10:30 A.M., 1:30 P.M., and 3:30 P.M. How long is it between the first two feedings?

2. If you arrive at the jellyfish exhibit at 1:45 P.M. and spend 35 minutes there, at what time will you leave?

For 3–4, use the sign.

3. How long is the aquarium open in the winter? In the summer and on holidays?

4. If you arrive at the aquarium at 11:30 A.M. and leave when it closes, how long will you be at the aquarium?

> *Open Daily* ..... 10 A.M. to 6 P.M.
> *Summer*
> *and Holidays* ..... 9:30 A.M. to 6 P.M.
>
> Monterey Bay
> **AQUARIUM**

▲ These marine animals are visible through the largest window on Earth.

For 5, use the calendar.

5. Suppose you buy your aquarium ticket on November 19 and visit the aquarium on December 5. How many days are there from the time you buy your ticket to the time you use it? Explain how you can use the calendar to help you find the answer.

November						
Sun	Mon	Tue	Wed	Thu	Fri	Sat
					1	2
3	4	5	6	7	8	9
10	11	12	13	14	15	16
17	18	19	20	21	22	23
24	25	26	27	28	29	30

THE AQUARIUM GIFT STORE

At the aquarium gift store, you can buy books, games, toys, and clothing related to the plants and animals in the exhibits.

1. Suzanne has five $1 bills, 3 quarters, 3 dimes, and 3 pennies. She wants to buy a book about jellyfish that costs $5.95. Does she have enough money? Explain.

2. Postcards of some of the marine animals cost 25¢ each. Joel has 6 quarters, 4 dimes, and 2 nickels. How many postcards can he buy? Draw a picture to explain.

3. Anna bought a green turtle beanbag for $5.95 and a shark puzzle for $8.50. How much did Anna spend in all?

4. Aaron bought a hermit crab finger puppet for $7.95 and a postcard for $0.25. He paid with a $10 bill. How much change did he get?

5. Justin has four $1 bills, 9 quarters, 7 dimes, 2 nickels, and 5 pennies. He wants to buy a poster that costs $5.29. Name three different money combinations Justin can use to pay for the poster.

▲ At the jellyfish exhibit, you can see dozens of varieties of jellyfish.

Understand Multiplication

Millions of pounds of fruit are grown all over the world each year. Knowing about how much different fruits weigh can help you decide how much fruit to buy. The pictograph shows the weights of some fruit. About how much more would 3 pears weigh than 3 peaches?

WEIGHT OF FRUITS		
Apple	🧺🧺🧺🧺🧺🧺	6
Orange	🧺🧺🧺🧺🧺	5
Peach	🧺🧺🧺	3
Banana	🧺🧺🧺🧺	4
Kiwi	🧺🧺🧺	3
Plum	🧺🧺	2
Pear	🧺🧺🧺🧺🧺🧺	6

Key: Each 🧺 = 1 ounce.

CHECK WHAT YOU KNOW ✓

Use this page to help you review and remember
important skills needed for Chapter 7.

✓ SKIP-COUNT (See p. H9.)

1. Skip-count by twos.

2, 4, 6, ■, ■, ■, ■

2. Skip-count by fives.

5, 10, 15, ■, ■, ■

Skip-count to find the missing numbers.

3. 2, 4, 6, ■, ■, ■, 14, 16, ■, ■

4. 3, 6, ■, 12, ■, ■, 21

5. 5, 10, ■, ■, 25, ■, ■, ■, ■, 50

6. 10, 20, ■, ■, 50, ■, ■, ■, 90

✓ EQUAL GROUPS (See p. H9.)

Write how many there are in all.

7.

3 groups of 3 = ■

8.

5 groups of 2 = ■

9.

3 groups of 4 = ■

Find how many in all. You may wish to draw a picture.

10. 2 groups of 5

11. 3 groups of 5

12. 4 groups of 2

13. 2 groups of 2

14. 1 group of 3

15. 2 groups of 3

✓ COLUMN ADDITION (See p. H10.)

Find the sum.

16.   1
           4
         +6
    ___

17.   2
           1
         +9
    ___

18.   6
           6
         +6
    ___

19.   5
           4
         +7
    ___

20.   8
           5
         +2
    ___

Algebra: Connect Addition and Multiplication

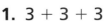
Quick Review

1. $3 + 3 + 3$

2. $5 + 5 + 5$

3. $2 + 2 + 2$

4. $2 + 2 + 2 + 2$

5. $4 + 4 + 4$

▶ Learn

SLURP! There are 3 juice boxes in a package. If Cara buys 5 packages, how many juice boxes will she have?

You can add to find how many in all.

VOCABULARY
multiply

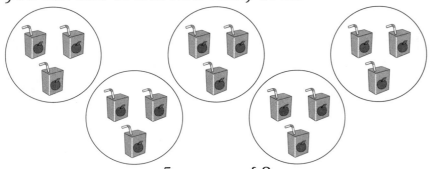

5 groups of 3

> **Remember**
> You can use a number sentence to show addition.
> $2 + 2 + 2 = 6$

Write: $3 + 3 + 3 + 3 + 3 = 15$

Say: 5 threes equal 15.

You can **multiply** to find how many in all.

Write: $5 \times 3 = 15$

Say: 5 times 3 equals 15.

So, Cara will have 15 juice boxes.

MATH IDEA When the groups have the same number, you can multiply to find how many in all.

REASONING Can you use multiplication to find $2 + 3 + 2$? Why or why not?

TECHNOLOGY LINK

More Practice:
Use E-Lab, *Exploring Multiplication.*
www.harcourtschool.com/elab2002

▶ Check

1. **Tell** two ways to find the total if the juice boxes come in packages of 4, and Cara buys 3 packages.

CALIFORNIA STANDARDS NS 2.0 Students calculate and solve problems involving addition, subtraction, multiplication, and division. ○━┓**NS 2.2** Memorize to automaticity the multiplication table for numbers between 1 and 10. *also* **NS 2.8, AF 1.0, MR 1.1, MR 2.3, MR 3.0, MR 3.2**

Copy and complete.

2.

3.

a. ▦ groups of ▦ = ▦
b. ▦ + ▦ + ▦ = ▦
c. ▦ × ▦ = ▦

a. ▦ groups of ▦ = ▦
b. ▦ + ▦ + ▦ + ▦ = ▦
c. ▦ × ▦ = ▦

▶ Practice and Problem Solving

Copy and complete.

4.

a. ▦ groups of ▦ = ▦
b. ▦ + ▦ + ▦ = ▦
c. ▦ × ▦ = ▦

5.

a. ▦ groups of ▦ = ▦
b. ▦ + ▦ = ▦
c. ▦ × ▦ = ▦

For 6–9, choose the letter of the number sentence
that matches.

a. $6 \times 2 = 12$	b. $3 \times 8 = 24$	c. $3 \times 4 = 12$	d. $6 \times 4 = 24$

6. $4 + 4 + 4 = 12$

7. $2 + 2 + 2 + 2 + 2 + 2 = 12$

8. $8 + 8 + 8 = 24$

9. $4 + 4 + 4 + 4 + 4 + 4 = 24$

10. **REASONING** Can you multiply to find $10 + 10 + 10 + 10$? Explain.

11. **REASONING** Yogurts come in packages of 6. How many packages are needed to give 23 students each a yogurt?

Mixed Review and Test Prep

12. 437 (p. 58) -229

13. 684 (p. 42) $+321$

14. $9,239$ (p. 46) $+1,605$

15. $3,030$ (p. 62) $-1,923$

16. **TEST PREP** What is the elapsed time from 3:15 P.M. until 4:30 P.M.? (p. 98)

A 15 minutes
B 45 minutes
C 1 hour 15 minutes
D 2 hours

Extra Practice page H38, Set A

Multiply with 2 and 5

Quick Review

How many are in all?

1. 1 row of 8
2. 3 rows of 2
3. 2 rows of 5
4. 4 rows of 2
5. 3 rows of 3

VOCABULARY

factors
product

▶ Learn

SMART ROCKS The chips that run computers are made from a mineral found in rocks. Mrs. Frank asked 5 students to bring in 2 rocks each for a science project. How many rocks does she need?

Use counters.

● ● ● ● ●
● ● ● ● ●

There are 5 groups with 2 in each group.

Since each group has the same number, you can multiply to find how many in all.

$$5 \times 2 = 10$$
↑   ↑   ↑
factor factor product

$$\begin{array}{r} 2 \\ \times 5 \\ \hline 10 \end{array}$$ ← factor
← factor
← product

So, Mrs. Frank needs 10 rocks in all.

MATH IDEA The numbers that you multiply are **factors**. The answer is the **product**.

• Name the factors and product in $3 \times 2 = 6$.

Computer chip

Crystal

▶ Check

1. **Find** the products 1×2 through 9×2. What do you notice about the products? Are they always even or odd numbers? Why?

Find the product.

2.

$$4 \times 2 = \blacksquare$$

3.

$$3 \times 5 = \blacksquare$$

4.

$$6 \times 2 = \blacksquare$$

CALIFORNIA STANDARDS NS 2.0 Students calculate and solve problems involving addition, subtraction, multiplication, and division. **O⊣ NS 2.2** Memorize to automaticity the multiplication table for numbers between 1 and 10. *also* **AF 1.0, O⊣ AF 2.1, MR 1.0, MR 2.2, MR 2.3, MR 2.4**

Practice and Problem Solving

Find the product.

5.

$2 \times 2 = $ ▨

6.

$5 \times 5 = $ ▨

7.

$4 \times 5 = $ ▨

Copy and complete the multiplication table.

×	1	2	3	4	5	6	7	8	9
8. 2	▨	▨	▨	▨	▨	▨	▨	▨	▨
9. 5	▨	▨	▨	▨	▨	▨	▨	▨	▨

Complete.

10. $8 \times 2 = $ ▨

11. ▨ $= 4 \times 5$

12. $6 \times 2 = $ ▨

13. ▨ $= 7 \times 5$

14. ▨ $= 9 \times 5$

15. $8 \times 5 = $ ▨

16. ▨ $= 7 \times 2$

17. $6 \times 5 = $ ▨

18.
$$\begin{array}{r} 2 \\ \times 8 \\ \hline \end{array}$$

19.
$$\begin{array}{r} 5 \\ \times 8 \\ \hline \end{array}$$

20.
$$\begin{array}{r} 2 \\ \times 9 \\ \hline \end{array}$$

21.
$$\begin{array}{r} 5 \\ \times 9 \\ \hline \end{array}$$

22. $3 + 3 = 2 \times$ ▨

23. $4 \times$ ▨ $= 4 + 4 + 4$

24. $2 \times 5 = $ ▨ $+ 5$

25. Sal bought 8 packages of rocks at the science center. There are 2 rocks in each package. How many rocks did Sal buy?

26. **REASONING** Drew has 5 pairs of white socks and 2 pairs of black socks. How many more white socks than black socks does Drew have?

27. **? What's the Error?** Midville School has 5 classrooms with 6 computers in each. Jan said that there are 11 computers in all. Describe Jan's error.

Mixed Review and Test Prep

28. 5 dimes = ▨ quarters (p. 80)

29. 9 nickels = ▨ pennies (p. 80)

30. Round 4,787 to the nearest hundred. (p. 28)

31.
$$\begin{array}{r} \$4.56 \\ +\$2.98 \\ \hline \end{array}$$ (p. 88)

32. **TEST PREP** Jim went to play basketball at 2:15 P.M. and returned at 4:45 P.M. How long was he gone? (p. 98)

A 2 hours

B 2 hours 15 minutes

C 2 hours 30 minutes

D 2 hours 45 minutes

HANDS ON
Arrays

Quick Review

Find how many in all.

1. 2 groups of 2

2. 1 group of 9

3. 4 groups of 2

4. 3 groups of 4

5. 2 groups of 3

▶ **Explore**

An **array** shows objects in rows and columns.

Make an array to find how many are in
3 rows of 5.

VOCABULARY
array
**Order Property
  of Multiplication**

MATERIALS
square tiles

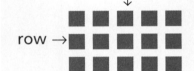

STEP 1	STEP 2
Make an array with 3 rows and 5 columns.	Count the tiles.

column
↓

row →

3 rows of 5 = ▉

$3 \times 5 = $ ▉

• How many tiles are in the 3 rows of 5?

• What multiplication sentence can you write to find
the number of tiles?

• Make an array with 3 rows of 3. What shape is
formed by this array? What multiplication sentence
can you write to find the number of tiles?

Try It

Copy and complete.

a.

▉ rows of ▉ = ▉
▉ × ▉ = ▉

b.

▉ rows of ▉ = ▉
▉ × ▉ = ▉

I have 2 rows of
4. How many
are there in all?

CALIFORNIA STANDARDS AF 1.5 Recognize and use the commutative and associative properties of
multiplication. **○┐NS 2.2** Memorize to automaticity the multiplication table for numbers between 1 and
10. *also* **NS 2.0, AF 1.0, MR 1.1, MR 2.0, MR 2.3, MR 3.0**

▶ Connect

The **Order Property of Multiplication** means that two numbers can be multiplied in any order. The product is the same.

Use arrays to show the Order Property of Multiplication.

Examples

A
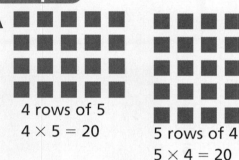

4 rows of 5
4 × 5 = 20

5 rows of 4
5 × 4 = 20

So, 4 × 5 = 5 × 4.

B
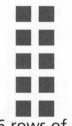

6 rows of 2
6 × 2 = 12

2 rows of 6
2 × 6 = 12

So, 6 × 2 = 2 × 6.

▶ Practice

Copy and complete.

1.

■ rows of ■ = ■
■ × ■ = ■

2.

■ rows of ■ = ■
■ × ■ = ■

3.

■ rows of ■ = ■
■ × ■ = ■

Find the product. You may wish to draw an array.

4. 2 × 3 = ■   **5.** 6 × 4 = ■   **6.** 8 × 3 = ■   **7.** 3 × 5 = ■

8. ✏ **Write About It** Tell how you would explain to a second grader the kinds of arrays you have used to help you understand multiplication. Give an example.

Mixed Review and Test Prep

9. 24
       46
     +93 (p. 36)

10. 446
       −267 (p. 58)

11.   34
        −15

12. Write the value of the 7 in 67,409.
(p. 10)

13. **TEST PREP** What is the total value of 3 dimes and 4 nickels?

A 34¢   C 55¢
B 50¢   D 70¢

Multiply with 3

Quick Review

1. $2 + 2 = 2 \times \blacksquare$

2. $3 \times \blacksquare = 5 + 5 + 5$

3. $2 \times 6 = 6 - \blacksquare$

4. $\blacksquare \times 5 = 10$

5. $2 \times 4 = \blacksquare$

▶ **Learn**

PRACTICE, PRACTICE, PRACTICE

Pat practiced soccer 2 hours each day for 3 days. How many hours did he practice in all?

Val practiced soccer 3 hours each day for 2 days. How many hours did she practice in all?

For 2 hours, move 2 spaces. For 3 days, make 3 jumps of 2 spaces.

For 3 hours, move 3 spaces. For 2 days, make 2 jumps of 3 spaces.

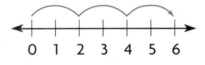

0 1 2 3 4 5 6

Multiply: $3 \times 2 = 6$

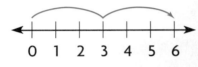

0 1 2 3 4 5 6

Multiply: $2 \times 3 = 6$

So, both Pat and Val practiced for 6 hours.

A number line can help you understand the Order Property of Multiplication.

Example

Multiply. $3 \times 5 = 15$

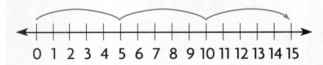

0 1 2 3 4 5 6 7 8 9 10 11 12 13 14 15

Multiply. $5 \times 3 = 15$

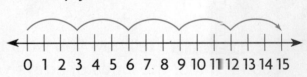

0 1 2 3 4 5 6 7 8 9 10 11 12 13 14 15

So, $3 \times 5 = 5 \times 3$.

• **REASONING** Use the factors 3 and 6 to explain the Order Property of Multiplication.

CALIFORNIA STANDARDS AF 1.5 Recognize and use the commutative and associative properties of multiplication. **NS 2.2** Memorize to automaticity the multiplication table for numbers between 1 and 10. *also* **NS 2.0, AF 1.0, AF 2.1, MR 1.0, MR 1.1, MR 2.0, MR 2.3**

Multiplication Practice

What if Pat scored 4 goals in each of 3 games and Val scored 3 goals in each of 4 games? Which number line shows Pat's goals? Val's goals? How many goals did each player score?

A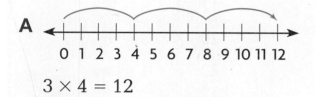

$3 \times 4 = 12$

B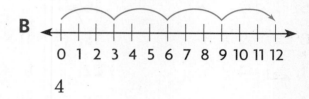

$$\begin{array}{r} 4 \\ \times 3 \\ \hline 12 \end{array}$$

Number line A shows Pat's goals.   Number line B shows Val's goals.

Each player scored 12 goals.

- **What if** Val scored 8 goals in each of 3 games and Pat scored 3 goals in each of 8 games? How many goals did each player score?

▶ **Check**

1. **Explain** how knowing the product 7×3 can help you find the product 3×7.

2. Is the product even or odd when you multiply 3 by an even number? by an odd number?

Use the number line to find the product.

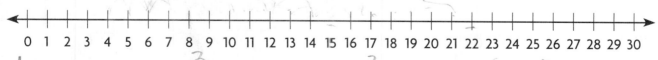

3. $4 \times 3 = \blacksquare\ 12$ 4. $12 = 3 \times 4$ 5. $7 \times 3 = \blacksquare\ 21$ 6. $21 = 3 \times 7$

7. $\blacksquare = 6 \times 3$ 8. $\blacksquare = 3 \times 6$ 9. $3 \times 8 = \blacksquare\ 24$ 10. $21 = 8 \times 3$

11. $\begin{array}{r} 5 \\ \times 3 \\ \hline 15 \end{array}$ 12. $\begin{array}{r} 3 \\ \times 5 \\ \hline 15 \end{array}$ 13. $\begin{array}{r} 3 \\ \times 9 \\ \hline 27 \end{array}$ 14. $\begin{array}{r} 9 \\ \times 3 \\ \hline 27 \end{array}$

LESSON CONTINUES ▶

Practice and Problem Solving

Use the number line to find the product.

0 1 2 3 4 5 6 7 8 9 10 11 12 13 14 15 16 17 18 19 20 21 22 23 24 25 26 27 28 29 30

15. $3 \times 6 =$ _18_

16. _25_ $= 5 \times 5$

17. $3 \times 9 =$ _27_

18. _27_ $= 9 \times 3$

19. _18_ $= 9 \times 2$

20. $7 \times 3 =$ _21_

21. _20_ $= 4 \times 5$

22. $2 \times 8 =$ _16_

23. $7 \times 2 =$ _14_

24. $\begin{array}{r} 3 \\ \times 8 \\ \hline 24 \end{array}$

25. $\begin{array}{r} 1 \\ \times 5 \\ \hline 5 \end{array}$

26. $\begin{array}{r} 3 \\ \times 9 \\ \hline 27 \end{array}$

Copy and complete the multiplication table.

27.

×	1	2	3	4	5	6	7	8	9
3	3	6	9	12	15	18	21	24	27

Write the missing factor.

28. $2 \times 3 = $ _3_ $\times 2$

29. $3 \times$ _7_ $= 7 \times 3$

30. $7 \times 3 = $ _3_ $\times 7$

31. $6 \times 3 = $ _9_ $\times 2$

32. $4 \times 3 = 6 \times$ _2_

33. $8 \times 3 = $ _3_ $\times 8$

TECHNOLOGY LINK

More Practice:
Use Mighty Math
Carnival Countdown,
Snap Clowns, Level P.

34. **? What's the Error?** Sam played soccer for 2 hours a day, 3 days a week, for 4 weeks. Sam said he played soccer for a total of 6 hours during the month. What is Sam's error?

She also needs to add 4 weeks

35. **REASONING** If you add 3 to an odd number, is the sum even or odd? If you multiply an odd number by 3, is the product even or odd? _A even number_

USE DATA For 36–39, use the bar graph.

36. How many more goals did Matt's soccer team score in Game 4 than in Game 1? _3._

37. How many goals did Matt's team score in the four games? _18_

38. If Matt's team scores twice as many goals in Game 5 as in Game 3, how many goals will it score? _4_

39. ✏ **Write a problem** using the information in the bar graph.

124

40. Last week, the team practiced 2 days for 2 hours each day. This week, they practiced 3 days for 2 hours each day. How many hours did they practice in all?

41. Tim needs 25 table tennis balls. There are 5 tennis balls in 1 package. If he buys 4 packages, will he have enough tennis balls? Explain.

Mixed Review and Test Prep

Write *A.M.* **or** *P.M.* **for each.** (p. 96)

42. see the moon: 10:00 ?

43. eat lunch: 11:45 ?

44. play at park: 4:30 ?

Find the product. (p. 118)

45. $6 \times 2 = \blacksquare$ **46.** $8 \times 2 = \blacksquare$

47. $4 \times 5 = \blacksquare$ **48.** $6 \times 5 = \blacksquare$

49. TEST PREP $35 + 14 + 26$ (p. 36)

A 65 **B** 75 **C** 76 **D** 85

50. TEST PREP Tina went to the mall at 3:00 P.M. She returned home at 4:15 P.M. How long was she gone? (p. 98)

F 15 minutes
G 1 hour
H 1 hour 15 minutes
J 1 hour 30 minutes

Thinker's Corner

SOLVE THE RIDDLE! Find the product. To answer the riddle, match the letters to the products below.

6 **I** ×3 = 18	6 **A** ×5 = 30	5 **H** ×3 = 15	3 **J** ×4 = 12
9 **Q** ×5 = 45	2 **S** ×7 = 14	8 **N** ×3 = 24	3 **L** ×2 = 6

16 = 2 × 8 **D**	2 × 2 = 4 **K**	8 = 4 × 2 **M**
7 × 3 = 21 **E**	27 = 9 × 3 **U**	40 = 8 × 5 **G**

What is a mouse's favorite game?

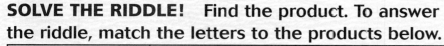

? ? ? ? — 15 18 16 21 — HIDe

? ? ? — 30 24 16 — and

? ? ? ? ? ? — 14 45 27 21 30 4 — SQueak

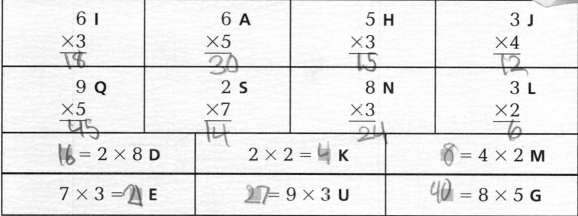

Extra Practice page H38, Set C

Problem Solving Skill
Too Much/Too Little Information

Understand ➤ Plan ➤ Solve ➤ Check

FIND THE FACTS   Three students walked 6 blocks to the craft store. Each one bought 5 pieces of poster board to make posters for the school book fair. They stayed at the store for 45 minutes. How many pieces of poster board in all did the students buy?

STEP 1 Find what the problem asks.
- How many pieces of poster board did the students buy in all?

STEP 2 Find what facts are needed to solve the problem.
- the number of students
- the number of pieces of poster board each one buys

STEP 3 Look for extra information.
- how far they walked
- how long they were at the store
Do you need this information to solve the problem?

STEP 4 Solve the problem.
- multiply
 3 students $\times$ 5 pieces $=$ 15 pieces
So, the students bought 15 pieces of poster board in all.

Talk About It

- Is there too much or too little information in the problem above?

- How do you know which information you need to solve a problem?

- Three students went to a restaurant. They each bought a sandwich. How much did they spend in all? Does this problem have too much, too little, or the right amount of information?

CALIFORNIA STANDARDS MR 1.1 Analyze problems by identifying relationships, distinguishing relevant from irrelevant information, sequencing and prioritizing information, and observing patterns. **NS 2.0** Students calculate and solve problems involving addition, subtraction, multiplication, and division. *also* **MR 1.0, MR 2.0, MR 2.3, MR 3.2**

▶ Problem Solving Practice

USE DATA For 1–4, use the table. Write *a*, *b*, or *c* to tell whether the problem has

a. too much information.

b. too little information.

c. the right amount of information.

Solve those with too much or the right amount of information.

SCHOOL SUPPLIES	
Pack of Paper	$1
Backpack	$9
Pack of Pencils	$3
Lunch Box	$4

1. Felix wants to buy a backpack and a box of crayons. How much will he spend?

2. Marisa bought 2 packs of pencils. She was second in line to pay for her supplies. How much did Marisa spend on school supplies?

3. Sam bought 2 backpacks and a lunch box. He received $3 change. How much money had he given the clerk?

4. Sally had $15. She bought 5 packs of paper and a lunch box. How much did she spend?

Mixed Applications

USE DATA For 5–6, use the table above.

You have $15 to spend on school supplies.

5. Which items can you buy?

   A a backpack, 3 packs of pencils

   B a backpack, a pack of pencils, a lunch box

   C a lunch box, 2 packs of paper, a backpack

   D 4 packs of pencils, a backpack

6. How much more money do you need if you choose to buy 2 packs of pencils, a lunch box, and a backpack?

   F $1   **H** $3

   G $2   **J** $4

7. Joe bought two tapes at the music store. They cost $7.28 and $7.71. How much change did he receive from $20.00?

8. **? What's the Question?** There are 4 people in the Tamura family. Movie tickets cost $6 each. The answer is $24.

Review/Test

✓ CHECK VOCABULARY AND CONCEPTS

Choose the best term from the box.

array
factors
multiply
product

1. When groups have the same number, you can __?__ to find how many in all. (p. 116)

2. The numbers you multiply are __?__. (p. 118)

3. The answer to a multiplication problem is the __?__. (p. 118)

Find the product. You may wish to draw an array. (pp. 120–121)

4. $4 \times 5 = $ ▧　　　**5.** $3 \times 2 = $ ▧　　　**6.** $3 \times 6 = $ ▧　　　**7.** $7 \times 3 = $ ▧

✓ CHECK SKILLS

For 8–9, choose the letter of the number sentence that matches. (pp. 116–117)

8. $3 + 3 + 3 + 3 = 12$

9. $4 + 4 + 4 + 4 + 4 = 20$

a. $5 \times 4 = 20$
b. $6 \times 2 = 12$
c. $4 \times 3 = 12$

Find the product. (pp. 118–125)

10. $3 \times 8 = $ ▧　　　**11.** $7 \times 3 = $ ▧　　　**12.** ▧ $= 5 \times 6$　　　**13.** $4 \times 2 = $ ▧

14. $\begin{array}{r} 3 \\ \times 9 \\ \hline \end{array}$　　**15.** $\begin{array}{r} 6 \\ \times 3 \\ \hline \end{array}$　　**16.** $\begin{array}{r} 8 \\ \times 5 \\ \hline \end{array}$　　**17.** $\begin{array}{r} 7 \\ \times 5 \\ \hline \end{array}$　　**18.** $\begin{array}{r} 5 \\ \times 9 \\ \hline \end{array}$

✓ CHECK PROBLEM SOLVING

Write *a*, *b*, or *c* to tell whether the problem has

a. too much information.　　**b.** too little information.　　**c.** the right amount of information.

Solve those with too much or the right amount of information. (pp. 126–127)

19. Pete practices piano 3 hours a day, Monday through Friday. On Saturday he plays soccer. How many hours does he practice each week?

20. Ramiro worked on his science project 3 hours longer than Sue. How much time did each of them spend on the project?

Cumulative Review

 Get the information you need.
See item **8**.

Don't be confused by information that is not needed. You need the cost of the packs and how many Ben bought to find how much he spent.

Also see problem 3, p. H63.

For 1–10, choose the best answer.

1. Which is another way to write
6 + 6 + 6 + 6 + 6 + 6 = 36?

   A 16 + 16 = 36
   B 6 × 6 = 36
   C 18 + 12 = 36
   D 12 × 4 = 36

2. Which multiplication sentence matches the array?

   F 3 × 5 = 15
   G 3 + 5 = 8
   H 5 × 5 = 25
   J 15 × 1 = 15

3. 3 × 9 = ■

   A 6   **C** 27
   B 12   **D** NOT HERE

4. Shelly made 6 glasses of grape drink. She used 2 scoops of powder in each glass. How many scoops did she use in all?

   F 12   **H** 4
   G 8   **J** NOT HERE

5. Libby bought 8 packages of fishing bait for $5 each. How much money did she spend on the fishing bait?

   A $30   **C** $38
   B $35   **D** $40

6. 3 × 8 = ■

   F 11   **H** 20
   G 16   **J** NOT HERE

7. Ramon has 64¢. A pack of gum costs 25¢ and a snack bar costs 35¢. How much money will Ramon have left if he buys a snack bar?

   A 60¢   **C** 29¢
   B 39¢   **D** 4¢

8. Ben had $25. He bought 7 packs of cards that cost $2 each. How much money did he spend?

   F $9   **H** $14
   G $11   **J** $25

9. What time does the clock show?

   A 6:08   **C** 7:08
   B 6:40   **D** 7:40

10. What is the value of the 1 in 17,405?

    F 10   **H** 1,000
    G 100   **J** 10,000

Multiplication Facts Through 5

There are about 175 roller coasters in the United States today. Some can reach speeds of 70 miles per hour! Compare the number of riders on the roller coasters named on the pictograph with the coaster in the photo. What is a quick way to find the number of riders on Mantis?

Mantis is a stand-up roller coaster in Sandusky, Ohio. Each train has eight rows of four riders.

ROLLER COASTER RIDERS

Hercules (Pennsylvania)	
The Psyclone (California)	
The Viper (Georgia)	
Mamba (Missouri)	

Key: Each 🚃 = 4 riders.

CHECK WHAT YOU KNOW ✔

Use this page to help you review and remember important skills needed for Chapter 8.

✔ VOCABULARY

Choose the best term from the box.

1. The numbers that you multiply are called ___?___ .

2. The answer to a multiplication problem is the ___?___ .

> addends
> factors
> product

✔ ADDITION (See p. H10.)

Complete.

3. $3 + 3 + 3 + 3 = \blacksquare$

4. $8 + \blacksquare = 8$

5. $1 + 1 + 1 + 1 + 1 = \blacksquare$

6. $4 + 4 + 4 + 4 = \blacksquare$

7. $\blacksquare = 1 + 9$

8. $6 + 0 = \blacksquare$

9. $0 + 0 + 0 = \blacksquare$

10. $4 + 4 + 4 = \blacksquare$

11. $\blacksquare + 7 = 15$

✔ ORDER PROPERTY OF ADDITION (See p. H11.)

Complete.

12. $4 + 7 = 7 + \blacksquare$

13. $8 + 5 = 5 + \blacksquare$

14. $9 + 6 = \blacksquare + 9$

15. $\blacksquare + 9 = 9 + 8$

16. $3 + \blacksquare = 7 + 3$

17. $\blacksquare + 4 = 4 + 9$

✔ MULTIPLICATION FACTS (See p. H11.)

Find the product.

18. $1 \times 2 = \blacksquare$

19. $4 \times 5 = \blacksquare$

20. $2 \times 5 = \blacksquare$

21. $\blacksquare = 3 \times 3$

22. $3 \times 6 = \blacksquare$

23. $\blacksquare = 2 \times 9$

24. $7 \times 3 = \blacksquare$

25. $4 \times 2 = \blacksquare$

26. $\blacksquare = 9 \times 3$

27. $8 \times 5 = \blacksquare$

28. $\blacksquare = 6 \times 2$

29. $8 \times 2 = \blacksquare$

30. $9 \times 5 = \blacksquare$

31. $\blacksquare = 5 \times 6$

32. $2 \times 7 = \blacksquare$

33. $\blacksquare = 3 \times 8$

1

Multiply with 0 and 1

► Learn

ALL OR NOTHING Tina saw 5 cars. One clown sat in each car. How many clowns were there in all?

STEP 1 Count the cars.

STEP 2 Count the clowns in the cars.

STEP 3 Write the multiplication sentence.

$$
\underset{\substack{\uparrow \\ \text{number of} \\ \text{groups}}}{5} \quad \times \quad \underset{\substack{\uparrow \\ \text{number in} \\ \text{each group}}}{1} \quad = \quad \underset{\substack{\uparrow \\ \text{number in} \\ \text{all}}}{5}
$$

So, there were 5 clowns in all.

Suppose Tina saw 3 cars with 0 clowns in each car. How many clowns were there in all?

$$
\underset{\substack{\uparrow \\ \text{number of} \\ \text{groups}}}{3} \quad \times \quad \underset{\substack{\uparrow \\ \text{number in} \\ \text{each group}}}{0} \quad = \quad \underset{\substack{\uparrow \\ \text{number in} \\ \text{all}}}{0}
$$

So, there were 0 clowns in all.

MATH IDEA The product of 1 and any number equals that number. The product of 0 and any number equals 0.

REASONING What is 498×1? 498×0? How do you know?

CALIFORNIA STANDARDS NS 2.6 Understand the special properties of 0 and 1 in multiplication and division.
O¬ NS 2.2 Memorize to automaticity the multiplication table for numbers between 1 and 10. *also* **NS 2.0, AF 1.0, MR 2.0, MR 2.3, MR 2.4, MR 3.0**

132

1. **Explain** what happens when you multiply by 1. What happens when you multiply by 0?

Find the product.

2. $4 \times 1 = \blacksquare$ 3. $5 \times 0 = \blacksquare$ 4. $1 \times 3 = \blacksquare$

► Practice and Problem Solving

Find the product.

5. $2 \times 1 = \blacksquare$ 6. $4 \times 0 = \blacksquare$ 7. $0 \times 5 = \blacksquare$ 8. $\blacksquare = 9 \times 1$

9. $1 \times 6 = \blacksquare$ 10. $\blacksquare = 0 \times 9$ 11. $7 \times 1 = \blacksquare$ 12. $2 \times 4 = \blacksquare$

13. $\blacksquare = 0 \times 7$ 14. $3 \times 3 = \blacksquare$ 15. $\blacksquare = 1 \times 8$ 16. $0 \times 0 = \blacksquare$

17. $\begin{array}{r} 9 \\ \times 0 \\ \hline \end{array}$ 18. $\begin{array}{r} 1 \\ \times 3 \\ \hline \end{array}$ 19. $\begin{array}{r} 5 \\ \times 4 \\ \hline \end{array}$ 20. $\begin{array}{r} 8 \\ \times 5 \\ \hline \end{array}$

21. Multiply 4 by 1.
22. Find the product of 0 and 8.
23. Find the product of 1 and 7.
24. What is 3 times 9?

Complete.

25. $\blacksquare = 9 \times 5$ 26. $3 + 9 = 3 \times \blacksquare$ 27. $3 \times 6 = \blacksquare \times 9$

28. $8 + 7 = \blacksquare \times 5$ 29. $0 \times 8 = \blacksquare \times 9$ 30. $9 \times \blacksquare = 3 \times 3$

31. **REASONING** Ann is younger than Rick. Rick is older than Tracy. Who is the oldest?

32. **Write About It** Which is less, the product of your age and 1 or the product of your age and 0? Explain.

33. Write a problem that has zero as the product.

Mixed Review and Test Prep

Find the value of the blue digit. (p. 10)

34. 92,348   35. 38,965

36. Find the sum of 652 and 738. (p. 42)

37. Put the numbers in order from least to greatest. 43, 24, 34, 42 (p. 24)

38. **TEST PREP** Carol bought a pencil for $0.32. She paid with a $1 bill. How much change did she get? (p. 86)

A $0.32   C $0.78

B $0.68   D $1.32

LESSON

2 Multiply with 4

▶ Learn

TWISTS AND TURNS There are 6 cars in the Twister ride at the amusement park. Each car holds 4 people. How many people does the Twister ride hold?

$$6 \times 4 = \blacksquare$$

- How could you use an array to find 6×4?

You can use a multiplication table to find the product.

The product is found where row 6 and column 4 meet.

Multiplication Table

column
↓

×	0	1	2	3	4	5	6	7	8	9
0	0	0	0	0	0	0	0	0	0	0
1	0	1	2	3	4	5	6	7	8	9
2	0	2	4	6	8	10	12	14	16	18
3	0	3	6	9	12	15	18	21	24	27
4	0	4	8	12	16	20	24	28	32	36
5	0	5	10	15	20	25	30	35	40	45
6	0	6	12	18	24	30	36	42	48	54
7	0	7	14	21	28	35	42	49	56	63
8	0	8	16	24	32	40	48	56	64	72
9	0	9	18	27	36	45	54	63	72	81

row →

$$6 \quad \times \quad 4 \quad = \quad 24 \qquad \begin{array}{r} 4 \\ \times 6 \\ \hline 24 \end{array} \begin{array}{l} \leftarrow \text{ factor} \\ \leftarrow \text{ factor} \\ \leftarrow \text{ product} \end{array}$$

factor   factor   product

So, the Twister ride holds 24 people.

MATH IDEA All multiplication facts that have 4 as a factor have an even product.

REASONING How do you know that 2×5 plus 2×5 is the same as finding 4×5?

▶ Check

1. **Explain** how you can use the multiplication table to find 4×8.

Find the product.

2. $2 \times 4 = \blacksquare 8$ 3. $9 \times 4 = 36$

4. $20 = 4 \times 5$ 5. $4 \times 3 = 12$

CALIFORNIA STANDARDS NS 2.0 Students calculate and solve problems involving addition, subtraction, multiplication, and division. **O─ NS 2.2** Memorize to automaticity the multiplication table for numbers between 1 and 10. *also* **NS 2.8, AF 2.0, O─ AF 2.1, MR 1.1, MR 2.0, QMR 2.3, MR 3.0**

134

▶ Practice and Problem Solving

Find the product.

6. $2 \times 4 = 8$ 7. $9 \times 0 = 0$ 8. $4 \times 5 = 20$

9. $40 = 5 \times 8$ 10. $4 \times 4 = 16$ 11. $1 \times 9 = 9$

12. $28 = 4 \times 7$ 13. $3 \times 0 = 0$ 14. $32 = 8 \times 4$

15. $\begin{array}{r} 2 \\ \times 3 \\ \hline 6 \end{array}$
16. $\begin{array}{r} 5 \\ \times 1 \\ \hline 5 \end{array}$
17. $\begin{array}{r} 4 \\ \times 6 \\ \hline 24 \end{array}$
18. $\begin{array}{r} 0 \\ \times 4 \\ \hline 0 \end{array}$
19. $\begin{array}{r} 7 \\ \times 4 \\ \hline 28 \end{array}$
20. $\begin{array}{r} 5 \\ \times 5 \\ \hline 25 \end{array}$

21. $\begin{array}{r} 3 \\ \times 6 \\ \hline 18 \end{array}$
22. $\begin{array}{r} 4 \\ \times 9 \\ \hline 36 \end{array}$
23. $\begin{array}{r} 8 \\ \times 2 \\ \hline 16 \end{array}$
24. $\begin{array}{r} 7 \\ \times 5 \\ \hline 35 \end{array}$
25. $\begin{array}{r} 4 \\ \times 8 \\ \hline 32 \end{array}$
26. $\begin{array}{r} 9 \\ \times 4 \\ \hline 36 \end{array}$

Copy and complete.

27.

×	0	1	2	3	4	5	6	7	8	9
4	■	■	■	■	■	■	■	■	■	■

28. $9 \times 0 = \blacksquare \times 4$ 29. $2 \times 9 = \blacksquare \times 3$ 30. $4 \times 4 = \blacksquare \times 2$

31. Multiply 5 by 4.   32. Find the product of 4 and 9.

33. Find the product of 7 and 0.   34. What is 1 times 12?

35. Each ride at the amusement park costs 4 tickets. If Tonya went on 7 different rides, how many tickets did she use?

36. Ahmed has 3 packs of 8 baseball cards and 11 extra cards. How many cards does he have in all?

37. Is the product 6×4 greater than, less than, or equal to 3×8? Explain.

38. Since $9 \times 4 = 36$ and $10 \times 4 = 40$, what is 11×4? How do you know?

Mixed Review and Test Prep

39. Ida has one $1 bill, 4 quarters, 2 dimes, and 1 nickel. How much money does she have? (p. 84)

40. $\begin{array}{r} \$4.89 \\ +\$7.77 \\ \hline \end{array}$ (p. 88)

41. $\begin{array}{r} \$9.42 \\ -\$5.54 \\ \hline \end{array}$ (p. 88)

42. Round 547 to the nearest ten. (p. 28)

43. **TEST PREP** Which shows the numbers in order from least to greatest? (p. 24)

   A 468, 486, 648

   B 468, 648, 486

   C 486, 468, 648

   D 648, 486, 468

Extra Practice page H39, Set B

Problem Solving Strategy
Find a Pattern

Understand → Plan → Solve → Check

Quick Review
1. $3 \times \blacksquare = 6$
2. $3 \times \blacksquare = 9$
3. $\blacksquare \times 4 = 12$
4. $3 \times \blacksquare = 15$
5. $3 \times \blacksquare = 18$

THE PROBLEM Emily is playing a number pattern game. She says the numbers 3, 5, 8, 10, 13, 15, 18, 20, and 23. What is the rule for her pattern? What are the next four numbers she will say?

• What are you asked to find?

• What information will you use?

• Is there information you will not use? If so, what?

• What strategy can you use to solve the problem?

You can *find a pattern*.

• How can you use the strategy to solve the problem?

Use a number line to find the pattern. Then write the rule and the next four numbers.

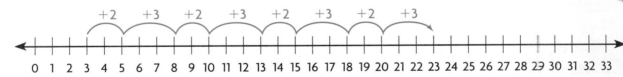

So, the rule is to add 2 and then add 3. The next four numbers in Emily's pattern will be 25, 28, 30, and 33.

• How do you know if your answer is correct?

CALIFORNIA STANDARDS MR 1.1 Analyze problems by identifying relationships, distinguishing relevant from irrelevant information, sequencing and prioritizing information, and observing patterns. **AF 2.2** Extend and recognize a linear pattern by its rules. *also* MR 1.0, MR 2.0, MR 2.3, MR 3.0

Problem Solving Practice

PROBLEM SOLVING STRATEGIES

Draw a Diagram or Picture
Make a Model or Act It Out
Make an Organized List
▶ **Find a Pattern**
Make a Table or Graph
Predict and Test
Work Backward
Solve a Simpler Problem
Write an Equation
Use Logical Reasoning

Use *find a pattern* to solve.

1. **What if** Emily's pattern is 5, 4, 9, 8, 13, 12, 17, 16, and 21? What are the rule and the next four numbers in her pattern?

2. Albert's pattern is 3, 6, 9, and 12. What are the rule and the next four numbers?

Karen is thinking of a number pattern. The first four numbers are 4, 8, 12, and 16.

3. What are the next three numbers in Karen's pattern?

   A 16, 20, 24   C 18, 20, 22

   B 17, 19, 21   D 20, 24, 28

4. Which number cannot be in Karen's pattern?

   F 20   H 32

   G 28   J 35

Mixed Strategy Practice

5. Bo bicycled 4 miles a day last week. He did not bicycle on Saturday or Sunday. How far did he bicycle last week?

6. ⟨? **What's the Error?**⟩ Look at this pattern. Describe the error and tell how to correct it.
   11, 21, 31, 41, 51, 60, 71

7. **REASONING** Write the greatest possible 4-digit number using 4 different digits. Write the smallest possible 4-digit number using 4 different digits.

8. If this is the time now, what time will it be in 2 hours 35 minutes?

CARL'S JUMPING JACKS	
Monday	𝅷 𝅷
Tuesday	𝅷 𝅷 𝅷
Wednesday	𝅷 𝅷 𝅷 𝅷

Key: Each 𝅷 = 3 jumping jacks.

9. **USE DATA** If Carl continues the pattern above, how many jumping jacks will he do on Saturday?

10. ✎ **Write a problem** about a number pattern. Tell how you would explain to a second grader how to find the next number in your pattern.

Practice Multiplication

Quick Review

1. $3 \times \blacksquare = 12$
2. $2 \times 6 = \blacksquare$
3. $0 \times 8 = \blacksquare$
4. $4 \times 5 = \blacksquare$
5. $9 \times \blacksquare = 9$

▶ Learn

FACTS IN FLIGHT   At the airport, Nicole saw 6 jets waiting to take off. Each jet had 3 engines. How many engines were there in all?

$$6 \times 3 = \blacksquare$$

There are many ways to find a product.

A. You can make equal groups or arrays.

$6 \times 3 = 18$

$6 \times 3 = 18$

B. You can skip-count on a number line.

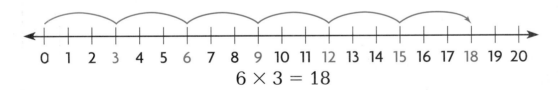

$6 \times 3 = 18$

C. You can double a fact that you already know.

Think: $3 \times 3 = 9$ and $9 + 9 = 18$, so,
$6 \times 3 = 18$.

D. You can use the Order Property of Multiplication.

Think: $3 \times 6 = 6 \times 3 = 18$.

CALIFORNIA STANDARDS NS 2.0 Students calculate and solve problems involving addition, subtraction, multiplication, and division. O━┓ NS 2.2 Memorize to automaticity the multiplication table for numbers between 1 and 10. *also* NS 2.8, AF 1.5, O━┓ AF 2.1, MR 1.1, MR 2.0, MR 2.3, MR 2.4

Ways to Find a Product

E. You can use a multiplication table.

column
↓

×	0	1	2	3	4	5	6	7	8	9
0	0	0	0	0	0	0	0	0	0	0
1	0	1	2	3	4	5	6	7	8	9
2	0	2	4	6	8	10	12	14	16	18
3	0	3	6	9	12	15	18	21	24	27
4	0	4	8	12	16	20	24	28	32	36
5	0	5	10	15	20	25	30	35	40	45
6	0	6	12	18	24	30	36	42	48	54
7	0	7	14	21	28	35	42	49	56	63
8	0	8	16	24	32	40	48	56	64	72
9	0	9	18	27	36	45	54	63	72	81

row→ **6**

Think: The product is found where row 6 and column 3 meet.

$$6 \times 3 = 18$$

So, there are 18 engines in all.

MATH IDEA You can use arrays, skip-counting, doubles, the Order Property of Multiplication, or a multiplication table to help you find products.

TECHNOLOGY LINK

More Practice:
Use Mighty Math
Carnival Countdown,
Snap Clowns, Level W.

▶ Check

1. Explain two ways to find 4×8.

Write a multiplication sentence for each.

2.

3.
✈✈✈ ✈✈✈
✈✈✈ ✈✈✈
✈✈✈ ✈✈✈
✈✈✈ ✈✈✈

4.

Find the product.

5. $2 \times 6 = \blacksquare$

6. $5 \times 3 = \blacksquare$

7. $\blacksquare = 1 \times 7$

8. $\blacksquare = 7 \times 3$

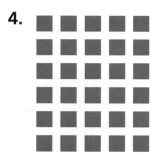

LESSON CONTINUES ▶

Find the product.

9. $7 \times 4 = 52$ 10. $2 \times 8 = 16$ 11. $7 \times 5 = 59$ 12 $4 \times 2 = 8$

13. $20 = 5 \times 4$ 14. $9 \times 4 = 36$ 15. $10 = 2 \times 5$ 16 $9 \times 1 = 9$

17. $9 = 1 \times 9$ 18. $6 \times 2 = 12$ 19. $5 \times 6 = 30$ 20 $32 = 8 \times 4$

21. $5 \times 1 = 5$ 22. $3 \times 0 = 0$ 23. $2 \times 9 = 18$ 24. $5 \times 6 = 30$

25. $2 \times 4 = 8$ 26. $21 = 3 \times 7$ 27. $8 \times 5 = 40$ 28. $27 = 9 \times 3$

29. 9   30. 0   31. 5   32. 8   33. 4   34. 4
   ×3   ×6   ×7   ×3   ×9   ×4
   ──   ──   ──   ──   ──   ──
   27   0   35   24   36   16

Copy and complete.

35.

×	2	4	7	8	9
2	4	8	14	16	18

36.

×	3	5	7	8	9
3	9	15	21	24	27

37.

×	2	7	5	3	8
4	8	28	20	12	32

38.

×	1	6	9	7	8
5	5	30	45	32	40

Compare. Write $<$ **,** $>$ **, or** $=$ **for each** ●**.**

39. 3×2 ● 4×1 40. 7×4 ● 4×8 41. 5×8 ● $35 + 6$

42. 4×6 ● 8×3 43. 3×6 ● 5×4 44. 7×5 ● 8×3

45. **REASONING** Jenny baked some cookies. She put 4 chocolate chips and 2 pecans on each cookie. If she used 24 chocolate chips in all, how many pecans did she use?

46. Pedro and Jon have 20 toy cars altogether. If Jon buys another toy car, he will have twice as many toy cars as Pedro. How many toy cars does Pedro have?

47. ❓ **What's the Error?** To find the product 5×6, Ellen made this array. What did Ellen do wrong?

48. **REASONING** Look at this number pattern. What is the rule? What are the missing numbers?

8, 11, 14, ■, ■, ■, 26

49. Three vans are going to the airport. There are 9 people in each van. How many people are going to the airport?

50. Marie has 4 loose stamps and 5 sheets of 8 stamps each. How many stamps does she have in all?

Mixed Review and Test Prep

51. How many minutes are between 12:25 P.M. and 12:30 P.M.? (p. 96)

52. $3,509$ (p. 46)
$+4,737$

53. $7,324$ (p. 62)
$-2,195$

54. What time will it be in 1 hour 45 minutes? (p. 98)

55. $17 + 88 + 71 = \blacksquare$
(p. 36)

56. $\blacksquare + 14 = 23$

57. $51 - \blacksquare = 39$

58. $872 + 208 = \blacksquare$ (p. 42)

59. (TEST PREP) Which shows the numbers in order from greatest to least? (p. 24)

A 998, 989, 999
B 989, 998, 999
C 999, 989, 998
D 999, 998, 989

60. (TEST PREP) Leo bought a comic book for $2.59. He paid with a $5 bill. How much change did he get back? (p. 86)

F $2.41   H $3.41
G $2.51   J $3.51

THINKER'S CORNER

USE DATA For 1–4, use the pictograph.

FLIGHTS FROM ORLANDO TO LOS ANGELES

Number of Flights per Day

Airline

Key: Each ✈ = 3 flights.

1. How many flights does Airline A have to Los Angeles each day? Airline B? Airline C?

2. Sara's mother is flying from Orlando to Los Angeles for business. How many flights does she have to choose from?

3. During a week, how many more flights does Airline C have than Airline A?

4. How many more flights does Airline B have to Los Angeles each day than Airline A? than Airline C?

Algebra: Find Missing Factors

Quick Review

1. $4 \times 6 = $
2. $2 \times 7 = $
3. $= 3 \times 0$
4. $6 \times 1 = $
5. $3 \times 6 = $

▶ **Learn**

BLUE RIBBON BAKING Mike's muffins won first prize at the county fair. Each plate held 5 muffins. He made 35 muffins. How many plates did he use?

A multiplication table can help you find the missing factor.

$$\blacksquare \qquad \times \qquad 5 \qquad = \qquad 35$$

↑ number of plates ↑ number of muffins on each plate ↑ total number of muffins

Start at the column for 5.
Look down to the product, 35.
Look left across the row from 35.
The missing factor is 7.

$$7 \qquad \times \qquad 5 \qquad = \qquad 35$$

↑ factor *row* ↑ factor *column* ↑ product *row 7 column 5*

So, Mike used 7 plates.

column
↓

×	0	1	2	3	4	5	6	7	8	9
0	0	0	0	0	0	0	0	0	0	0
1	0	1	2	3	4	5	6	7	8	9
2	0	2	4	6	8	10	12	14	16	18
3	0	3	6	9	12	15	18	21	24	27
4	0	4	8	12	16	20	24	28	32	36
5	0	5	10	15	20	25	30	35	40	45
row 6	0	6	12	18	24	30	36	42	48	54
→ 7	0	7	14	21	28	35	42	49	56	63
8	0	8	16	24	32	40	48	56	64	72
9	0	9	18	27	36	45	54	63	72	81

EXAMPLE

Find the missing factor.
$4 \times \blacksquare = 12$
Start at the row for 4.
Look right, to the product, 12.
Look up the column from 12.
The missing factor is 3.

So, $4 \times 3 = 12$.

column
↓

×	0	1	2	3	4	5	6	7	8	9
0	0	0	0	0	0	0	0	0	0	0
1	0	1	2	3	4	5	6	7	8	9
2	0	2	4	6	8	10	12	14	16	18
row 3	0	3	6	9	12	15	18	21	24	27
→ 4	0	4	8	12	16	20	24	28	32	36
5	0	5	10	15	20	25	30	35	40	45
6	0	6	12	18	24	30	36	42	48	54
7	0	7	14	21	28	35	42	49	56	63
8	0	8	16	24	32	40	48	56	64	72
9	0	9	18	27	36	45	54	63	72	81

MATH IDEA When you know the product and one factor, a multiplication table can help you find the missing factor.

CALIFORNIA STANDARDS AF 1.0 Students select appropriate symbols, operations, and properties to represent, describe, simplify, and solve simple number relationships. **O—n AF 1.1** Represent relationships of quantities in the form of mathematical expressions, equations, or inequalities. *also* **NS 2.0, O—n NS 2.2, O—n NS 2.3, NS 2.8, AF 1.2, MR 1.0, MR 1.1, MR 2.2, MR 2.3**

1. **Explain** how to use the table to find the missing factor in ■ × 6 = 18.

Find the missing factor.

2. ■ × 2 = 8   **3.** 3 × ■ = 9   **4.** 5 × ■ = 20

► **Practice and Problem Solving**

Find the missing factor.

5. ■ × 4 = 12   **6.** ■ × 3 = 21   **7.** 5 × ■ = 0   **8.** 2 × ■ = 12

9. 1 × ■ = 9   **10.** 8 × ■ = 24   **11.** ■ × 6 = 30   **12.** ■ × 4 = 32

13. ■ × 2 = 18   **14.** 4 × ■ = 16   **15.** 5 × ■ = 15   **16.** ■ × 6 = 24

17. ■ × 4 = 36   **18.** ■ × 3 = 27   **19.** 6 × ■ = 18   **20.** 8 × ■ = 40

21. 4 × 6 = ■ × 3   **22.** 9 × ■ = 50 − 5   **23.** 7 × ■ = 32 − 4

24. The product of 4 and another factor is 28. What is the other factor?

25. If you multiply 9 by a number, the product is 27. What is the number?

26. There are 2 chairs at each table. If there are 14 chairs, how many tables are there? Write a multiplication sentence to solve.

27. There are 4 oatmeal cookies and 3 sugar cookies on each plate. How many cookies are on 5 plates?

28. Isabel has $1.29 and Luke has $5.95. How much more does Luke have than Isabel?

29. **? What's the Question?** Pies are on sale for $3 each. Carly spent $12 on pies. The answer is 4 pies.

─ **Mixed Review and Test Prep** ─

30. Tony has two $1 bills and 1 dime. Benita has one $1 bill and 4 quarters. Who has the greater amount of money? (p. 84)

31. 400 (p. 58)
      −137

32. 453 (p. 42)
      +487

33. Write <, >, or = . (p. 20)

5,450 ● 5,405

34. **TEST PREP** Which number means 10,000 + 1,000 + 10 + 1? (p. 10)

A 10,111   C 11,101

B 11,011   D 11,110

Review/Test

✔ CHECK VOCABULARY

Choose the best term from the box.

1. The product of _?_ and any number equals that number. (p. 132)

2. The product of _?_ and any number equals zero. (p. 132)

3. On a multiplication table, the _?_ is found where the factor row and the factor column meet. (p. 134)

> array
> zero
> product
> one
> factor

✔ CHECK SKILLS

Find the product. (pp. 132-133)

4. $5 \times 1 = \blacksquare$
5. $0 \times 6 = \blacksquare$
6. $\blacksquare = 1 \times 8$
7. $\blacksquare = 9 \times 0$

Find the product. (pp. 134-135)

8. $\blacksquare = 4 \times 3$
9. $2 \times 4 = \blacksquare$
10. $5 \times 4 = \blacksquare$
11. $\blacksquare = 8 \times 4$
12. $4 \times 9 = \blacksquare$
13. $4 \times 4 = \blacksquare$
14. $\blacksquare = 6 \times 4$
15. $4 \times 0 = \blacksquare$

Find the missing factor. (pp. 142-143)

16. $2 \times \blacksquare = 8$
17. $\blacksquare \times 5 = 30$
18. $1 \times \blacksquare = 9$
19. $\blacksquare \times 6 = 18$
20. $\blacksquare \times 1 = 9$
21. $4 \times \blacksquare = 32$
22. $\blacksquare \times 3 = 18$
23. $7 \times \blacksquare = 0$

✔ CHECK PROBLEM SOLVING

Solve. (pp. 136-137)

24. The first four numbers in the pattern are 4, 8, 12, 16. What is the rule? What are the next three numbers?

25. Lin saw this number pattern: 10, 13, 16, 19, 22, 25, and 28. What is the rule? What are the next three numbers?

Cumulative Review

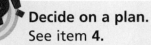

Decide on a plan.
See item **4.**

The problem tells you to multiply a number by 4. You should start by writing a number sentence from the words. Then solve the number sentence.

Also see problem **4,** p. H63.

For 1–10, choose the best answer.

1. Maria had 8 plates. She put 1 cookie on each plate. How many cookies did she use?

A 0	**C** 6
B 4	**D** 8

2. $9 \times 0 = \blacksquare$

F 0	**H** 9
G 1	**J** NOT HERE

3. Matt has 7 folders for his school work. Each folder has 4 assignments in it. How many assignments are there altogether?

A 28	**C** 21
B 24	**D** 11

4. A number multiplied by 4 is 36. What is the number?

F 40	**H** 9
G 32	**J** 8

5. Justin put 5 jelly beans in each of 8 cups. How many jelly beans are in the cups altogether?

A 13	**C** 33
B 32	**D** 40

6. These numbers follow a pattern.

20, 17, 14, 11

Which numbers continue the pattern?

F 8, 5, 2	**H** 10, 9, 8
G 9, 7, 5	**J** 12, 13, 14

7. These numbers follow a pattern.

5, 9, 13, 17

Which numbers continue the pattern?

A 18, 19, 20	**C** 19, 21, 23
B 21, 25, 29	**D** 20, 23, 26

8. Bradley has 4 each of 3 kinds of marbles. How many marbles does he have in all?

F 7	**H** 15
G 12	**J** NOT HERE

9. Which number makes this equation true?

$$\blacksquare \times 1 = 5$$

A 6	**C** 4
B 5	**D** 2

10. Pete is setting the table for 9 people. Each person will get 1 fork, 1 knife, and 1 spoon. How many forks, knives and spoons will Pete put on the table in all?

F 45	**H** 27
G 42	**J** 14

CHAPTER 9

Multiplication Facts and Strategies

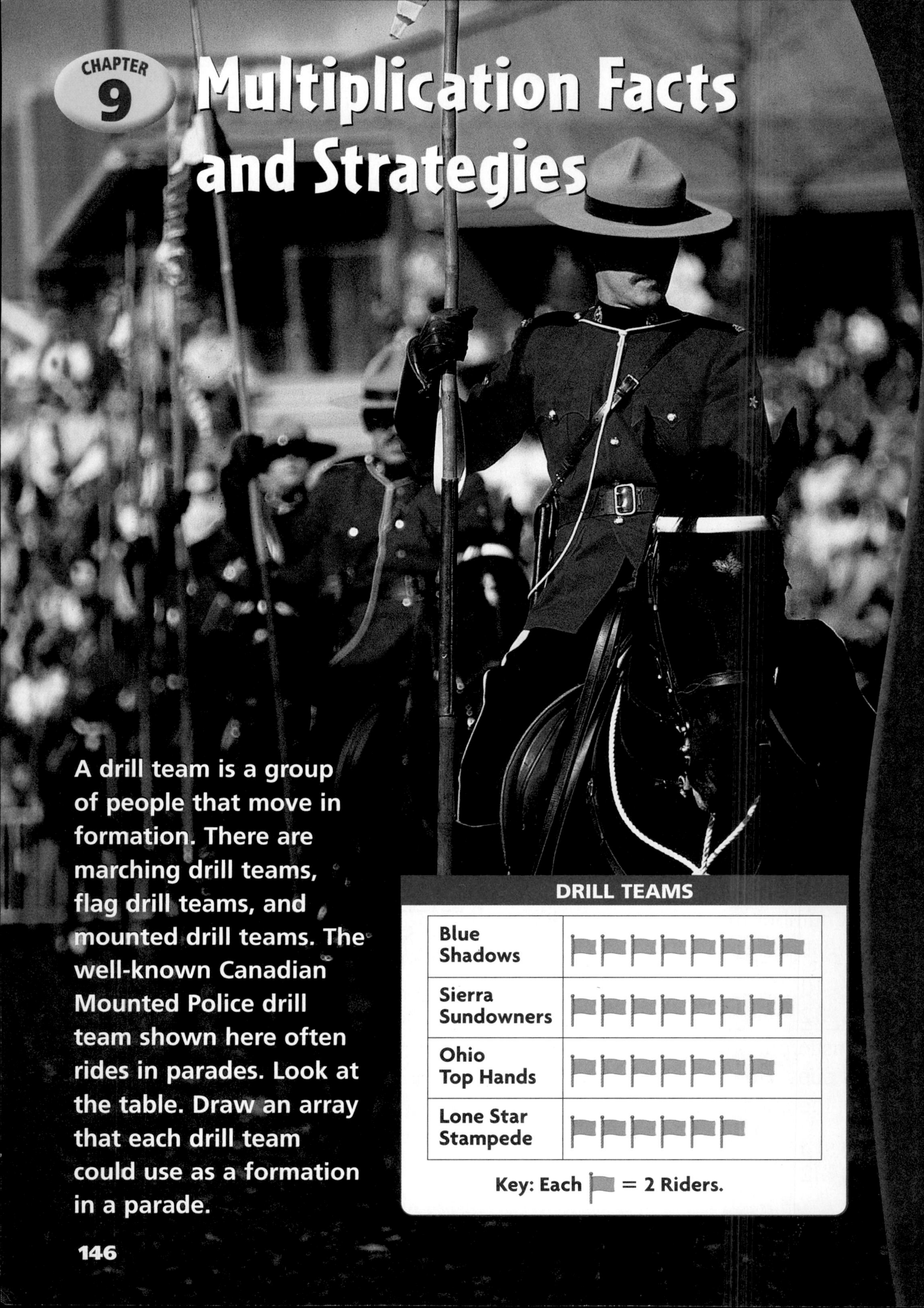

A drill team is a group of people that move in formation. There are marching drill teams, flag drill teams, and mounted drill teams. The well-known Canadian Mounted Police drill team shown here often rides in parades. Look at the table. Draw an array that each drill team could use as a formation in a parade.

DRILL TEAMS

Blue Shadows	⚑⚑⚑⚑⚑⚑⚑⚑
Sierra Sundowners	⚑⚑⚑⚑⚑⚑⚑ and half of ⚑
Ohio Top Hands	⚑⚑⚑⚑⚑⚑⚑
Lone Star Stampede	⚑⚑⚑⚑⚑⚑

Key: Each ⚑ = 2 Riders.

CHECK WHAT YOU KNOW ✓

Use this page to help you review and remember
important skills needed for Chapter 9.

✓ VOCABULARY

Choose the best term from the box to
describe the example $2 \times 5 = 10$.

> difference
> factor
> product

1. In the example, the number 2 is called a __?__ .

2. In the example, the number 10 is called the __?__ .

✓ ADD EQUAL GROUPS (See p. H12.)

Write how many there are in all.

3.

2 groups of 4 = ■

4.

3 groups of 6 = ■

5.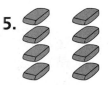

3 groups of 4 = ■

6.

5 groups of 2 = ■

7.

1 group of 5 = ■

8.

4 groups of 2 = ■

✓ MULTIPLICATION FACTS THROUGH 5 (See p. H12.)

Find the product.

9. $7 \times 3 = $ ■

10. $5 \times 5 = $ ■

11. ■ $ = 7 \times 4$

12. $1 \times 2 = $ ■

13. $4 \times 1 = $ ■

14. $6 \times 3 = $ ■

15. $9 \times 5 = $ ■

16. ■ $ = 6 \times 1$

17. $5 \times 3 = $ ■

18. ■ $ = 9 \times 4$

19. $5 \times 0 = $ ■

20. $7 \times 5 = $ ■

21.   6
       $\times 4$

22.   4
       $\times 2$

23.   2
       $\times 3$

24.   8
       $\times 4$

25.   3
       $\times 1$

26.   5
       $\times 2$

27.   1
       $\times 7$

28.   4
       $\times 5$

29.   0
       $\times 3$

30.   2
       $\times 8$

Multiply with 6

Quick Review

1. $5 \times 3 = \blacksquare$
2. $5 \times \blacksquare = 25$ 3. $5 \times 9 = \blacksquare$
4. $\blacksquare \times 5 = 30$ 5. $40 = 5 \times \blacksquare$

▶ Learn

MARCHING MULTIPLES   The school band has 6 rows, with 6 students in each row. How many students are in the band?

$$6 \times 6 = \blacksquare$$

One Way   Break apart an array to find the product.

STEP 1
Make an array that shows 6 rows of 6.

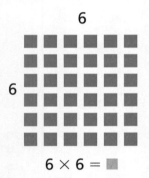

$6 \times 6 = \blacksquare$

STEP 2
Break the array into two smaller arrays.

$1 \times 6 = 6$

$5 \times 6 = 30$

STEP 3
Add the products of the two arrays.

$$\begin{array}{r} 6 \\ +30 \\ \hline 36 \end{array}$$

$6 \times 6 = 36$
So, there are 36 students in the band.

Another Way   When one of the factors is an even number, you can use doubles. The product of each 6's fact is double the product of each 3's fact.

To find 6×6
- First find the 3's fact.
  Think: $6 \times 3 = 18$
- Double the product.
  $18 + 18 = 36$
- So, $6 \times 6 = 36$.

$0 \times 3 = 0$	$0 \times 6 = 0$
$1 \times 3 = 3$	$1 \times 6 = 6$
$2 \times 3 = 6$	$2 \times 6 = 12$
$3 \times 3 = 9$	$3 \times 6 = 18$
$4 \times 3 = 12$	$4 \times 6 = 24$
$5 \times 3 = 15$	$5 \times 6 = 30$
$6 \times 3 = 18$	$6 \times 6 = \blacksquare$
$7 \times 3 = 21$	$7 \times 6 = 42$
$8 \times 3 = 24$	$8 \times 6 = 48$
$9 \times 3 = 27$	$9 \times 6 = 54$

CALIFORNIA STANDARDS O⎯ⁿ **NS 2.2** Memorize to automaticity the multiplication table for numbers between 1 and 10. O⎯ⁿ **AF 2.1** Solve simple problems involving a functional relationship between two quantities. *also* **NS 2.0, NS 2.8, AF 1.0, AF 1.2, MR 1.2, MR 2.2, MR 3.2**

► Check

1. Explain how you can use 8×3 to find 8×6.

Find each product.

2. $2 \times 6 = $ ■   **3.** $4 \times 6 = $ ■   **4.** $5 \times 6 = $ ■   **5.** $6 \times 9 = $ ■

► Practice and Problem Solving

Find each product.

6. $3 \times 6 = $ ■   **7.** $6 \times 5 = $ ■   **8.** $5 \times 9 = $ ■   **9.** ■ $= 8 \times 6$

10. $4 \times 7 = $ ■   **11.** ■ $= 3 \times 4$ **12.** $4 \times 9 = $ ■   **13.** $6 \times 0 = $ ■

14. ■ $= 2 \times 9$ **15.** ■ $= 8 \times 4$ **16.** $3 \times 5 = $ ■   **17.** $9 \times 6 = $ ■

18.	**19.**	**20.**	**21.**	**22.**	**23.**
5	6	8	6	5	6
×7	×7	×3	×1	×8	×6

Copy and complete the multiplication table.

24.

×	0	1	2	3	4	5	6	7	8	9
6	■	■	■	■	■	■	■	■	■	■

Complete.

25. ■ $\times 4 = 12$ **26.** ■ $\times 6 = 42$ **27.** $48 = 8 \times$ ■

28. ■ $\times 4 = 4 \times 3$ **29.** $3 \times 6 = $ ■ $\times 2$ **30.** ■ $\times 6 = 40 + 8$

31. A guitar has 6 strings. A banjo has 5 strings. How many strings are on 4 guitars and 2 banjos?

32. **Write About It** Draw arrays to show that 6×4 is the same as 1×4 plus 5×4.

Mixed Review and Test Prep

33. 327 (p. 58)
$- \ 82$

34. 600 (p. 58)
-346

Write the value of the blue digit. (p. 10)

35. $57,899$ **36.** $98,365$

37. **TEST PREP** Find the sum of 452 and 678. (p. 42)

A 226   **C** 1,120

B 1,030   **D** 1,130

Multiply with 7

Quick Review

1. $6 \times 5 = $
2. $6 \times 4 = $
3. $3 \times 6 = $
4. $2 \times 6 = $
5. $5 \times 6 = $

▶ **Learn**

PARADE! PARADE! Students from the local high school built a float for a parade. They worked on the float for 8 weeks. How many days did they work on the float?

$$8 \times 7 = \blacksquare$$

Break apart an array to find the product.

STEP 1

Make an array that shows 8 rows of 7.

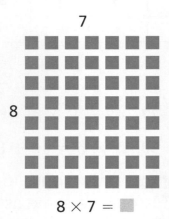

$$8 \times 7 = \blacksquare$$

STEP 2

Break the array into two smaller arrays.

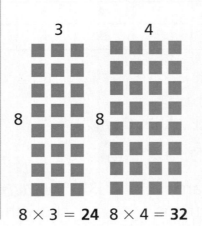

$$8 \times 3 = \mathbf{24} \quad 8 \times 4 = \mathbf{32}$$

STEP 3

Add the products of the two arrays.

$$\begin{array}{r} 24 \\ +32 \\ \hline 56 \end{array}$$

$$8 \times 7 = 56$$

So, the students worked 56 days on the float.

▶ **Check**

1. Tell how you could break apart 7×5 into two facts to help you find the product.

TECHNOLOGY LINK

More Practice: Use E-Lab, Multiplication Arrays.

www.harcourtschool.com/elab2002

Find each product.

2. $7 \times 4 = \blacksquare$ **3.** $0 \times 7 = \blacksquare$ **4.** $7 \times 5 = \blacksquare$ **5.** $7 \times 5 = \blacksquare$

CALIFORNIA STANDARDS O━ᅮNS 2.2 Memorize to automaticity the multiplication table for numbers between 1 and 10. O━ᅮAF 2.1 Solve simple problems involving a functional relationship between two quantities. *also* NS 2.0, NS 2.8, AF 1.0, AF 1.2, MR 1.2, MR 2.2, MR 3.2

Find each product.

6. $2 \times 7 = \blacksquare$ **7.** $2 \times 9 = \blacksquare$ **8.** $3 \times 7 = \blacksquare$

9. $6 \times 6 = \blacksquare$ **10.** $7 \times 6 = \blacksquare$ **11.** $\blacksquare = 6 \times 9$

12. $\blacksquare = 5 \times 9$ **13.** $7 \times 7 = \blacksquare$ **14.** $4 \times 7 = \blacksquare$

15. $3 \times 6 = \blacksquare$ **16.** $\blacksquare = 4 \times 9$ **17.** $\blacksquare = 5 \times 5$

18. $\begin{array}{r} 1 \\ \times 7 \\ \hline \end{array}$ **19.** $\begin{array}{r} 7 \\ \times 9 \\ \hline \end{array}$ **20.** $\begin{array}{r} 6 \\ \times 8 \\ \hline \end{array}$ **21.** $\begin{array}{r} 5 \\ \times 8 \\ \hline \end{array}$ **22.** $\begin{array}{r} 4 \\ \times 5 \\ \hline \end{array}$ **23.** $\begin{array}{r} 5 \\ \times 7 \\ \hline \end{array}$

Copy and complete the multiplication table.

24.

×	0	1	2	3	4	5	6	7	8	9
7	■	■	■	■	■	■	■	■	■	■

Complete.

25. $7 \times 6 = \blacksquare + 21$ **26.** $\blacksquare \times 4 = 30 - 2$ **27.** $8 + 6 = 7 \times \blacksquare$

28. REASONING How can you tell without multiplying that 7×9 is less than 9×8?

29. REASONING Explain how you can use $6 \times 2 = 12$ to find 7×2.

30. Shayla was on vacation for 7 weeks. She spent 3 weeks at band camp and the rest of the time at home. How many days did she spend at home?

31. Break apart the array. Then write the multiplication fact.

32. (a+b over c) **Algebra** Write a one-digit number to make this number sentence true.
$\blacksquare \times 7 + 10 > 67 - 9$

Mixed Review and Test Prep

33. $\begin{array}{r} 3,458 \text{ (p. 46)} \\ +1,679 \\ \hline \end{array}$ **34.** $\begin{array}{r} 2,008 \text{ (p. 46)} \\ +1,256 \\ \hline \end{array}$ **35.** $\begin{array}{r} 2,814 \text{ (p. 62)} \\ -1,680 \\ \hline \end{array}$ **36.** $\begin{array}{r} 8,093 \text{ (p. 62)} \\ -5,934 \\ \hline \end{array}$

37. TEST PREP Subtract 49 from 201. (p. 58)

A 152   **B** 162   **C** 250   **D** 252

Multiply with 8

Quick Review

1. $35 = \blacksquare \times 7$

2. $2 \times 7 = \blacksquare$

3. $7 \times 3 = \blacksquare$

4. $7 \times \blacksquare = 49$

5. $6 \times 7 = \blacksquare$

▶ Learn

BAKE-OFF Mr. Lee baked 6 peach pies for the state fair. He used 8 peaches in each pie. How many peaches did he use in all?

$$6 \times 8 = \blacksquare$$

One Way Break apart an array to find the product.

STEP 1	**STEP 2**	**STEP 3**
Make an array that shows 6 rows of 8.	Break the array into two smaller arrays.	Add the products of the two arrays.

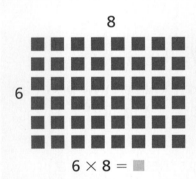

$6 \times 8 = \blacksquare$

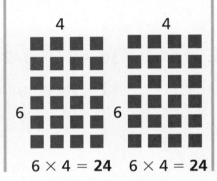

$6 \times 4 = 24$ $6 \times 4 = 24$

$$\begin{array}{r} 24 \\ +24 \\ \hline 48 \end{array}$$

$6 \times 8 = 48$

So, Mr. Lee used 48 peaches in all.

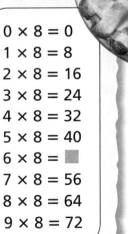

Another Way When one of the factors is an even number, you can use doubles. The product of each 8's fact is double the product of each 4's fact.

To find 6×8

- First find the 4's fact.
  Think: $6 \times 4 = 24$

- Double the product.
  $24 + 24 = 48$

- So, $6 \times 8 = 48$.

$0 \times 4 = 0$	$0 \times 8 = 0$
$1 \times 4 = 4$	$1 \times 8 = 8$
$2 \times 4 = 8$	$2 \times 8 = 16$
$3 \times 4 = 12$	$3 \times 8 = 24$
$4 \times 4 = 16$	$4 \times 8 = 32$
$5 \times 4 = 20$	$5 \times 8 = 40$
$6 \times 4 = 24$	$6 \times 8 = \blacksquare$
$7 \times 4 = 28$	$7 \times 8 = 56$
$8 \times 4 = 32$	$8 \times 8 = 64$
$9 \times 4 = 36$	$9 \times 8 = 72$

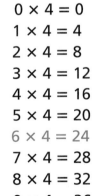

CALIFORNIA STANDARDS O¬NS 2.2 Memorize to automaticity the multiplication table for numbers between 1 and 10. O¬AF 2.1 Solve simple problems involving a functional relationship between two quantities. *also* NS 2.0, NS 2.8, AF 1.0, AF 1.2, MR 1.2, MR 2.2, MR 3.2

▶ Check

1. Explain how you can use $4 \times 5 = 20$ to find 8×5.

Find each product.

2. $8 \times 4 = $ ▢   **3.** $8 \times 2 = $ ▢   **4.** $5 \times 8 = $ ▢   **5.** $1 \times 8 = $ ▢

▶ Practice and Problem Solving

Find each product.

6. $5 \times 4 = $ ▢   **7.** $8 \times 3 = $ ▢   **8.** $9 \times 8 = $ ▢   **9.** $4 \times 7 = $ ▢

10. $8 \times 6 = $ ▢   **11.** $3 \times 4 = $ ▢   **12.** $5 \times 7 = $ ▢   **13.** $8 \times 8 = $ ▢

14. $2 \times 9 = $ ▢   **15.** $7 \times 8 = $ ▢   **16.** $5 \times 9 = $ ▢   **17.** $6 \times 6 = $ ▢

18. $\begin{array}{r} 4 \\ \times 8 \\ \hline \end{array}$ **19.** $\begin{array}{r} 7 \\ \times 7 \\ \hline \end{array}$ **20.** $\begin{array}{r} 9 \\ \times 6 \\ \hline \end{array}$ **21.** $\begin{array}{r} 7 \\ \times 9 \\ \hline \end{array}$ **22.** $\begin{array}{r} 3 \\ \times 7 \\ \hline \end{array}$ **23.** $\begin{array}{r} 6 \\ \times 8 \\ \hline \end{array}$

Complete the multiplication table.

24.

×	0	1	2	3	4	5	6	7	8	9
8	▢	▢	▢	▢	▢	▢	▢	▢	▢	▢

Compare. Write $<$, $>$, or $=$ for each ⬤.

25. 2×3 ⬤ 2×4 **26.** 5×8 ⬤ 8×5 **27.** 5×5 ⬤ 4×6

28. $\frac{a+b}{c}$ **Algebra** Hal has 7 bags of 8 green apples and 1 bag of red apples. Altogether he has 60 apples. How many red apples does he have?

29. **? What's the Error?** Robin says, "I can find 8×7 by thinking of $3 \times 7 = 21$ and doubling it."

30. REASONING If you know $9 \times 4 = 36$, how can you find 8×4?

31. **? What's the Question?** Joanna has 9 boxes of pears. She has 72 pears in all. The answer is 8 pears.

Mixed Review and Test Prep

32. $\begin{array}{r} 172 \text{ (p. 42)} \\ +781 \\ \hline \end{array}$ **33.** $\begin{array}{r} 399 \text{ (p. 42)} \\ +421 \\ \hline \end{array}$ **34.** $\begin{array}{r} 2,523 \text{ (p. 46)} \\ +1,607 \\ \hline \end{array}$ **35.** $\begin{array}{r} 3,627 \text{ (p. 46)} \\ +4,482 \\ \hline \end{array}$

36. **TEST PREP** What is another way to show $2 + 2 + 2 + 2$? (p. 116)

A 2×2 **B** 4×2 **C** 2×8 **D** 8×4

Problem Solving Strategy

Draw a Picture

 Understand → Plan → Solve → Check

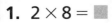

PROBLEM Jacob's grandmother has made 16 squares for a quilt. How can she arrange the squares so the quilt is the same number of squares wide as it is long?

Understand

• What are you asked to find?

• What information will you use?

• Is there information you will not use?

Plan

• What strategy can you use to solve the problem?

Draw a picture to show how to arrange the squares.

Solve

• How can you use the strategy to solve the problem?

Try 3 squares wide and 3 squares long.

$3 \times 3 = 9$

This quilt is 9 squares in all. Try again.

Try 4 squares wide and 4 squares long.

$4 \times 4 = 16$

This quilt is 16 squares in all.

So, her quilt should be 4 squares wide and 4 squares long.

Check

• Look at the Problem again. Explain why this answer makes sense.

CALIFORNIA STANDARDS MR 2.0 Students use strategies, skills, and concepts in finding solutions. **MR 2.3** Use a variety of methods, such as words, numbers, symbols, charts, graphs, tables, diagrams, and models, to explain mathematical reasoning. *also* **O—πNS 2.2, NS 2.8, MR 1.1, MR 2.4, MR 3.1, MR 3.2**

Problem Solving Practice

PROBLEM SOLVING STRATEGIES

▶ Draw a Diagram or Picture
Make a Model or Act It Out
Make an Organized List
Find a Pattern
Make a Table or Graph
Predict and Test
Work Backward
Solve a Simpler Problem
Write an Equation
Use Logical Reasoning

Use *draw a picture* to solve.

1. **What if** Jacob's grandmother had a total of 25 squares? How could she arrange the squares so the quilt is the same number of squares long as it is wide?

2. Amrita has completed 32 squares. How many more does she have to make to be able to arrange them so that the quilt is the same number of squares long as it is wide?

Mrs. Adams is hanging her students' pictures. She has 24 pictures to arrange.

3. Which arrangement can Mrs. Adams **not** use?

   A 3 rows of 8   **C** 4 rows of 6
   B 6 rows of 4   **D** 4 rows of 8

4. Mrs. Adams decides to put 4 pictures in a row. How many rows does she need?

   F 3   **G** 4   **H** 6   **J** 8

Mixed Strategy Practice

5. **REASONING** I am a one-digit number. If you multiply me by 3 the product has a 1 in the ones place. What number am I?

6. List all the ways you can use coins to make 21¢.

USE DATA For 7–9, use the graph.

7. If 3 frogs leave the pond, how many animals in all are in the pond?

8. If 5 ducks come to the pond and 2 geese leave the pond, how many ducks and geese are in the pond?

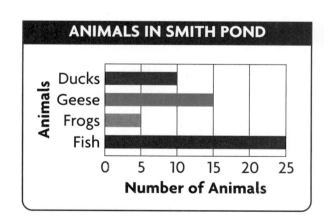

9. If 5 more frogs come to the pond, how should the bar graph change?

Algebra:
Practice the Facts

▶ **Learn**

Splash! Each instructor teaches a group of 6 children. If there are 7 instructors, how many children are taking swimming lessons?

$$7 \times 6 = \blacksquare$$

You have learned many ways to find 7×6.

A. Break an array into known facts.

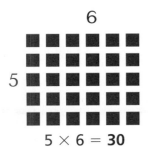

$2 \times 6 = 12$

6

5

$5 \times 6 = 30$

$12 + 30 = 42$, so $7 \times 6 = 42$.

B. Use a multiplication table.

✕	0	1	2	3	4	5	6	7	8	9
0	0	0	0	0	0	0	0	0	0	0
1	0	1	2	3	4	5	6	7	8	9
2	0	2	4	6	8	10	12	14	16	18
3	0	3	6	9	12	15	18	21	24	27
4	0	4	8	12	16	20	24	28	32	36
5	0	5	10	15	20	25	30	35	40	45
6	0	6	12	18	24	30	36	42	48	54
7	0	7	14	21	28	35	42	49	56	63
8	0	8	16	24	32	40	48	56	64	72
9	0	9	18	27	36	45	54	63	72	81

$7 \times 6 = 42$

C. Use the Order Property of Multiplication.

Try changing the order of the factors:
Think: If $6 \times 7 = 42$, then $7 \times 6 = 42$.

D. When one of the factors is an even number, you can use doubles.

To find a 6's fact, you can double a 3's fact.

- First find the 3's fact.
  Think: $7 \times 3 = 21$
- Double the product. $21 + 21 = 42$
  $7 \times 6 = 42$.

So, 42 children are taking lessons.

CALIFORNIA STANDARDS O⊸ **NS 2.2** Memorize to automaticity the multiplication table for numbers between 1 and 10. **AF 1.5** Recognize and use the commutative and associative properties of multiplication. *also* **NS 2.0, AF 1.0, AF 1.2,** O⊸**AF 2.1, MR 1.2, MR 2.2, MR 2.3, MR 2.4, MR 3.2**

Ways to Find a Product

What if there are 8 instructors with 5 swimmers each? How many children are taking lessons?

$$8 \times 5 = \blacksquare$$

David and Niam use different ways to find 8×5.

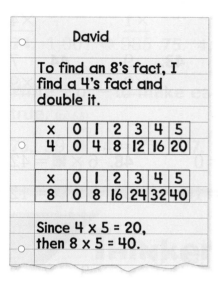

David

To find an 8's fact, I find a 4's fact and double it.

x	0	1	2	3	4	5
4	0	4	8	12	16	20

x	0	1	2	3	4	5
8	0	8	16	24	32	40

Since $4 \times 5 = 20$, then $8 \times 5 = 40$.

Niam

I can use the Order Property of Multiplication.

I know that 8×5 is the same as 5×8.

$5 \times 8 = 8 \times 5 = 40$

- What is another way that David or Niam could find 8×5?

- If you can't remember the fact 7×8, what strategy can you use?

▶ Check

1. **Explain** how you could use $9 \times 5 = 45$ to find 8×5.

2. **Describe** how you could use doubles to find 6×9.

Find each product.

3. $4 \times 5 = \blacksquare$

4. $3 \times 7 = \blacksquare$

5. $6 \times 4 = \blacksquare$

6. $7 \times 6 = \blacksquare$

7. $2 \times 5 = \blacksquare$

8. $6 \times 6 = \blacksquare$

9. $\blacksquare = 4 \times 3$

10. $2 \times 2 = \blacksquare$

11. $1 \times 8 = \blacksquare$

12. $\blacksquare = 5 \times 3$

13. $9 \times 2 = \blacksquare$

14. $5 \times 5 = \blacksquare$

15. $\begin{array}{r} 5 \\ \times 9 \\ \hline \end{array}$

16. $\begin{array}{r} 6 \\ \times 3 \\ \hline \end{array}$

17. $\begin{array}{r} 7 \\ \times 7 \\ \hline \end{array}$

18. $\begin{array}{r} 4 \\ \times 8 \\ \hline \end{array}$

19. $\begin{array}{r} 0 \\ \times 4 \\ \hline \end{array}$

20. $\begin{array}{r} 3 \\ \times 3 \\ \hline \end{array}$

Review/Test

✓ CHECK CONCEPTS

Name a way to break apart each array.
Then write the product. (pp. 148–153)

1. 3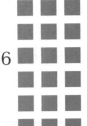

2. 7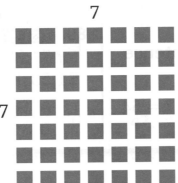

3. 5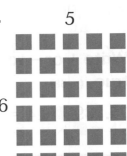

✓ CHECK SKILLS

Find each product. (pp. 148–153, 156–159)

4. $6 \times 7 = \blacksquare$ 5. $\blacksquare = 8 \times 7$ 6. $4 \times 6 = \blacksquare$ 7. $9 \times 7 = \blacksquare$

8. $\blacksquare = 5 \times 8$ 9. $6 \times 0 = \blacksquare$ 10. $3 \times 8 = \blacksquare$ 11. $5 \times 7 = \blacksquare$

12. $6 \times 6 = \blacksquare$ 13. $8 \times 9 = \blacksquare$ 14. $8 \times 6 = \blacksquare$ 15. $6 \times 9 = \blacksquare$

16. $\begin{array}{r} 8 \\ \times 8 \\ \hline \end{array}$ 17. $\begin{array}{r} 8 \\ \times 4 \\ \hline \end{array}$ 18. $\begin{array}{r} 2 \\ \times 7 \\ \hline \end{array}$ 19. $\begin{array}{r} 7 \\ \times 3 \\ \hline \end{array}$ 20. $\begin{array}{r} 6 \\ \times 2 \\ \hline \end{array}$ 21. $\begin{array}{r} 7 \\ \times 4 \\ \hline \end{array}$

✓ CHECK PROBLEM SOLVING

Solve. (pp. 154–155)

22. Hisako has 2 rows of 6 cookies. What is another way she could arrange the cookies in equal rows?

23. Calvin had 3 rows of 7 cookies. He gave away 8 cookies. How many were left?

24. Rita has 49 squares for a quilt. How can she arrange the squares so the quilt is the same number of squares long as it is wide?

25. Brad had 6 equal rows of stamps. He then bought 3 more stamps. If he now has 33 stamps, how many were in each row?

Cumulative Review

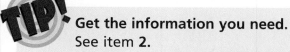

Get the information you need.
See item **2.**

You need to remember how many days are in a week to solve this problem.

Also see problem **3,** p. H63.

For 1–10, choose the best answer.

1. Betty baked some cookies. She put 6 rows of cookies, with 5 cookies in each row, on each pan. How many cookies did she put on each pan?

A 30   **C** 24
B 25   **D** 11

2. Evan attended camp for 4 weeks. How many days are in 4 weeks?

F 30   **H** 24
G 28   **J** 20

3. Which number makes this number sentence true? $7 \times \blacksquare = 42$

A 5   **C** 7
B 6   **D** 8

4. There are 8 rows of seats on a bus. There is room for 4 people in each row. How many people can ride on the bus?

F 8   **H** 24
G 16   **J** NOT HERE

5. What is another way to show $7 + 7 + 7$?

A 7×7 **C** 3×7
B 21×3 **D** $3 + 7$

6. $\begin{array}{r} 8 \\ \times 8 \\ \hline \end{array}$

F 56   **H** 72
G 64   **J** NOT HERE

7. Marcie bought 3 colors of yarn to knit a sweater. She bought 6 packages of each color. Which number sentence tells how many packages of yarn she bought?

A $6 \times 1 = 6$
B $3 \times 3 = 9$
C $3 \times 6 = 18$
D $6 \times 6 = 36$

8. Mr. Perez made a display with 4 rows of 9 apples. Ted's mother bought 15 apples. How many apples were left in the display?

F 36   **H** 24
G 28   **J** 21

9. Mrs. Kahn chose 7 students to make bookmarks for the book fair. Each student made 9 bookmarks. How many bookmarks did they make in all?

A 16   **C** 63
B 56   **D** 72

10. Andy wants to buy colored pencils for $0.59, paper for $0.85, and an eraser for $0.29. How much money does Andy need?

F $1.73   **H** $3.27
G $2.27   **J** $4.73

Multiplication Facts and Patterns

WEIGHT OF A NEWBORN PUPPY

Siberian Husky	🐾🐾🐾🐾🐾🐾
Alaskan Malamute	🐾🐾🐾🐾🐾🐾🐾🐾
American Eskimo	🐾🐾🐾🐾🐾
Samoyed	🐾🐾🐾🐾🐾

Key: Each 🐾 = 2 ounces.

For more than 1,000 years, sled dogs were used for transportation, protection, and companionship. Dogsled teams were used to deliver mail in places such as Michigan, Minnesota, Wisconsin, and Alaska. Large teams of dogs could pull 400 to 500 pounds of mail. A full-grown Samoyed sled dog weighs about 50 pounds. A team of 10 Samoyeds would weigh 500 pounds. What would 10 newborn Samoyed puppies weigh?

CHECK WHAT YOU KNOW

Use this page to help you review and remember
important skills needed for Chapter 10.

SKIP-COUNT BY TENS (See p. H13.)

Continue the pattern.

1. 10, 20, 30, 40, ▪, ▪

2. 30, 40, 50, 60, 70, ▪, ▪

Skip-count by tens to find the missing numbers.

3. 3, 13, 23, ▪, ▪, 53, ▪, ▪, 83, ▪

4. 7, 17, 27, ▪, ▪, 57, ▪, 77, ▪, ▪

5. 5, 15, 25, ▪, ▪, 55, ▪, ▪, ▪

6. 64, 54, 44, ▪, ▪, ▪, ▪

MULTIPLICATION FACTS THROUGH 8 (See p. H13.)

Find the product.

7. $5 \times 1 = $ ▪

8. ▪ $= 6 \times 3$

9. ▪ $= 5 \times 2$

10. $3 \times 7 = $ ▪

11. ▪ $= 8 \times 4$

12. $5 \times 6 = $ ▪

13. $7 \times 4 = $ ▪

14. ▪ $= 8 \times 2$

15. $4 \times 5 = $ ▪

16. $8 \times 6 = $ ▪

17. $5 \times 8 = $ ▪

18. $0 \times 6 = $ ▪

19. $3 \times 8 = $ ▪

20. ▪ $= 6 \times 7$

21. ▪ $= 1 \times 4$

22. ▪ $= 3 \times 2$

23. ▪ $= 6 \times 6$

24. $5 \times 3 = $ ▪

25. $2 \times 7 = $ ▪

26. $3 \times 1 = $ ▪

27.
$$\begin{array}{r} 9 \\ \times 6 \\ \hline \end{array}$$

28.
$$\begin{array}{r} 7 \\ \times 8 \\ \hline \end{array}$$

29.
$$\begin{array}{r} 8 \\ \times 8 \\ \hline \end{array}$$

30.
$$\begin{array}{r} 9 \\ \times 0 \\ \hline \end{array}$$

31.
$$\begin{array}{r} 8 \\ \times 9 \\ \hline \end{array}$$

32.
$$\begin{array}{r} 4 \\ \times 2 \\ \hline \end{array}$$

33.
$$\begin{array}{r} 5 \\ \times 7 \\ \hline \end{array}$$

34.
$$\begin{array}{r} 5 \\ \times 5 \\ \hline \end{array}$$

35.
$$\begin{array}{r} 1 \\ \times 7 \\ \hline \end{array}$$

36.
$$\begin{array}{r} 3 \\ \times 3 \\ \hline \end{array}$$

▶ Practice and Problem Solving

Find the product.

19. ■ = 10 × 4 **20.** 4 × 8 = ■ **21.** 9 × 8 = ■ **22.** ■ = 8 × 6

23. 9 × 9 = ■ **24.** ■ = 7 × 10 **25.** 10 × 10 = ■ **26.** 2 × 8 = ■

27. ■ = 5 × 5 **28.** 5 × 9 = ■ **29.** ■ = 6 × 6 **30.** ■ = 10 × 0

31. 10
  × 8

32. 9
  ×4

33. 5
  ×10

34. 8
  ×8

35. 9
  ×3

36. 7
  ×8

37. 4
  ×3

38. 10
  × 1

39. 8
  ×3

40. 6
  ×9

41. 9
  ×7

42. 10
  × 6

Find the missing factor.

43. ■ × 6 = 0 **44.** 10 × ■ = 20 **45.** ■ × 7 = 28

46. ■ × 5 = 50 **47.** 6 × ■ = 54 **48.** 7 × ■ = 42

49. ■ × 3 = 9 **50.** 8 × ■ = 64 **51.** 3 × 6 = ■ × 9

52. 5 × ■ = 4 × 10 **53.** ■ × 2 = 12 + 8 **54.** 6 × 6 = ■ × 4

Compare. Write < , >, or = for each ●.

55. 10 × 6 ● 75 − 15 **56.** 9 × 9 ● 10 × 8 **57.** 7 × 9 ● 10 × 7

58. 8 × 9 ● 9 × 8 **59.** 16 + 40 ● 9 × 6 **60.** 10 × 10 ● 50 + 50

61. **Write About It** Sydney says, "The problem 3 × 10 is the same as 10 + 10 + 10." Do you agree or disagree? Explain.

62. Malcolm cut 3 pies into 10 pieces each and 2 pies into 8 pieces each. How many pieces of pie did he have in all?

63. **? What's the Error?** Describe Mike's error. Then solve the problem correctly.

> Mike
> 9 x 4 = ■
> Think: 10 x 4 = 40
> 40 − 9 = 31
> So, 9 x 4 = 31.

64. Emiko had 4 sheets with 10 animal stickers on each. After she gave some stickers away, she had 37 left. How many stickers did she give away?

Order each group of numbers from least to greatest. (p. 24)

65. 243, 536, 144

66. 1,390; 1,039; 1,930

67. 6,321; 6,967; 5,644

Find the sum. (p. 46)

68. $1,568 + 3,215 = \blacksquare$

69. $2,513 + 874 = \blacksquare$

70. Meg went to the store at 11:15 A.M. She got home 2 hours later. At what time did she get home? (p. 98)

A 9:15 A.M. **C** 1:15 P.M.
B 12:15 P.M. **D** 2:15 P.M.

71. **TEST PREP** Jesse has 3 rows of 9 stamps. Lou has 2 rows of 6 stamps. How many stamps do they have in all? (p. 122)

F 27 **G** 39 **H** 56 **J** 60

LiNKUP
to Reading

Strategy • Analyze Information The information in a problem can offer clues about how to solve it. *Analyze,* or look carefully, at each part of the problem. Read the following problems carefully.

USE DATA For 1–4, use the graph.

1. A beaver community is made up of several beaver families. If two of the families have 4 members and the rest of the families have 5 members, how many beavers are there in all?

2. Deer mice eat berries, leaves, nuts, seeds and insects. If there are 10 deer mice in each family, how many are there in all?

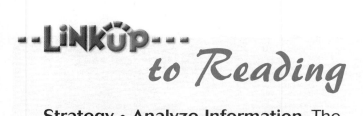

ANIMAL FAMILIES AT THE POND	
Ducks	🐾🐾🐾🐾🐾 🐾🐾🐾🐾🐾
Beavers	🐾🐾🐾🐾🐾
Turtles	🐾🐾🐾🐾 🐾🐾🐾
Deer Mice	🐾🐾🐾🐾

Key: Each 🐾 = 1 family.

3. Ducks build homes in clumps of grass or reeds. If each duck family has 8 ducklings, how many ducklings are there in all?

4. Otters make their homes in burrows near water or under rocky ledges. If a family of otters moves to the pond, how many animal families live at the pond in all?

Algebra: Find a Rule

Quick Review

1. $5 \times 4 = \blacksquare$ 2. $7 \times 5 = \blacksquare$

3. $6 \times 8 = \blacksquare$ 4. $7 \times 2 = \blacksquare$

5. $8 \times 6 = \blacksquare$

▶ Learn

CLIP CLOP   Horses wear a horseshoe on each of their 4 hooves. How many horseshoes are needed for 6 horses?

Think: 1 horse needs 4 horseshoes.
2 horses need 8 horseshoes.
3 horses need 12 horseshoes, and so on.

Look for a pattern. Write a rule.

Horses	1	2	3	4	5	6
Horseshoes	4	8	12	16	20	■

Pattern: The number of horseshoes equals the number of horses times 4.

Rule: Multiply the number of horses by 4.

Since $6 \times 4 = 24$, then 24 horseshoes are needed for 6 horses.

MATH IDEA You can write a rule to describe a number pattern in a table.

Example

Write a rule to find the cost of the bread.

Loaves of bread	1	2	4	5	7	9
Cost	$3	$6	$12	$15	$21	$27

Rule: Multiply the number of loaves of bread by $3.

• **Explain** how to use the rule to find the cost of 3 loaves of bread.

▶ Check

1. **Describe** how you could use the rule to find the cost of 10 loaves of bread.

  CALIFORNIA STANDARDS ○━┓**AF 2.1** Solve simple problems involving a functional relationship between two quantities (e.g., find the total cost of multiple items given the cost per unit). **AF 2.2** Extend and recognize a linear pattern by its rules. *also* **NS 2.0,** ○━┓**NS 2.2, AF 1.0, AF 1.1, AF 2.0, MR 1.1, MR 2.3, MR 2.4, MR 3.2**

2. Write a rule for the table. Then copy and complete the table.

Nickels	1	2	3	4	5	6	7	8	9	10
Pennies	5	10	15	▨	▨	▨	▨	▨	▨	▨

▶ Practice and Problem Solving

Write a rule for each table. Then copy and complete the table.

3.

Spiders	1	2	3	4	5	6
Legs	8	16	24	▨	▨	▨

4.

Toy cars	1	2	3	4	5	6
Cost	$2	$4	$6	▨	▨	▨

5.

Tables	3	4	5	7	8	9
Legs	12	16	20	▨	▨	▨

6.

Guitars	2	3	5	6	7	8
Strings	12	18	30	▨	▨	▨

For 7-9, use the table below.

Dimes	1	2	3	4	5	6	7	8	9	10
Nickels	2	4	6	▨	▨	▨	▨	▨	▨	▨

7. Write a rule to find the number of nickels. Copy and complete the table.

8. How many nickels can you trade for 8 dimes?

9. REASONING How many dimes can you trade for 18 nickels?

10. Each pudding pack costs $4. How much would 5 packs cost? Make a table and write a rule to find your answer.

Mixed Review and Test Prep

For 11–14, find the elapsed time. (p. 98)

11. 8:00 A.M. to 3:00 P.M.

12. 11:30 A.M. to 1:05 P.M.

13. 3:00 A.M. to 5:45 A.M.

14. 11:00 A.M. to midnight

15. TEST PREP Nieta had 314 stickers. She gave 54 away and collected 32 more. How many does she have now? (p. 58)

A 400   **C** 260

B 292   **D** 228

Algebra: Multiply with 3 Factors

▶ **Learn**

VOCABULARY

Grouping Property of Multiplication

PRACTICE, PRACTICE . . .   Julia has been taking horseback riding lessons for 3 months. At each lesson she rides for 2 hours. If she has 4 lessons each month, for how many hours has she ridden?

$$3 \times 2 \times 4 = \blacksquare$$

MATH IDEA   The **Grouping Property of Multiplication** states that when the grouping of factors is changed, the product remains the same.

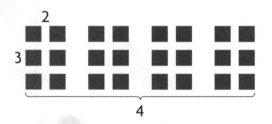

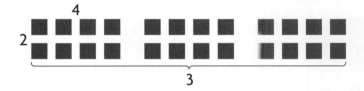

$(3 \times 2) \times 4 = \blacksquare$ $3 \times (2 \times 4) = \blacksquare$ Multiply the numbers in ( ) first.

↓   ↓

$6 \quad \times 4 = 24$ $3 \times \quad 8 \quad = 24$

So, Julia has ridden for 24 hours.

▶ **Check**

1. **Tell** which numbers you would multiply first to find $7 \times 2 \times 3$ mentally.

Find each product.

2. $(2 \times 4) \times 1 = \blacksquare$ 3. $2 \times (1 \times 3) = \blacksquare$

4. $2 \times (4 \times 2) = \blacksquare$ 5. $(3 \times 3) \times 2 = \blacksquare$

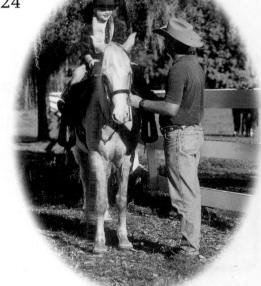

CALIFORNIA STANDARDS AF 1.5 Recognize and use the commutative and associative properties of multiplication. **NS 2.2** Memorize to automaticity the multiplication table for numbers between 1 and 10. *also* **NS 2.0, NS 2.8, AF 1.0, AF 1.2, AF 2.1, MR 2.3, MR 2.4, MR 3.2**

Find each product.

6. $(4 \times 2) \times 5 = $ ■

7. $(3 \times 3) \times 6 = $ ■

8. $8 \times (2 \times 2) = $ ■

9. $(6 \times 1) \times 2 = $ ■

10. ■ $= 5 \times (7 \times 1)$

11. ■ $= 3 \times (3 \times 3)$

12. ■ $= (2 \times 3) \times 5$

13. ■ $= (5 \times 2) \times 7$

14. $(4 \times 2) \times 4 = $ ■

Use the Grouping Property to find the product.

15. $6 \times 1 \times 8 = $ ■

16. $9 \times 2 \times 1 = $ ■

17. ■ $= 7 \times 4 \times 2$

18. ■ $= 9 \times 8 \times 0$

19. $6 \times 5 \times 2 = $ ■

20. $4 \times 2 \times 9 = $ ■

Find the missing factor.

21. $(1 \times $ ■$) \times 8 = 64$

22. $(2 \times 4) \times $ ■ $= 24$

23. $42 = 7 \times ($ ■ $\times 2)$

24. $2 \times 4 \times $ ■ $= 8$

25. $2 \times 4 \times $ ■ $= 40$

26. $14 = $ ■ $\times 2 \times 7$

27. Ross made 2 pies for each of 3 friends. In each pie he used 3 apples. How many apples did he use?

28. **REASONING** Explain why 18×2 is the same as $9 \times (2 \times 2)$.

29. **REASONING** Explain why $6 \times 1 \times 8$ has the same product as $2 \times 3 \times 8$.

30. $\frac{a+b}{c}$ **Algebra** Darla had 2 singing lessons a month for 2 months. She learned the same number of songs at each lesson. She learned 12 songs. How many songs did she learn at each lesson?

Mixed Review and Test Prep

Choose $<$, $>$, or $=$ for each ●. (p. 148)

31. 6×8 ● $50 - 2$

32. 7×6 ● 8×5

33. 9×6 ● 6×9

34. Jed has three $1 bills, 5 quarters, and 2 nickels. How much money does he have? (p. 84)

35. **TEST PREP** Irene bought 100 apples. She made 7 pies. Each pie had 8 apples in it. How many apples were left over? (p. 150)

A 56   C 34

B 44   D 16

Problem Solving Skill
Multistep Problems

Understand → Plan → Solve → Check

Quick Review

1. $(2 \times 3) \times 4 = \blacksquare$

2. $(5 \times 1) \times 7 = \blacksquare$

3. $4 + 16 + 5 = \blacksquare$

4. $7 + 8 + 2 = \blacksquare$

5. $3 \times (4 \times 2) = \blacksquare$

KNOW THE SCORE   A football team scores 6 points for a touchdown and 3 points for a field goal. The high school team made 1 touchdown and 4 field goals. How many points did they score?

To find how many points in all, you must solve a **multistep problem**, or a problem with more than one step.

VOCABULARY
multistep problem

STEP 1 Find how many points were scored by touchdowns.

1 touchdown was scored. Each touchdown = 6 points.

$1 \times 6 = 6$

6 points were scored by touchdowns.

STEP 2 Find how many points were scored by field goals.

4 field goals were scored. Each field goal = 3 points.

$4 \times 3 = 12$

12 points were scored by field goals.

STEP 3 Find how many points were scored in all.

Add the points scored by touchdowns and field goals.

$6 + 12 = 18$

So, 18 points were scored in all.

Talk About It

• Does it matter if you find the points scored for touchdowns first or the points scored for field goals first? Explain.

CALIFORNIA STANDARDS NS 2.8 Solve problems that require two or more of the skills (addition, subtraction, multiplication, and division) mentioned above. **MR 2.0** Students use strategies, skills, and concepts in finding solutions. *also* Oⁿ AF 2.1, MR 1.1, MR 2.3, MR 2.4, MR 3.0, MR 3.1, MR 3.2,

▶ Problem Solving Practice

Solve.

1. To raise money for the school, Lucia sold 9 boxes of cards. Ginger sold 7 boxes. Each box cost $3. How much money did they raise in all?

2. The Wilsons drove 598 miles in 3 days. They drove 230 miles the first day and 175 miles the second day. How far did they go the third day?

Kelsey bought 3 boxes of tacos. Each box had 6 tacos. Then she gave 4 tacos away.

3. Which shows the first step you take to find how many tacos Kelsey had left?

   A $3 + 6 = 9$
   B $6 - 3 = 3$
   C $3 \times 6 = 18$
   D $3 \times 4 = 12$

4. How many tacos did Kelsey have left?

   F 12
   G 14
   H 18
   J 22

Mixed Applications

USE DATA For 5–7, use the pictograph.

5. How many students did NOT vote for hot dogs?

6. How many students voted in all?

7. ✎ **Write a problem** about the graph. Exchange with a partner and solve.

8. **? What's the Question?** Rob spent $30 for 4 tickets. He bought 3 children's tickets for $7 each and 1 adult ticket. The answer is $9.

FAVORITE HOT LUNCHES	
Tacos	🍪 🍪 🍪
Hot Dogs	🍪 🍪 🍪 🍪
Hamburgers	🍪 🍪
Pizza	🍪 🍪 🍪

Key: Each 🍪 = 5 votes.

Review/Test

✓ CHECK VOCABULARY AND CONCEPTS

Choose the best term from the box.

> factor
> Grouping Property
> multistep problem

1. A problem with more than one step is a __?__. (p. 172)

2. The __?__ of Multiplication states that when the grouping of factors is changed, the product remains the same. (p. 170)

Solve. (pp. 164–167)

3. Explain how to use $6 \times 10 = 60$ to find 6×9.

✓ CHECK SKILLS

Find the product. (pp. 164–167)

4. $9 \times 7 = $ ■

5. ■ $= 9 \times 4$

6. $6 \times 9 = $ ■

7. $8 \times 9 = $ ■

8. $10 \times 5 = $ ■

9. $3 \times 10 = $ ■

10. ■ $= 9 \times 9$

11. $10 \times 7 = $ ■

Write a rule for the table. Then copy and complete the table. (pp. 168–169)

12.

Insects	1	2	3	4	5	6	7
Legs	6	12	18	■	■	■	■

Find each product. (pp. 170–171)

13. $(3 \times 1) \times 6 = $ ■

14. ■ $= 5 \times (2 \times 2)$

15. $(3 \times 3) \times 9 = $ ■

16. $4 \times (2 \times 5) = $ ■

17. ■ $= (2 \times 4) \times 8$

18. $9 \times (4 \times 1) = $ ■

✓ CHECK PROBLEM SOLVING

Solve. (pp. 172–173)

19. In March, Mr. Holly's class raised $176. In April, they raised $209. How much do they still need in order to raise $500?

20. Joe bought 5 guppies for $3 each and 8 goldfish for $2 each. How much did he spend?

Cumulative Review

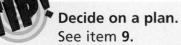

Decide on a plan.
See item **9**.
Look at the table to see the relationship of pints to cups. Choose the rule that describes this relationship.

Also see problem **4**, p. H63.

For 1–9, choose the best answer.

1. There are 9 girls in Mary Jo's scout troop. Their leader gave each girl 6 tickets for rides at the county fair. How many tickets did the scouts have?

A 15   **C** 45
B 36   **D** 54

2. The scout leader bought 4 pizzas. Each pizza was cut into 10 pieces. She saved 2 pieces for herself. How many pieces were left for the scouts?

F 16   **H** 38
G 20   **J** 42

3. The scout leader spent $6.78 on drinks. How much change did she get back from $10.00?

A $3.22   **C** $3.42
B $3.32   **D** $3.78

4. $(2 \times 5) \times 5 = \blacksquare$

F 12   **H** 25
G 15   **J** NOT HERE

5. $9 \times (3 \times 2) = \blacksquare$

A 15   **C** 54
B 30   **D** NOT HERE

6. Katie has some coins worth 85 cents in her pocket. Which coins could she have?

F 3 quarters
G 2 quarters, 3 dimes, 1 nickel
H 8 dimes, 2 nickels
J 1 quarter, 5 dimes

7. To raise money for school, Mona sold 3 boxes of wrapping paper. Jenny sold 4 boxes. They collected $6 for each box. How much money did they collect altogether?

A $49   **C** $13
B $42   **D** $7

8. Which number completes the table?

Dollars	1	2	3	4
Quarters	4	8	12	■

F 13   **H** 15
G 14   **J** 16

9. What is the rule for this table?

Pints	1	2	3	4	5	6
Cups	2	4	6	8	10	12

A Multiply the number of pints by 2.
B Add 1 to the number of pints.
C Multiply the number of pints by 4.
D Subtract 2 from the number of cups.

MATH DETECTIVE

Follow the Leader

Solve the case. Then on your paper, put the letter for each case above the answer below to solve the riddle.

Case 1

Start with 4.   **O**
Add 2.
Subtract 3.
Multiply by 5.
Subtract 9.
Multiply by 4.
End Number ■

Case 2

Start with 6.   **U**
Multiply by 2.
Subtract 5.
Multiply by 3.
Add 9.
Multiply by 0.
End Number ■

Case 3

Start with 2.   **P**
Multiply by 3.
Multiply by 4.
Subtract 4.
Add 7.
Multiply by 1.
End Number ■

Case 4

Start with 5.   **y**
Multiply by 5.
Subtract 15.
Add 5.
Subtract 9.
Multiply by 6.
End Number ■

Case 5

Start with 3.   **N**
Add 6.
Multiply by 2.
Subtract 10.
Multiply by 3.
Subtract 4.
End Number ■

Case 6

Start with 7.   **H**
Multiply by 5.
Subtract 30.
Multiply by 4.
Add 15.
Subtract 5.
End Number ■

Why doesn't a frog jump when it's sad?

It's too __?__ __?__ __?__   __?__ __?__ __?__   __?__ .
        0   20   30   24   27   27   36

CASE CLOSED

Challenge

Square Numbers

When both factors are the same, the product is called a **square number**.

Activity

Copy and complete the table.

$5 \times 5 = 25$

25 is a square number.

Number	Square Number	Number Added to Square Number to Get the Next Square Number
1	$1 \times 1 = 1$	3
2	$2 \times 2 = 4$	
3	$3 \times 3 = $	
4	$4 \times 4 = $	
5	$5 \times 5 = $	
6	$6 \times 6 = $	
7	$7 \times 7 = $	
8	$8 \times 8 = $	
9	$9 \times 9 = $	
10	$10 \times 10 = $	

Talk About It

• What pattern do you see in the numbers added to get the next square number?

Try It

1. How can you use this pattern to predict the square numbers for 11×11 and 12×12?

2. Check your prediction. Show another way to find 11×11 and 12×12.

Study Guide and Review

VOCABULARY

Choose the best term from the box.

1. The _?_ means that two numbers can be multiplied in any order. The product is the same. (p. 122)

2. The _?_ means that when the grouping of factors is changed, the product remains the same. (p. 170)

> factors
> **Grouping Property of Multiplication**
> **Order Property of Multiplication**

STUDY AND SOLVE

Chapter 7

Use the Order Property of Multiplication.

Numbers can be multiplied in any order. The product is the same.

$3 \times 4 = 12$

$4 \times 3 = 12$

Find the product. (pp. 122–125)

3. $2 \times 3 = \blacksquare$ $3 \times 2 = \blacksquare$

4. $3 \times 7 = \blacksquare$ $7 \times 3 = \blacksquare$

5. $4 \times 5 = \blacksquare$ $5 \times 4 = \blacksquare$

6. $3 \times 5 = \blacksquare$ $5 \times 3 = \blacksquare$

7. $2 \times 7 = \blacksquare$ $7 \times 2 = \blacksquare$

Chapter 8

Find missing factors.

$\blacksquare \times 7 = 28$

The multiplication table on page 142 can help you. Look down the column for 7 to the product 28. Look across the row from 28. The factor in that row is 4. So, $4 \times 7 = 28$.

Find the missing factor. (pp. 142–143)

8. $\blacksquare \times 8 = 16$ 9. $5 \times \blacksquare = 30$

10. $9 \times \blacksquare = 45$ 11. $3 \times \blacksquare = 3$

12. $4 \times \blacksquare = 0$ 13. $\blacksquare \times 3 = 15$

Chapter 9

Write mutiplication facts with factors 6, 7, and 8.

You can double products of facts you already know to help you find products you don't know.

6 × 8 = ■
Think: 6 × 4 = 24
24 + 24 = 48, so 6 × 8 = 48.

Use the Order Property of Multiplication.
7 × 5 = ■
5 × 7 = 35, so 7 × 5 = 35.

Find the product. (pp. 148–153)

14. 6 × 8 = ■   **15.** 7 × 6 = ■

16. 8 × 4 = ■   **17.** 6 × 6 = ■

18. 6 × 3 = ■   **19.** 4 × 8 = ■

20. 8 × 7 = ■   **21.** 8 × 8 = ■

22. 7 × 9 = ■   **23.** 7 × 7 = ■

24. 8 × 5 = ■   **25.** 7 × 4 = ■

Chapter 10

Find a rule for the pattern.

Write a rule for the pattern in the table.

Cars	1	2	3	4	5	6	7	8	9	10
Tires	4	8	12	16	20	■	■	■	■	■

Think: The number of tires is 4 times the number of cars.
Rule: Multiply by 4.

For 26–27, use the table below.
(pp. 168–169)

Spiders	1	2	4	5	6	8
Legs	8	16	32	40	■	■

26. Write a rule for the table.

27. Use the rule to complete the table.

PROBLEM SOLVING PRACTICE

Solve. (pp. 126–127, 172–173)

28. Pencils are in packages of 4. Erasers are in packages of 7. Marian bought 16 pencils. How many packages of pencils did she buy? Is there too much or too little information? Explain.

29. A box of cookies costs $3. A bag of nuts costs $2. Aimee bought 4 boxes of cookies and 7 bags of nuts for her friends. How much did she spend?

California Connections

CENTRAL VALLEY

The Central Valley region of California is about 430 miles long and contains some of the richest farmland in the world.

1. Orange trees are planted in arrays. Suppose there are 4 rows of 6 trees.

   a. How many trees are planted in all? Draw an array to show your answer.

   b. Draw a different array that has the same number of trees.

USE DATA For 2–4, use the table.

bags	1	2	3	4	5	6	7	8	9	10
oranges	8	16	24	▪	▪	▪	▪	▪	▪	▪

2. Write a rule to find the number of oranges. Copy the table and complete it, using your rule.

3. How many oranges are in 9 bags?

4. **REASONING** How many oranges are in 19 bags? Explain how the table can help you find the answer.

Oranges are one of over 200 different crops grown in the Central Valley. ▼

OPEN-AIR MARKETS

Some of California's fresh fruits, nuts, and vegetables are sold at open-air markets throughout the state.

The Grand Central Market is the oldest and largest open-air market in Los Angeles. ▼

USE DATA For 1–5, use the signs.

1. How much will it cost to buy 18 lemons? Explain how you know.

2. Maria spent $4 for grapes. How many pounds of grapes did she buy? Explain how you know.

3. How much will it cost to buy 2 cantaloupes and 6 pounds of walnuts?

4. Suppose you pay for the 2 cantaloupes and 6 pounds of walnuts with a $10 bill. How much change should you receive?

5. ✎ **Write a problem** using the data in the signs. Challenge a classmate to solve it.

Walnuts
2 lb for $1

Grapes
3 lb for $2

Cantaloupe
50¢ each

Lemons
6 for $1

Celery
2 for $1

Understand Division

Trading cards first became popular in the 1930s. Today, people collect many different kinds of trading cards. To protect and display their cards, collectors put them into plastic sleeves. These sleeves can hold one card or as many as 12 cards. Look at the collections listed. If each collection is divided into groups of 4 to a plastic sleeve, how many sleeves will each collector need?

COLLECTIONS

Aaron	
Jameel	
Lee	
Jenny	
Danielle	

Key: Each ▨ = 2 Cards.

CHECK WHAT YOU KNOW

Use this page to help you review and remember
important skills needed for Chapter 11.

✓ VOCABULARY

Choose the best term from the box.

| factor |
| product |
| sum |

1. In the multiplication sentence $6 \times 4 = 24$,
   the number 24 is the ? .

2. In the multiplication sentence $5 \times 3 = 15$,
   the number 5 is a ? .

✓ MEANING OF MULTIPLICATION (See p. H14.)

Copy and complete.

3.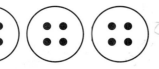

   a. ■ groups of ■
   b. ■ + ■ + ■ = ■
   c. ■ × ■ = ■

4.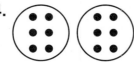

   a. ■ groups of ■
   b. ■ + ■ = ■
   c. ■ × ■ = ■

Write an addition and a multiplication sentence for each.

5.

6.

✓ MULTIPLICATION FACTS THROUGH 10 (See p. H14.)

Find the product.

7. $6 \times 7 = $ ■
8. $3 \times 8 = $ ■
9. ■ $= 8 \times 9$
10. $9 \times 0 = $ ■

11. $5 \times 7 = $ ■
12. ■ $= 10 \times 4$
13. $4 \times 4 = $ ■
14. $2 \times 4 = $ ■

15. ■ $= 9 \times 9$
16. ■ $= 6 \times 8$
17. $8 \times 7 = $ ■
18. $4 \times 3 = $ ■

19. $\begin{array}{r} 1 \\ \times 5 \\ \hline \end{array}$
20. $\begin{array}{r} 2 \\ \times 6 \\ \hline \end{array}$
21. $\begin{array}{r} 3 \\ \times 0 \\ \hline \end{array}$
22. $\begin{array}{r} 5 \\ \times 2 \\ \hline \end{array}$
23. $\begin{array}{r} 0 \\ \times 7 \\ \hline \end{array}$

HANDS ON
The Meaning of Division

Quick Review

1. $6 \times 4 = $ ▪
2. $3 \times 5 = $ ▪   3. $2 \times 6 = $ ▪
4. $7 \times 3 = $ ▪   5. $4 \times 3 = $ ▪

VOCABULARY
divide

MATERIALS
counters

▶ **Explore**

When you multiply, you put equal groups together. When you **divide**, you separate into equal groups.

Divide 14 counters into 2 equal groups. How many counters are in each group?

STEP 1 Use 14 counters. Show 2 groups.

STEP 2 Place a counter in each group.

STEP 3 Continue until all counters are used.

So, there are 7 counters in each group.

Try It

Use counters to make equal groups.

a. Divide 20 counters into 5 equal groups. How many are in each group?

b. Divide 12 counters into equal groups in different ways.

• How did you find different equal groups using 12 counters?

We are putting 20 counters in 5 equal groups. How many should be in each group?

CALIFORNIA STANDARDS MR 2.3 Use a variety of methods, such as words, numbers, symbols, charts, graphs, tables, diagrams, and models, to explain mathematical reasoning. **NS 2.0** Students calculate and solve problems involving addition, subtraction, multiplication, and division. *also* **NS 2.8 MR 1.0, MR 3.2**

▶ Connect

MATH IDEA Use division to find how many items are in each group or how many equal groups there are.

Four friends share 20 marbles equally. How many marbles will each person get?

Put one marble in each group until all marbles are used.

Each person will get 5 marbles.

Each person wants 4 marbles. How many people can share 20 marbles?

Make equal groups of 4 marbles until all marbles are used.

Five people can share 20 marbles.

▶ Practice

Copy and complete the table. Use counters to help.

	COUNTERS	NUMBER OF EQUAL GROUPS	NUMBER IN EACH GROUP
1.	15	5	▪
2.	21	▪	3
3.	24	3	▪
4.	28	▪	7

For 5–8, use counters.

5. Five friends share 30 stickers equally. How many will each person get?

6. Hanae puts 18 crackers in groups of 3. How many equal groups can she make?

7. **REASONING** Three friends share some grapes equally. If each gets 9 grapes, how many grapes are there all together?

8. **Write About It** Explain how to divide 32 counters into 4 equal groups.

Mixed Review and Test Prep

9. 463 (p. 42)
  +297

10. 805 (p. 58)
  −176

11. 2 (p. 118)
  ×9

12. 3 (p. 122)
  ×8

13. **TEST PREP** The Boy Scouts charged $4 to wash each car. How much money did they make for washing 8 cars? (p. 134)

  A $2   **C** $12

  B $4   **D** $32

Relate Subtraction and Division

Quick Review

How many are there in all?

1. 2 groups of 3

2. 4 groups of 4

3. 5 groups of 2

4. 3 groups of 9

5. 1 group of 8

▶ **Learn**

GET IN THE GAME   Ana has 16 game pieces for a game. Each player gets 4 pieces. How many people can play?

$$16 \quad \div \quad 4 \quad = \quad \blacksquare$$

number of   number for   number of
pieces   each player   players

Start with 16. Take away groups of 4 until you reach 0. Count the number of times you subtract 4.

$$\begin{array}{cccc} 16 & 12 & 8 & 4 \\ -\ 4 & -\ 4 & -4 & -4 \\ \hline 12 & 8 & 4 & 0 \end{array}$$

Number of times
you subtract 4:   **1**   **2**   **3**   **4**

Since you subtract 4 from 16 four times, there are 4 groups of 4 in 16.

So, 4 people can play.

Write: $16 \div 4 = 4$ or $4\overline{)16}$ with quotient 4

Read: Sixteen divided by four equals four.

MATH IDEA   You can use repeated subtraction to find how many groups when you know how many in all and how many in each group.

• **Discuss** how to skip-count to find $15 \div 5$.

CALIFORNIA STANDARDS MR 2.3 Use a variety of methods, such as words, numbers, symbols, charts, graphs, tables, diagrams, and models, to explain mathematical reasoning. **NS 2.0** Students calculate and solve problems involving addition, subtraction, multiplication, and division. *also* **NS 2.8, AF 1.0, O⌐ AF 1.1, AF 1.3, MR 1.0, MR 2.0, MR 2.4, MR 3.0, MR 3.2**

▶ Check

1. Explain how to use repeated subtraction to prove that $18 \div 6 = 3$.

Write the division sentence for each.

2.
$$
\begin{array}{cc}
12 & 6 \\
-6 & -6 \\
\hline
6 & 0
\end{array}
$$

3.
$$
\begin{array}{cccc}
8 & 6 & 4 & 2 \\
-2 & -2 & -2 & -2 \\
\hline
6 & 4 & 2 & 0
\end{array}
$$

▶ Practice and Problem Solving

Write a division sentence for each.

4.
$$
\begin{array}{cccc}
20 & 15 & 10 & 5 \\
-5 & -5 & -5 & -5 \\
\hline
15 & 10 & 5 & 0
\end{array}
$$

5.
$$
\begin{array}{ccc}
24 & 16 & 8 \\
-8 & -8 & -8 \\
\hline
16 & 8 & 0
\end{array}
$$

Use subtraction to solve.

6. $15 \div 3 = \blacksquare$ **7.** $21 \div 7 = \blacksquare$ **8.** $30 \div 5 = \blacksquare$ **9.** $36 \div 6 = \blacksquare$

10. $2\overline{)10}$ **11.** $8\overline{)16}$ **12.** $7\overline{)35}$ **13.** $5\overline{)25}$

Algebra Complete. Write $+$, $-$, $\times$, or $\div$ for each ●.

14. $20 - 5 = 5$ ● 3 **15.** $24 \div 6 = 18$ ● 14

16. 32 ● $8 = 4 \times 10$ **17.** 8 ● $2 = 2 + 2$

18. Scott buys 22 baseball cards. He keeps 10 cards and divides the rest equally between 2 friends. How many cards will each friend get?

19. REASONING Nora says that $8 \div 4 = 0$ because $8 - 4 = 4$ and $4 - 4 = 0$. Is Nora correct? Explain.

20. Explain how to use repeated subtraction to find $100 \div 10$.

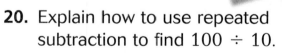

Mixed Review and Test Prep

21. $\begin{array}{r} 3 \\ \times 6 \\ \hline \end{array}$ (p. 122)   **22.** $\begin{array}{r} 4 \\ \times 7 \\ \hline \end{array}$ (p. 150)

23. $\begin{array}{r} \$3.57 \\ +\$7.94 \\ \hline \end{array}$ (p. 88)   **24.** $\begin{array}{r} \$9.26 \\ -\$5.83 \\ \hline \end{array}$ (p. 88)

25. **TEST PREP** Felipe buys a notebook for $1.39. He pays with a $5 bill. How much change should he get? (p. 86)

A $3.61   **C** $4.61

B $3.71   **D** $6.39

Extra Practice page H42, Set A

Algebra: Relate Multiplication and Division

▶ **Learn**

STICK WITH STAMPS   Use what you know about arrays and multiplication to understand division.

VOCABULARY

dividend   divisor
quotient
inverse operations

Mark is putting stamps into his stamp album. Each page holds 18 stamps in 3 equal rows. How many stamps are in each row?

$$18 \quad \div \quad 3 \quad = \quad \blacksquare$$

   ↑   ↑   ↑

number of   number of   number in
 stamps   rows   each row

Show an array with 18 in 3 equal rows. Find how many are in each row.

Since $3 \times 6 = 18$, then $18 \div 3 = 6$.

3 rows of 6 = 18

$$18 \quad \div \quad 3 \quad = \quad 6$$

   ↑   ↑   ↑

dividend   **divisor**   **quotient**

$$\text{divisor} \longrightarrow 3\overline{)18} \quad \begin{array}{l} 6 \leftarrow \text{quotient} \end{array}$$

            ↑

       dividend

So, there are 6 stamps in each row.

MATH IDEA Multiplication and division are opposite or **inverse operations**.

TECHNOLOGY LINK

More Practice:
Use E-Lab, *Exploring Division.*

www.harcourtschool.com/
elab2002

▶ **Check**

1. **Explain** how you use equal groups when you multiply and divide.

CALIFORNIA STANDARDS ⊶ **NS 2.3** Use the inverse relationship of multiplication and division to compute and check results. **NS 2.0** Students calculate and solve problems involving addition, subtraction, multiplication, and division. *also* ⊶ **NS 2.2, NS 2.8,** ⊶ **AF 1.1, AF 1.2, MR 1.0, MR 3.2**

Copy and complete.

2.

3 rows of ■ = 24

24 ÷ 3 = ■

3. 4 rows of ■ = 20

20 ÷ 4 = ■

4.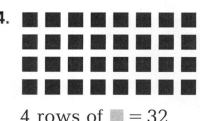

4 rows of ■ = 32

32 ÷ 4 = ■

▶ Practice and Problem Solving

Copy and complete.

5.

3 rows of ■ = 21

21 ÷ 3 = ■

6.

5 rows of ■ = 30

30 ÷ 5 = ■

7.

5 rows of ■ = 40

40 ÷ 5 = ■

Complete each number sentence. Draw an array to help.

8. 3 × ■ = 18   18 ÷ 3 = ■   **9.** 5 × ■ = 25   25 ÷ 5 = ■

10. 6 × ■ = 24   24 ÷ 6 = ■   **11.** 3 × ■ = 24   24 ÷ 3 = ■

a+b/c Algebra   Complete.

12. 4 × 2 = 24 ÷ ■   **13.** ■ × 3 = 30 ÷ 5   **14.** 4 × 1 = ■ ÷ 4

15. REASONING Mark bakes 14 muffins. He eats 2 muffins and divides the rest equally among 6 friends. What division sentence shows how many muffins each friend gets?

16. **? What's the Question?** Christy puts 36 pennies into 4 equal piles. The answer is 9 pennies.

Mixed Review and Test Prep

Write + or − for each ●. (p. 186)

17. 20 ● 5 = 5 × 3

18. 24 ÷ 3 = 6 ● 2

19. 4 × 9 = 38 ● 2

20. 7 ● 2 = 36 ÷ 4

21. **TEST PREP** Hector puts 1 ice cube in each of 7 cups. How many ice cubes are there in all? (p. 132)

A 1   C 7
B 6   D 8

Extra Practice page H42, Set B

Algebra: Fact Families

Quick Review

1. $3 \times \blacksquare = 18$

2. $\blacksquare \times 6 = 18$

3. $18 \div 6 = \blacksquare$

4. $5 \times \blacksquare = 25$

5. $25 \div 5 = \blacksquare$

▶ Learn

FUN FACTS   A set of related multiplication and division number sentences is called a **fact family**.

Fact Family for 3, 5, and 15

factor		factor		product		dividend		divisor		quotient
3	×	5	=	15		15	÷	5	=	3
5	×	3	=	15		15	÷	3	=	5

VOCABULARY

fact family

Activity

Materials: square pieces of paper, scissors

Use this triangle fact card to think of the fact family for 3, 5, and 15.

Make a set of triangle fact cards. Use them to write fact families.

Product
15

Factor **3**   **5** Factor

A

Fold each paper in half three times. Open up the paper and cut along the folds to make triangle cards.

Fold

Fold   Fold

B

Make triangle fact cards for each of these products: 12, 15, 18, 20, 24, 25, and 30.

C

Write fact families for at least 3 triangle fact cards.

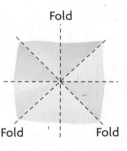

• **REASONING** How many triangle fact cards can you make for the product 12? Explain.

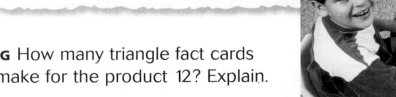

CALIFORNIA STANDARDS ⊶ **NS 2.3** Use the inverse relationship of multiplication and division to compute and check results. **NS 2.0** Students calculate and solve problems involving addition, subtraction, multiplication, and division. *also* ⊶**NS 2.2, NS 2.8, AF 1.0,** ⊶**AF 1.1, AF 1.2, MR 1.0, MR 2.0, MR 2.3, MR 3.0, MR 3.2**

Using a Multiplication Table

MATH IDEA Use related multiplication facts to find quotients or missing divisors in division sentences.

Examples

A

Find the quotient.

$12 \div 3 = $ ▨

Think: $3 \times$ ▨ $= 12$

Find row 3. Look across to find the product 12. Look up to find the missing factor, 4.

$3 \times 4 = 12$

So, $12 \div 3 = 4$.

B

Find the missing divisor.

$30 \div$ ▨ $= 5$

Think: ▨ $\times 5 = 30$

Find the factor 5 in the top row. Look down to find the product 30. Look left to find the missing factor, 6.

$6 \times 5 = 30$

So, $30 \div 6 = 5$.

×	0	1	2	3	4	5	6
0	0	0	0	0	0	0	0
1	0	1	2	3	4	5	6
2	0	2	4	6	8	10	12
3	0	3	6	9	12	15	18
4	0	4	8	12	16	20	24
5	0	5	10	15	20	25	30
6	0	6	12	18	24	30	36

- **REASONING** How can you use multiplication to check $20 \div 5 = 4$?

Check

1. **Explain** how you can use a multiplication table to show how $5 \times 2 = 10$ and $10 \div 2 = 5$ are related.

Write the missing number for each triangle fact card.

2.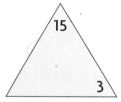

3.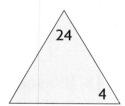

4.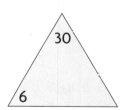

5.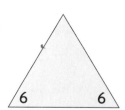

Write the fact family.

6. 3, 6, 18   7. 4, 4, 16   8. 4, 5, 20   9. 3, 7, 21

LESSON CONTINUES ▶

▶ Practice and Problem Solving

Write the missing number for each triangle fact card.

10.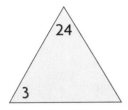

11.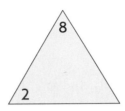

12.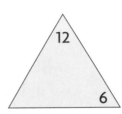

13.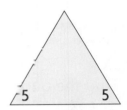

Write the fact family.

14. 5, 6, 30

15. 2, 8, 16

16. 4, 7, 28

17. 5, 5, 25

Find the quotient or product.

18. $3 \times 6 =$ ■

19. $6 \times 3 =$ ■

20. $18 \div 3 =$ ■

21. $18 \div 6 =$ ■

22. $4 \times 9 =$ ■

23. $9 \times 4 =$ ■

24. $36 \div 4 =$ ■

25. $36 \div 9 =$ ■

26. $8 \times 5 =$ ■

27. $5 \times 8 =$ ■

28. $40 \div 8 =$ ■

29. $40 \div 5 =$ ■

Write the other three sentences in the fact family.

30. $3 \times 7 = 21$

31. $1 \times 5 = 5$

32. $4 \times 3 = 12$

33. $5 \times 3 = 15$

34. $6 \times 4 = 24$

35. $9 \times 2 = 18$

Find the quotient or the missing divisor.

36. $8 \div 4 =$ ■

37. $16 \div 2 =$ ■

38. $7 = 21 \div$ ■

39. $2 = 12 \div$ ■

40. $24 \div 8 =$ ■

41. $10 \div$ ■ $= 2$

42. $30 \div$ ■ $= 6$

43. $28 \div 4 =$ ■

$\frac{a+b}{c}$ Algebra   Complete.

44. ■ $\div 5 = 6 + 3$

45. $6 \times$ ■ $= 54 \div 9$

46. $42 - 6 =$ ■ $\times 9$

47. What do you notice about the fact family for 6, 6, and 36?

48. **REASONING** How are $20 \div 5 = 4$ and $20 \div 4 = 5$ alike? How are they different?

49. Geri made 20 bookmarks. She kept 2 and then put an equal number in each of 3 gift boxes. How many bookmarks are in each box?

50. **REASONING** Kendra says, "There are 3 teaspoons in 1 tablespoon, so there are 15 teaspoons in 5 tablespoons." Do you agree or disagree? Explain.

51. Mr. Tapia uses 14 bowls for the dogs in his pet store. Each dog has a water bowl and a food bowl. How many dogs are in the store?

52. **? What's the Error?** John says that since $4 + 4 = 8$, then $8 \div 4 = 4$. Describe his error and give the correct quotient.

Mixed Review and Test Prep

Find the sum or difference. (p. 88)

53. $4.57
  +$0.82

54. $3.19
  −$1.35

Complete. Write +, −, ×, or = for each ●. (p. 186)

55. $14 \; ● \; 7 = 7$

56. $9 \; ● \; 5 = 45$

57. $28 \; ● \; 4 = 7$

58. $9 \; ● \; 9 = 18$

59. **TEST PREP** Denise spent 3 hours at math camp each day for 1 week. How many hours did she spend at math camp? (p. 122)

A 3 hours   **C** 14 hours

B 4 hours   **D** 21 hours

60. **TEST PREP** Mr. Li picked up Frank from soccer practice 20 minutes before 5:00. What time was it? (p. 94)

F 4:20   **H** 5:20

G 4:40   **J** 5:40

Thinker's Corner

TRICKY TRIANGLES

Materials: triangle fact cards, paper

Players: 2

a. One player chooses a triangle fact card and holds it so that one number is hidden.

b. The other player names the hidden number and writes the fact family.

c. Players take turns until all cards are used. The player with the most correct number sentences wins.

1. How can you make sure your fact families are complete?

2. What card did you hope to choose for your opponent? Explain.

Problem Solving Strategy
Write a Number Sentence

Understand → Plan → Solve → Check

Quick Review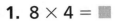

1. $8 \times 4 = \blacksquare$

2. $30 = 6 \times \blacksquare$

3. $10 \times 5 = \blacksquare$

4. $18 = \blacksquare \times 6$

5. $27 = 9 \times \blacksquare$

PROBLEM Megan puts 36 animal trading cards in her binder. She puts 9 cards on each page. How many pages will Megan need for her cards?

- What are you asked to find?

- What information will you use?

- Is there information you will not use? If so, what?

- What strategy can you use?

Write a number sentence to find the number of pages Megan will need.

- How can you use the strategy to solve it?

Write a number sentence and solve.

$$36 \div 9 = 4$$

$\uparrow$ $\uparrow$ $\uparrow$

number of   number on   number of
trading cards   each page   pages

So, Megan needs 4 pages for her cards.

- How can you decide if your answer is correct?

- What other strategy could you use?

CALIFORNIA STANDARDS ⊶ **AF 1.1** Represent relationships of quantities in the form of mathematical expressions, equations, or inequalities. **NS 2.0** Students calculate and solve problems involving addition, subtraction, multiplication, and division. *also* **AF 1.0, AF 1.2, MR 1.0, MR 1.1, MR 2.0, MR 2.3, MR 2.4, MR 2.6**

Problem Solving Practice

PROBLEM SOLVING STRATEGIES

Draw a Diagram or Picture
Make a Model or Act It Out
Make an Organized List
Find a Pattern
Make a Table or Graph
Predict and Test
Work Backward
Solve a Simpler Problem
► **Write a Number Sentence**
Use Logical Reasoning

Write a number sentence to solve.

1. **What if** Megan buys 27 trading cards to add to her collection? How many pages will she need for the new cards?

2. Rosita has 28 cards. She wants to keep 4 cards and divide the rest equally among 4 friends. How many cards will each friend get?

Jorge has 45 trading cards in his collection. His binder holds 10 pages. Each page holds 9 trading cards.

3. How many pages will Jorge use for the cards he has?

   A 5   **C** 45
   B 10   **D** 90

4. Which number sentence shows how to find how many trading cards fit in Jorge's binder?

   F $45 + 10 = 55$ **H** $10 + 9 = 19$
   G $9 \times 10 = 90$ **J** $45 \div 9 = 5$

Mixed Strategy Practice

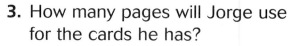

USE DATA For 5–8, use the graph.

5. How many coins in all are in Sebastian's coin collection?

6. Sebastian put the same type of coins on separate pages. The coins from Japan are on page 2. The piasters are just before the coins from France. Which coins are on page 4?

7. Sebastian adds some coins from Mexico to his collection. There are 5 more coins from Japan than from Mexico. How many coins are there from Mexico?

SEBASTIAN'S COIN COLLECTION

francs (France)	🪙 🪙 🪙
yen (Japan)	🪙 🪙 🪙 🪙
colones (Costa Rica)	🪙 🪙
piasters (Egypt)	🪙

Key: Each 🪙 = 5 coins.

8. ✏️ **Write a problem** using the data in the pictograph.

Review/Test

✓ CHECK VOCABULARY AND CONCEPTS

Choose the best term from the box.

| divide |
| divisor |
| quotient |
| inverse operations |

1. Multiplication and division are opposite operations, or __?__ . (p. 188)

2. In 18 ÷ 6 = 3, the number 3 is called the __?__ . (p. 188)

3. When you separate into equal groups, you __?__ . (p. 184)

Use subtraction to solve. (pp. 186–187)

4. $18 \div 6 = \blacksquare$ 5. $10 \div 5 = \blacksquare$ 6. $27 \div 9 = \blacksquare$ 7. $12 \div 4 = \blacksquare$

8. $3\overline{)6}$ 9. $3\overline{)15}$ 10. $4\overline{)20}$ 11. $8\overline{)32}$

✓ CHECK SKILLS

Complete each number sentence.
Draw an array to help. (pp. 188–189)

12. $2 \times \blacksquare = 6$ $6 \div 2 = \blacksquare$ 13. $3 \times \blacksquare = 15$ $15 \div 3 = \blacksquare$

14. $4 \times \blacksquare = 4$ $4 \div 4 = \blacksquare$ 15. $5 \times \blacksquare = 30$ $30 \div 5 = \blacksquare$

Write the fact family. (pp. 190–193)

16. 4, 5, 20   17. 2, 7, 14   18. 3, 8, 24

✓ CHECK PROBLEM SOLVING

Write a number sentence to solve. (pp. 194–195)

19. Ms. Kraft has 20 pencils to divide equally among 5 groups of students. How many pencils does each group get?

20. Fernando has 24 rocks in his collection. If a box holds 6 rocks, how many boxes will Fernando need for his collection?

Cumulative Review

TIP!
Decide on a plan.
See item **9.**

Every 20 minutes means there is a pattern of times. Write this pattern beginning at 8:00

Also see problem **4,** p. H63.

For 1–9, choose the best answer.

1. Joanne has 40 candy hearts to put into 8 cups. How many hearts will she put in each cup so each cup has the same number?

   A 3　　　　　**C** 5
   B 4　　　　　**D** 6

2. Kevin gets paid $6 when he mows one lawn. How much money is he paid if he mows 4 lawns?

   F $18　　　　**H** $24
   G $20　　　　**J** NOT HERE

3. Which multiplication fact could help solve $6 \div 2 = \blacksquare$?

   A $2 \times 2 = 4$
   B $2 \times 6 = 12$
   C $2 \times 4 = 8$
   D $2 \times 3 = 6$

4. $20 \div 4 = \blacksquare$

   F 6　　　　　**H** 4
   G 5　　　　　**J** NOT HERE

5. Cari took 24 ballet lessons. She took the same number of lessons each week for 8 weeks. Which number sentence tells how many lessons she took each week?

   A $24 + 8 = 32$
   B $24 - 8 = 16$
   C $6 \times 4 = 24$
   D $24 \div 8 = 3$

6. Which sentence does **not** belong in the fact family for 7, 6, and 42?

   F $7 \times 6 = 42$
   G $42 - 6 = 36$
   H $6 \times 7 = 42$
   J $42 \div 7 = 6$

7. Jeremy has 2 pizzas. Each pizza is cut into 8 pieces. He wants to share the pizzas equally among 4 people. How many pieces will each person get?

   A 4　　　　　**C** 10
   B 6　　　　　**D** 14

8. $30 \div 5 = \blacksquare$

   F 5　　　　　**H** 25
   G 6　　　　　**J** 30

9. A train leaves the station every 20 minutes starting at 8:00. Marcel gets to the station at 9:10. When will the next train leave?

   A 9:20　　　　**C** 9:35
   B 9:30　　　　**D** 9:45

Division Facts Through 5

This farmer is shearing a sheep.

AMOUNT OF YARN NEEDED

scarf	🧶🧶🧶🧶
adult sweater	🧶🧶🧶🧶🧶🧶🧶🧶🧶🧶 🧶🧶🧶🧶🧶🧶🧶🧶
pair of socks	🧶
dog sweater (medium)	🧶🧶🧶🧶

Key: Each 🧶 = 2 ounces.

Sheep are an important source of wool for clothing. The white, fluffy fur on a sheep is called *fleece*. An average fleece makes about 48 ounces of yarn. Look at the pictograph. How many scarves could be made from the fleece of one sheep? Can 6 dog sweaters be made from the fleece of one sheep?

CHECK WHAT YOU KNOW ✓

Use this page to help you review and remember
important skills needed for Chapter 12.

✓ VOCABULARY

Choose the best term from the box.

1. $15 + 26 = 41$ is called a ? .

> expression
>
> number sentence

✓ SUBTRACTION (See p. H15.)

Find each difference.

2. $\begin{array}{r} 14 \\ -\ 2 \\ \hline \end{array}$
3. $\begin{array}{r} 10 \\ -\ 0 \\ \hline \end{array}$
4. $\begin{array}{r} 20 \\ -\ 4 \\ \hline \end{array}$
5. $\begin{array}{r} 18 \\ -\ 6 \\ \hline \end{array}$
6. $\begin{array}{r} 27 \\ -\ 9 \\ \hline \end{array}$

7. $\begin{array}{r} 45 \\ -\ 9 \\ \hline \end{array}$
8. $\begin{array}{r} 21 \\ -\ 3 \\ \hline \end{array}$
9. $\begin{array}{r} 15 \\ -\ 5 \\ \hline \end{array}$
10. $\begin{array}{r} 30 \\ -10 \\ \hline \end{array}$
11. $\begin{array}{r} 16 \\ -\ 2 \\ \hline \end{array}$

Find each missing number.

12. $24 - \blacksquare = 18$
13. $30 - \blacksquare = 25$
14. $27 - \blacksquare = 24$

15. $32 - \blacksquare = 24$
16. $18 - \blacksquare = 16$
17. $36 = 42 - \blacksquare$

✓ MULTIPLICATION FACTS THROUGH 10 (See p. H14.)

Find each product.

18. $6 \times 3 = \blacksquare$
19. $3 \times 10 = \blacksquare$
20. $\blacksquare = 9 \times 2$
21. $7 \times 4 = \blacksquare$

22. $\blacksquare = 0 \times 7$
23. $\blacksquare = 10 \times 1$
24. $5 \times 8 = \blacksquare$
25. $3 \times 9 = \blacksquare$

26. $1 \times 1 = \blacksquare$
27. $6 \times 5 = \blacksquare$
28. $4 \times 6 = \blacksquare$
29. $\blacksquare = 2 \times 10$

30. $\begin{array}{r} 9 \\ \times 6 \\ \hline \end{array}$
31. $\begin{array}{r} 8 \\ \times 9 \\ \hline \end{array}$
32. $\begin{array}{r} 0 \\ \times 4 \\ \hline \end{array}$
33. $\begin{array}{r} 7 \\ \times 2 \\ \hline \end{array}$

Divide by 2 and 5

Quick Review

1. $2 \times \blacksquare = 10$

2. $4 \times \blacksquare = \varepsilon$

3. $5 \times \blacksquare = \bar{2}0$

4. $\blacksquare \times 5 = \bar{3}5$

5. $\blacksquare \times 2 = 18$

▶ Learn

CRAFTY MATH Mrs. Jackson knit 12 hats. She put an equal number of hats on each of 2 shelves in the craft shop. How many hats are on each shelf?

$12 \div 2 = \blacksquare$

Use a related multiplication fact to find the quotient.

Think: $2 \times \blacksquare = 12$

$$2 \times 6 = 12$$

$$12 \div 2 = 6, \text{ or } 2\overline{)12}^{\,6}$$

So, there are 6 hats on each shelf.

Remember

$$\underset{\underset{\text{dividend}}{\uparrow}}{16} \div \underset{\underset{\text{divisor}}{\uparrow}}{2} = \underset{\underset{\text{quotient}}{\uparrow}}{8}$$

What if Mrs. Jackson knits 15 hats and puts an equal number of hats on each of 5 shelves? How many hats are on each shelf?

$15 \div 5 = \blacksquare$

Think: $5 \times \blacksquare = 15$ $5 \times 3 = 15$ $15 \div 5 = 3, \text{ or } 5\overline{)15}^{\,3}$

So, there are 3 hats on each shelf.

$\times$	0	1	2	3	4	5
0	0	0	0	0	0	0
1	0	1	2	3	4	5
2	0	2	4	6	8	10
3	0	3	6	9	12	15
4	0	4	8	12	16	20
5	0	5	10	15	20	25
6	0	6	12	18	24	30

MATH IDEA You can find missing factors in related multiplication facts to help you divide.

• How can you use $2 + 2 + 2 = 6$ to help you find $6 \div 2$?

▶ Check

1. **Explain** how you can use multiplication to check $20 \div 5 = 4$.

CALIFORNIA STANDARDS O—π**NS 2.3** Use the inverse relationship of multiplication and division to compute and check results. **NS 2.0** Students calculate and solve problems involving addition, subtraction, multiplication, and division. *also* O—π**NS 2.2, NS 2.8, AF 1.0,** O—π**AF 1.1, AF 1.2, AF 2.2, MR 1.0, MR 2.3, MR 2.4, MR 3.2**

Copy and complete each table.

2.

÷	2	4	6	8
2	■	■	■	■

3.

÷	10	15	20	25
5	■	■	■	■

▶ **Practice and Problem Solving**

Copy and complete each table.

4.

÷	10	12	14	16
2	■	■	■	■

5.

÷	30	35	40	45
5	■	■	■	■

Find each missing factor and quotient.

6. $2 \times ■ = 4$ $4 \div 2 = ■$

7. $5 \times ■ = 20$ $20 \div 5 = ■$

8. $5 \times ■ = 35$ $35 \div 5 = ■$

9. $2 \times ■ = 16$ $16 \div 2 = ■$

Find each quotient.

10. $6 \div 2 = ■$

11. $10 \div 2 = ■$

12. $■ = 10 \div 5$

13. $5 \div 5 = ■$

14. $25 \div 5 = ■$

15. $■ = 14 \div 2$

16. $40 \div 5 = ■$

17. $20 \div 2 = ■$

18. $2\overline{)2}$

19. $5\overline{)15}$

20. $5\overline{)35}$

21. $2\overline{)16}$

a+b/c Algebra Complete.

22. $10 \div 2 = ■ \times 1$

23. $40 \div 5 = 4 \times ■$

24. $■ \div 2 = 3 + 4$

25. REASONING What multiplication sentence can you use to check $14 \div 2 = 7$?

26. Mrs. Jackson sells hats for $5. She has $15. How many more hats must she sell to have $35 in all?

27. ? What's the Error? Philip used the multiplication fact $2 \times 8 = 16$ to find $8 \div 2 = ■$. Describe his error. What is the correct quotient?

Mixed Review and Test Prep

28. $\begin{array}{r} 7 \\ \times 8 \\ \hline \end{array}$ (p. 150)

29. $\begin{array}{r} 8 \\ \times 6 \\ \hline \end{array}$ (p. 152)

30. $76 + 67 + 22 = ■$ (p. 36)

31. $3 \times 2 \times ■ = 48$ (p. 170)

32. TEST PREP James made an array with 3 rows of 9 tiles. Choose the number sentence that shows how many tiles are in the array. (p. 120)

A $9 - 3 = 6$ **C** $9 \div 3 = 3$

B $3 + 9 = 12$ **D** $3 \times 9 = 27$

Divide by 3 and 4

Quick Review

1. ■ × 3 = ¯8
2. 4 × ■ = ¹2
3. ■ × 4 = ¹6
4. 4 × ■ = ³2
5. 3 × ■ = ²1

▶ Learn

PADDLE POWER The Traveler Scouts want to rent canoes. There are 24 people in the group. A canoe can hold 3 people. How many canoes should the group rent?

$24 \div 3 = ■$

Use a related multiplication fact.

Think: $3 × ■ = 24$

$3 × 8 = 24$ $24 \div 3 = 8$, or $3\overline{)24}^{\,8}$

So, the group should rent 8 canoes.

What if the group wants to rent rowboats instead? If each rowboat holds 4 people, how many rowboats should they rent?

$24 \div 4 = ■$

Think: $4 × ■ = 24$

$4 × 6 = 24$ $24 \div 4 = 6$, or $4\overline{)24}^{\,6}$

So, the group should rent 6 rowboats.

- **REASONING** How can you use $21 \div 3 = 7$ to find $24 \div 3$?

×	0	1	2	3	4	5	6	7	8	9
0	0	0	0	0	0	0	0	0	0	0
1	0	1	2	3	4	5	6	7	8	9
2	0	2	4	6	8	10	12	14	16	18
3	0	3	6	9	12	15	18	21	24	27
4	0	4	8	12	16	20	24	28	32	36
5	0	5	10	15	20	25	30	35	40	45

▶ Check

1. **Explain** how you can use multiplication to find $12 \div 4$.

Write the multiplication fact you can use to find the quotient. Then write the quotient.

2. $12 \div 3 = ■$ 3. $8 \div 4 = ■$

4. $15 \div 3 = ■$ 5. $28 \div 4 = ■$

CALIFORNIA STANDARDS O—n NS 2.3 Use the inverse relationship of multiplication and division to compute and check results. **NS 2.0** Students calculate and solve problems involving addition, subtraction, multiplication, and division. also O—n NS 2.2, NS 2.8, AF 1.0, O—n AF 1.1, AF 1.2, AF 2.2, MR 1.0, MR 2.3, MR 2.4, MR 3.2

▶ Practice and Problem Solving

TECHNOLOGY LINK

More Practice:
Use Mighty Math
Carnival Countdown,
Snap Clowns,
Levels **R** and **Y**.

Write the multiplication fact you can use to find
the quotient. Then write the quotient.

6. $27 \div 3 = \blacksquare$ **7.** $\blacksquare = 4 \div 4$ **8.** $30 \div 3 = \blacksquare$

9. $16 \div 4 = \blacksquare$ **10.** $18 \div 3 = \blacksquare$ **11.** $\blacksquare = 20 \div 4$

Copy and complete each table.

12.

÷	9	12	15	18
3	■	■	■	■

13.

÷	16	20	24	28
4	■	■	■	■

Find each quotient.

14. $12 \div 4 = \blacksquare$ **15.** $\blacksquare = 6 \div 3$ **16.** $\blacksquare = 14 \div 2$ **17.** $12 \div 2 = \blacksquare$

18. $15 \div 3 = \blacksquare$ **19.** $25 \div 5 = \blacksquare$ **20.** $24 \div 4 = \blacksquare$ **21.** $\blacksquare = 40 \div 4$

22. $\blacksquare = 18 \div 2$ **23.** $32 \div 4 = \blacksquare$ **24.** $\blacksquare = 9 \div 3$ **25.** $30 \div 5 = \blacksquare$

26. $3\overline{)3}$ **27.** $3\overline{)18}$ **28.** $4\overline{)36}$ **29.** $5\overline{)20}$

a+b/c Algebra Complete.

30. $20 \div 4 = 8 - \blacksquare$ **31.** $24 \div 3 = \blacksquare \times 2$ **32.** $36 \div \blacksquare = 18 \div 2$

33. Yusef collected 38 pine cones. He kept 11 pine cones for himself and divided the rest equally among 3 friends. How many pine cones did each friend get?

34. The scouts saw squirrels and birds in the woods. If there were 4 animals and 12 legs, how many squirrels and birds were there?

35. REASONING Two numbers have a product of 16 and a quotient of 4. What are they?

36. ✎ **Write About It** Explain how to solve $32 \div 4$ in at least 2 different ways.

Mixed Review and Test Prep

37.
$$\begin{array}{r} 3 \\ \times 5 \\ \hline \end{array}$$ (p. 122)

38.
$$\begin{array}{r} 5 \\ \times 7 \\ \hline \end{array}$$ (p. 118)

39.
$$\begin{array}{r} 4 \\ \times 7 \\ \hline \end{array}$$ (p. 134)

40.
$$\begin{array}{r} 9 \\ \times 8 \\ \hline \end{array}$$ (p. 164)

41. **TEST PREP** Paula had $1.36. Her aunt gave her $5.25 for her birthday. How much money does Paula have now? (p. 88)

  A $3.89   **C** $6.51

  B $5.61   **D** $6.61

Extra Practice page H43, Set B

Divide with 0 and 1

▶ **Learn**

MOO . . . VE OVER Here are some rules for dividing with 0 and 1.

RULE A Any number divided by 1 equals that number.

3	÷	1	=	3
↑		↑		↑
number of cows		number of stalls		number in each stall

If there is only 1 stall, then all of the cows must be in that stall.

RULE B Any number (except 0) divided by itself equals 1.

3	÷	3	=	1
↑		↑		↑
number of cows		number of stalls		number in each stall

If there are the same number of cows and stalls, then one cow goes in each stall.

RULE C Zero divided by any number (except 0) equals 0.

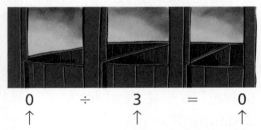

0	÷	3	=	0
↑		↑		↑
number of cows		number of stalls		number in each stall

If there are no cows, then no matter how many stalls you have, there won't be any cows in the stalls.

RULE D You cannot divide by 0.

If there are no stalls, then you aren't separating cows into equal groups. So, using division doesn't make sense.

• **REASONING** How can you use multiplication to show that $3 \div 0 = \blacksquare$ doesn't make sense?

CALIFORNIA STANDARDS NS 2.6 Understand the special properties of 0 and 1 in multiplication and division.
NS 2.0 Students calculate and solve problems involving addition, subtraction, multiplication and division.
also **AF 1.0, AF 1.2, MR 1.0, MR 2.3, MR 2.4, MR 3.2**

1. **Explain** how you can use multiplication to check $0 \div 9 = 0$.

Find each quotient.

2. $3 \div 3 = \blacksquare$
3. $\blacksquare = 5 \div 1$
4. $0 \div 2 = \blacksquare$
5. $\blacksquare = 6 \div 6$

6. $7 \div 1 = \blacksquare$
7. $0 \div 6 = \blacksquare$
8. $\blacksquare = 4 \div 4$
9. $10 \div 1 = \blacksquare$

► **Practice and Problem Solving**

Find each quotient.

10. $2 \div 1 = \blacksquare$
11. $8 \div 8 = \blacksquare$
12. $\blacksquare = 6 \div 3$
13. $1 \div 1 = \blacksquare$

14. $20 \div 5 = \blacksquare$
15. $\blacksquare = 0 \div 4$
16. $\blacksquare = 5 \div 5$
17. $10 \div 2 = \blacksquare$

18. $3 \div 1 = \blacksquare$
19. $21 \div 3 = \blacksquare$
20. $32 \div 4 = \blacksquare$
21. $\blacksquare = 0 \div 7$

22. $\blacksquare = 0 \div 8$
23. $18 \div 2 = \blacksquare$
24. $\blacksquare = 9 \div 1$
25. $24 \div 4 = \blacksquare$

26. Divide 2 by 2.
27. Divide 4 by 1.
28. Divide 0 by 3.
29. Divide 14 by 2.

30. $5\overline{)35}$
31. $9\overline{)9}$
32. $2\overline{)16}$
33. $5\overline{)0}$
34. $2\overline{)14}$

35. $3\overline{)18}$
36. $1\overline{)8}$
37. $4\overline{)36}$
38. $7\overline{)7}$
39. $1\overline{)0}$

$\frac{a+b}{c}$ **Algebra** **Compare. Write** $<$, $>$, **or** $=$ **for each** ●.

40. $4 \div 1 \; ● \; 4 \div 4$
41. $0 \div 9 \; ● \; 9 \div 1$
42. $6 + 4 \; ● \; 5 \div 1$

43. A farmer has 6 bales of hay. He feeds 2 bales to his cows. He divides the rest equally among 4 stalls. How many bales are in each stall?

44. **REASONING** Chelsea says, "Ask me to divide any number by 1, and I'll give you the quotient." What is her strategy?

45. Use what you know about 0 and 1 to find each quotient.
    a. $398 \div 398 = \blacksquare$
    b. $971 \div 1 = \blacksquare$
    c. $0 \div 426 = \blacksquare$

Mixed Review and Test Prep

46. $\begin{array}{r} 0 \\ \times 4 \\ \hline \end{array}$ (p. 132)

47. $\begin{array}{r} 5 \\ \times 1 \\ \hline \end{array}$ (p. 132)

48. $387 + 132 + 155 = \blacksquare$ (p. 42)

49. $\$10.00 - \$5.69 = \blacksquare$ (p. 58)

50. **TEST PREP** Akiko bought 3 sandwiches. Each sandwich cost $4. How much did she spend for the sandwiches? (p. 122)

    A $4   **B** $7   **C** $12   **D** $15

Algebra: Write Expressions

▶ Learn

HAPPY CAMPERS   The camp counselor divided 21 campers into 3 equal groups. How many campers are in each group?

Write an expression to show how many campers are in each group.

21 campers	divided into	3 groups
↓	↓	↓
21	÷	3

Remember

An *expression* is part of a number sentence. Examples:

$3 + 4$ 1×2

$25 - 12$ $12 \div 6$

$21 \div 3 = 7$ is a number sentence. 7 is the number of campers in each group.

You know a number sentence can be true or false.

$$12 \div 3 = 4 \text{ is true.}$$

$$12 \times 3 = 4 \text{ is false.}$$

Mr. Gonzales is lining up 4 rows of 5 campers for relay races. There are 20 campers lined up.

$$4 \bullet 5 = 20$$

Which symbol will make the sentence true?

Try +	$4 + 5 = 20$	*Not* true
Try −	$4 - 5 = 20$	*Not* true
Try ÷	$4 \div 5 = 20$	*Not* true
Try ×	$4 \times 5 = 20$	True

So, the correct operation symbol is ×.

▶ Check

1. **Write** a different expression to describe the relay race problem above.

CALIFORNIA STANDARDS ⊶ **AF 1.1** Represent relationships of quantities in the form of mathematical expressions, equations, or inequalities. **AF 1.0** Students select appropriate symbols, operations, and properties to represent, describe, simplify, and solve simple number relationships. *also* **NS 2.0, NS 2.8, AF 1.3, MR 1.0, MR 2.3, MR 2.4, MR 3.2**

Write an expression to describe each problem.

2. Nine campers each ate 7 carrot sticks. How many carrot sticks did they eat in all?

3. Jo had 9 carrot sticks. She ate 7 of them. How many carrot sticks does she have left?

Write +, −, ×, or ÷ to make the number sentence true.

4. $9 = 18 \bullet 9$

5. $6 \times 6 = 4 \bullet 9$

6. $72 \bullet 8 = 3 \times 3$

▶ Practice and Problem Solving

Write an expression to describe each problem.

7. Antoine had 24 grapes. His mother gave him 6 more. How many grapes does he have now?

8. Six students share 30 crayons equally. How many crayons does each student get?

9. Beth bought 6 pieces of drawing paper, 8 crayons, and 1 eraser. How many items did she buy?

10. Alison had $0.45. She lost a dime. How much money does she have now?

Write +, −, ×, or ÷ to make the number sentence true.

11. $13 \bullet 7 = 2 \times 3$

12. $12 + 5 = 9 \bullet 8$

13. $6 \times 4 = 8 \bullet 3$

For 14–15, write two different expressions you can use to describe each problem.

14. Carl and 3 of his friends each ate 5 roasted marshmallows. How many marshmallows did Carl and his friends eat in all?

15. Bruce has a photo album with 8 pages. He can fit 6 photos on each page. How many photos can he put in the album?

16. **Write a problem** for each expression.

   a. $35 \div 5$ b. 7×3 c. $15 - 3$

Mixed Review and Test Prep

17. $125 + 291 = \blacksquare$ (p. 42)

18. $3,538 + 1,896 = \blacksquare$ (p. 46)

19. $345 - 77 = \blacksquare$ (p. 58)

20. $622 - 404 = \blacksquare$ (p. 58)

21. **TEST PREP** Anna has a music lesson at 4:30 P.M. The lesson will last 45 minutes. At what time will Anna's lesson end? (p. 98)

   A 4:15 P.M.   C 5:15 P.M.
   B 4:45 P.M.   D 5:45 P.M.

Extra Practice page H43, Set D

Problem Solving Skill
Choose the Operation

Understand ➤ Plan ➤ Solve ➤ Check

NATURE WALK On a hike, the campers saw 6 chipmunks, 4 deer, and 8 butterflies. How many more butterflies did they see than chipmunks?

This chart can help you decide when to use each operation.

ADD	• Join groups of different sizes.
SUBTRACT	• Take away. • Compare amounts.
MULTIPLY	• Join equal groups.
DIVIDE	• Separate into equal groups. • Find the number in each group.

MATH IDEA Before you solve a problem, decide what operation to use. Write a number sentence to solve the problem.

Since you are comparing amounts, subtract.

8	−	6	=	2
↓		↓		↓
number of butterflies		number of chipmunks		how many more butterflies than chipmunks

So, they saw 2 more butterflies than chipmunks.

- **REASONING** When would you use division to solve a problem?

- Write a number sentence to find how many animals they saw in all.

CALIFORNIA STANDARDS ⊶AF 1.1 Represent relationships of quantities in the form of mathematical expressions, equations, or inequalities. **AF 1.3** Select appropriate operational and relational symbols to make an expression true. *also* **NS 2.0, NS 2.8, AF 1.0, AF 1.2, MR 1.0, MR 1.1, MR 2.0, MR 2.3, MR 2.4**

▶ Problem Solving Practice

Choose the operation. Write a number sentence. Then solve.

1. David collected 8 acorns and 16 wildflowers. He put the same number of wildflowers in each of 8 vases. How many wildflowers were in each vase?

2. The camp counselor put 4 pears, 5 apples, and 7 bananas in a basket. If 3 pieces of fruit were taken, how many pieces of fruit were left?

3. Beth has a scrap book. Each page can hold 8 small postcards or 6 large postcards. How many small postcards fit on 4 pages?

4. Shawn took 8 photos of birds, 9 photos of wildflowers, and 15 photos of campers. How many photos did he take?

Thirty students went on an amusement park ride. Five students sat in each car. How many cars did they fill?

5. Which number sentence can you use to solve the problem?

   A $30 + 5 =$ ▮   **C** $30 \div 5 =$ ▮
   B $30 - 5 =$ ▮   **D** $30 \times 5 =$ ▮

6. What is the answer to the question?

   F 25 students   **H** 6 students
   G 6 cars   **J** 1 car

Mixed Applications

USE DATA For 7–9, use the graph.

7. Gina took 5 packs of soda to her club meeting. Were there enough bottles of soda for 28 people? Explain.

8. Khar bought 3 packs of water. He gave 4 bottles to friends. How many bottles of water did he have left?

9. **? What's the Question?** The answer is 30 bottles of juice.

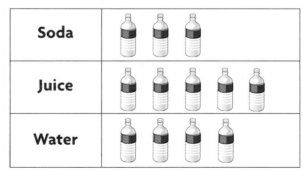

NUMBER OF BOTTLES IN A PACK

Soda	🍾 🍾 🍾
Juice	🍾 🍾 🍾 🍾 🍾
Water	🍾 🍾 🍾 🍾

Key: Each 🍾 = 2 bottles.

Review/Test

✅ CHECK VOCABULARY

Choose the best term from the box.

1. You cannot divide by ___?___ . (p. 204)

2. Any number (except 0) ___?___ by itself equals 1. (p. 204)

3. A number phrase like 12 ÷ 2 or 3 × 5 is called an ___?___ . (p. 206)

> expression
> divided
> multiplied
> zero

✅ CHECK SKILLS

Find each quotient. (pp. 200–205)

4. $16 \div 4 = \blacksquare$
5. $\blacksquare = 21 \div 3$
6. $6 \div 1 = \blacksquare$
7. $\blacksquare = 25 \div 5$

8. $8 \div 2 = \blacksquare$
9. $0 \div 5 = \blacksquare$
10. $\blacksquare = 9 \div 3$
11. $\blacksquare = 8 \div 1$

12. $4\overline{)32}$
13. $1\overline{)10}$
14. $2\overline{)18}$
15. $3\overline{)0}$
16. $5\overline{)15}$

17. $2\overline{)14}$
18. $3\overline{)15}$
19. $5\overline{)0}$
20. $4\overline{)24}$
21. $5\overline{)40}$

Write an expression to describe each problem. (pp. 206–207)

22. Lila made 18 muffins. She put 3 muffins in each bag. How many bags did Lila fill?

23. Kyle had 32 shells. He gave 8 to his friend. How many shells does Kyle have now?

✅ CHECK PROBLEM SOLVING

Choose the operation. Write a number sentence. Then solve. (pp. 208–209)

24. Chiang has 24 trading cards. She puts the cards into piles of 6. How many piles does she make?

25. Casey has 5 packs of stickers. Each pack has 8 stickers. How many stickers does he have?

Cumulative Review

TIP! Check your work.
See item **4.**
Count by 5 and keep track of the numbers you say until you get to 45.
Also see problem **7**, p. H65.

For 1–10, choose the best answer.

1. On Friday, 25 students will go to the zoo. They will be in 5 equal groups. How many students will be in each group?

A 30   C 6
B 10   D 5

2. Which sentence does **not** belong in the fact family for 2, 8, and 16?

F $16 - 8 = 8$
G $16 \div 2 = 8$
H $8 \times 2 = 16$
J $2 \times 8 = 16$

3. There are 8 cookies for 4 people to share equally. Which number sentence tells how many cookies each person can have?

A $8 + 4 = 12$
B $8 \div 4 = 2$
C $8 \times 4 = 32$
D $8 - 4 = 4$

4. Farmer Mac has 45 pigs and wants to put 5 pigs in each pen. How many pens will he use?

F 6   H 8
G 7   J 9

5.

$$1\overline{)6}$$

A 6   C 1
B 3   D NOT HERE

6. Joey has 27 stamps to put in his stamp book. He wants to put 9 stamps on each page. Which expression tells how many pages he will need?

F $27 - 9$ H $27 \div 9$
G $27 + 9$ J 9×3

7.   509
       $+493$

A 1,002   C 902
B 996   D 196

8. $15 \div 3 = $ ▓

F 4   H 6
G 5   J NOT HERE

9. Mr. Sato bought a television that cost $287. What is that amount rounded to the nearest hundred dollars?

A $200   C $300
B $250   D $400

10. Samantha bought a sandwich for $1.68, apple chips for $0.79, and a drink for $1.29. How much change did she get back from $5.00?

F $3.76   H $1.34
G $2.24   J $1.24

Division Facts Through 10

There are more than 61,000 pizza restaurants in the United States! America's favorite pizza topping is pepperoni. Suppose each pizza in the chart is cut into 8 slices, and the pieces of pepperoni are divided equally among the 8 slices. How many pieces of pepperoni are on 1 slice of pizza from each restaurant?

PEPPERONI PIZZA	
Restaurant	Pieces of Pepperoni
Mario's Pizza	32
Broadway Pizza	56
Mamma Mia's	40
Joanne's Pizza	48

CHECK WHAT YOU KNOW ✓

Use this page to help you review and remember
important skills needed for Chapter 13.

✓ VOCABULARY

Choose the best term from the box.

Example:
$18 \div 3 = 6$

| quotient |
| dividend |
| divisor |

1. In the example, the number 6 is called the __?__.

2. In the example, the number 3 is called the __?__.

✓ DIVISION FACTS THROUGH 5 (See p. H15.)

Find each quotient.

3. $35 \div 5 = \blacksquare$ 4. $8 \div 2 = \blacksquare$ 5. $20 \div 4 = \blacksquare$ 6. $0 \div 5 = \blacksquare$

7. $\blacksquare = 18 \div 3$ 8. $12 \div 4 = \blacksquare$ 9. $\blacksquare = 9 \div 1$ 10. $16 \div 2 = \blacksquare$

11. $4\overline{)16}$ 12. $3\overline{)24}$ 13. $1\overline{)6}$ 14. $8\overline{)0}$ 15. $2\overline{)14}$

16. $1\overline{)10}$ 17. $6\overline{)0}$ 18. $3\overline{)15}$ 19. $4\overline{)28}$ 20. $2\overline{)10}$

✓ ORDER PROPERTY OF MULTIPLICATION (See p. H16.)

Use the Order Property of Multiplication to help you
find each product.

21. $3 \times 5 = \blacksquare$ $5 \times 3 = \blacksquare$ 22. $7 \times 8 = \blacksquare$ $8 \times 7 = \blacksquare$

23. $6 \times 8 = \blacksquare$ $8 \times 6 = \blacksquare$ 24. $3 \times 6 = \blacksquare$ $6 \times 3 = \blacksquare$

25. $7 \times 9 = \blacksquare$ $9 \times 7 = \blacksquare$ 26. $6 \times 10 = \blacksquare$ $10 \times 6 = \blacksquare$

✓ MISSING FACTORS (See p. H16.)

Find the missing factor.

27. $3 \times \blacksquare = 27$ 28. $25 = \blacksquare \times 5$ 29. $\blacksquare \times 6 = 12$ 30. $45 = 9 \times \blacksquare$

31. $6 \times \blacksquare = 48$ 32. $\blacksquare \times 8 = 32$ 33. $35 = 5 \times \blacksquare$ 34. $20 = \blacksquare \times 4$

35. $\blacksquare \times 2 = 18$ 36. $80 = \blacksquare \times 10$ 37. $10 \times \blacksquare = 10$ 38. $56 = 8 \times \blacksquare$

Divide by 6, 7, and 8

Quick Review

1. $6 \times 3 = \blacksquare$

2. $6 \times 7 = \blacksquare$ 3. $7 \times 4 = \blacksquare$

4. $7 \times 8 = \blacksquare$ 5. $8 \times 4 = \blacksquare$

▶ **Learn**

IT'S IN THE BAG   The Bagel Stop sells bagels in bags of 6. Ramona has 24 fresh bagels to put in bags. How many bags does she need?

$24 \div 6 = \blacksquare$

Use a related multiplication fact to find the quotient.

Think: $6 \times \blacksquare = 24$
$6 \times 4 = 24$

$24 \div 6 = 4$, or $6)\overline{24}$ with quotient 4

So, Ramona needs 4 bags.

Examples

A $63 \div 7 = \blacksquare$
Think: $7 \times \blacksquare = 63$
$7 \times 9 = 63$

$63 \div 7 = 9$, or $7)\overline{63}$ with quotient 9

B $56 \div 8 = \blacksquare$
Think: $8 \times \blacksquare = 56$
$8 \times 7 = 56$

$56 \div 8 = 7$, or $8)\overline{56}$ with quotient 7

MATH IDEA Think of related multiplication facts to help you divide.

- What multiplication fact can you use to find $42 \div 7$? What is the quotient?

TECHNOLOGY LINK

To learn more about division, watch the Harcourt Math Newsroom Video *Giant Panda Bears.*

CALIFORNIA STANDARDS ○━ⁿNS 2.3 Use the inverse relationship of multiplication and division to compute and check results. ○━ⁿAF 1.1 Represent relationships of quantities in the form of mathematical expressions, equations, or inequalities. *also* NS 2.0, ○━ⁿNS 2.2, NS 2.8, AF 1.2, AF 2.0, MR 1.0, MR 1.1, MR 2.0, MR 2.3, MR 3.2

Equal Groups

Remember, you can also use equal groups and arrays to help you find a quotient.

Here are two different ways to find $28 \div 7$.

Jane
I used counters to model equal groups.

$28 \div 7 = \blacksquare$

number of counters   number in each group   number of groups

So, $28 \div 7 = 4$.

Kevin
I modeled the problem with an array.

$28 \div 7 = \blacksquare$

number in array   number in each row   number of rows

So, $28 \div 7 = 4$.

- **REASONING** You have used equal groups and arrays to help you find products. Why can you also use them to help you find quotients?

▶ Check

1. **Explain** how you would use a related multiplication fact to find $18 \div 6$.

Find the missing factor and quotient.

2. $8 \times \blacksquare = 16$ $16 \div 8 = \blacksquare$ 3. $7 \times \blacksquare = 28$ $28 \div 7 = \blacksquare$

4. $6 \times \blacksquare = 36$ $36 \div 6 = \blacksquare$ 5. $6 \times \blacksquare = 30$ $30 \div 6 = \blacksquare$

Copy and complete each table.

6.

÷	14	21	28	35
7	▨	▨	▨	▨

7.

÷	6	12	18	24
6	▨	▨	▨	▨

Find the quotient.

8. $21 \div 7 = \blacksquare$ 9. $42 \div 6 = \blacksquare$ 10. $\blacksquare = 56 \div 7$ 11. $32 \div 8 = \blacksquare$

12. $8\overline{)40}$ 13. $7\overline{)42}$ 14. $8\overline{)24}$ 15. $6\overline{)24}$

LESSON CONTINUES ▶

▶ Practice and Problem Solving

Find the missing factor and quotient.

16. $7 \times \blacksquare = 14$ $14 \div 7 = \blacksquare$ **17.** $6 \times \blacksquare = 60$ $60 \div 6 = \blacksquare$

18. $6 \times \blacksquare = 48$ $48 \div 6 = \blacksquare$ **19.** $8 \times \blacksquare = 72$ $72 \div 8 = \blacksquare$

Copy and complete each table.

20.

÷	42	63	56	49
7	■	■	■	■

21.

÷	56	40	48	32
8	■	■	■	■

Find the quotient.

22. $36 \div 6 = \blacksquare$ **23.** $80 \div 8 = \blacksquare$ **24.** $\blacksquare = 0 \div 7$ **25.** $8 \div 1 = \blacksquare$

26. $\blacksquare = 15 \div 3$ **27.** $\blacksquare = 18 \div 6$ **28.** $45 \div 5 = \blacksquare$ **29.** $24 \div 8 = \blacksquare$

30. $8\overline{)64}$ **31.** $2\overline{)14}$ **32.** $7\overline{)28}$ **33.** $6\overline{)0}$

34. $5\overline{)10}$ **35.** $7\overline{)7}$ **36.** $8\overline{)0}$ **37.** $3\overline{)30}$

38. Divide 42 by 6. **39.** Divide 8 by 8. **40.** Divide 35 by 5.

Write a division sentence for each.

41.

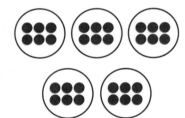

42.

a+b/c Algebra Complete.

43. $3 + \blacksquare = 49 \div 7$ **44.** $8 \times 5 = \blacksquare \times 10$ **45.** $\blacksquare - 4 = 24 \div 6$

46. $\blacksquare \times 4 = 8 \times 3$ **47.** $6 \div 6 = 0 + \blacksquare$ **48.** $5 + 3 = 16 \div \blacksquare$

49. REASONING Is the quotient $24 \div 6$ greater than or less than the quotient $24 \div 4$? How do you know?

50. **? What's the Question?** Hikara bought 35 fruit chews. Fruit chews come in packs of 5. The answer is 7 packs.

216

51. Jessica has 48 marbles. She puts an equal number of marbles in each of 6 bags. How many marbles are in 2 bags?

52. Asha had 24 pictures of her friends. She put 8 pictures on each page in a photo album. Her album has 20 pages. How many album pages do not have pictures?

Mixed Review and Test Prep

Write the time. (p. 94)

53.

54.

Find the missing factor. (p. 142)

55. $3 \times \blacksquare = 24$ **56.** $4 \times \blacksquare = 28$

57. $\blacksquare \times 8 = 72$ **58.** $\blacksquare \times 1 = 8$

59. **TEST PREP** Luther read 54 pages of his book on Saturday. He read 39 pages on Sunday. How many pages did he read in all?

  A 15   **C** 93
  B 83   **D** 94

60. **TEST PREP** Patricia collected 205 stickers. She gave 28 stickers to her sister. How many stickers does Patricia have left? (p. 58)

  F 177   **H** 233
  G 187   **J** 277

--LiNKUP--- to Reading

STRATEGY • CHOOSE IMPORTANT INFORMATION

Some word problems have more information than you need. Before you solve a problem, find the facts you need to solve the problem.

Mrs. Taylor baked 8 batches of muffins. She had a total of 48 muffins. She also baked 2 cakes. How many muffins were in each batch?

Facts You Need: baked 8 batches of muffins, total of 48 muffins

Fact You Don't Need: baked 2 cakes

$48 \div 8 = 6$

So, there were 6 muffins in each batch.

Write the important facts. Solve the problem.

1. Bonnie made 4 batches of cookies and 3 pies in the morning. She made 3 batches of cookies in the afternoon. She made 63 cookies in all. How many cookies were in each batch?

Divide by 9 and 10

Quick Review

1. $9 \times 3 = \blacksquare$

2. $9 \times 5 = \blacksquare$

3. $10 \times 4 = \blacksquare$

4. $8 \times 10 = \blacksquare$

5. $9 \times 10 = \blacksquare$

▶ Learn

PLENTY OF PINS Katie collects different kinds of pins. She has boxes that hold 9 or 10 pins. Help Katie organize her collection.

Examples

A Katie puts her 45 state flag pins in boxes that hold 9 pins each. How many boxes does she need?

$45 \div 9 = \blacksquare$

Think: $9 \times \blacksquare = 45$

$9 \times 5 = 45$

$45 \div 9 = 5$, or $9\overline{)45}$ with quotient 5

So, Katie needs 5 boxes for her state flag pins.

B Katie puts her 60 flower pins in boxes that hold 10 pins each. How many boxes does she need?

$60 \div 10 = \blacksquare$

Think: $10 \times \blacksquare = 60$

$10 \times 6 = 60$

$60 \div 10 = 6$, or $10\overline{)60}$ with quotient 6

So, Katie needs 6 boxes for her flower pins.

Katie's Pin Collection

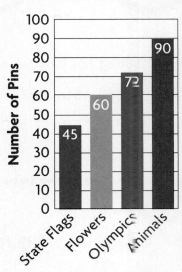

▶ Check

1. **Explain** how to use a related multiplication fact to find $36 \div 9$. What is the quotient?

Copy and complete each table.

2.

÷	9	18	27	36
9	■	■	■	■

3.

÷	20	30	40	50
10	■	■	■	■

CALIFORNIA STANDARDS O—┓NS 2.3 Use the inverse relationship of multiplication and division to compute and check results. O—┓AF 1.1 Represent relationships of quantities in the form of mathematical expressions, equations, or inequalities. *also* NS 2.0, O—┓NS 2.2, NS 2.8, AF 1.0, AF 1.2, AF 2.0, MR 1.0, MR 2.0, MR 2.3, MR 2.4

Copy and complete each table.

4.

÷	54	72	63	81
9	▦	▦	▦	▦

5.

÷	70	90	80	100
10	▦	▦	▦	▦

Find the quotient.

6. $45 \div 9 = $ ▦

7. ▦ $= 10 \div 10$

8. ▦ $= 0 \div 9$

9. $60 \div 10 = $ ▦

10. $9 \div 1 = $ ▦

11. $12 \div 6 = $ ▦

12. $18 \div 3 = $ ▦

13. ▦ $= 36 \div 9$

14. ▦ $= 50 \div 10$

15. $14 \div 2 = $ ▦

16. ▦ $= 12 \div 4$

17. $40 \div 5 = $ ▦

18. $10\overline{)0}$

19. $6\overline{)42}$

20. $9\overline{)90}$

21. $9\overline{)63}$

22. $8\overline{)64}$

23. $5\overline{)25}$

24. $7\overline{)28}$

25. $3\overline{)24}$

26. Divide 72 by 8.

27. Divide 42 by 7.

28. Divide 70 by 10.

$\frac{a+b}{c}$ **Algebra** Write +, −, ×, or ÷ for each ●.

29. $10 ● 10 = 2 - 1$

30. $8 \times 3 = 20 ● 4$

31. $12 ● 7 = 50 \div 10$

32. $3 \times 6 = 2 ● 9$

33. $81 ● 9 = 3 \times 3$

34. $6 \times 7 = 35 ● 7$

35. REASONING Ken has 89 patches in his collection. He puts 8 patches on his vest and puts the rest in boxes of 9 patches each. How many boxes does Ken need?

36. REASONING Boxes for 9 pins cost $4 each. Boxes for 10 pins cost $5 each. Janine has 90 pins. If she wants to spend the least amount of money, what type of boxes should she buy?

37. Trish says, "I can find the quotient of 70 divided by 10 by taking the zero off the end of 70. So, $70 \div 10 = 7$." Do you agree or disagree? Explain.

38. **Write About It** Make a table showing the 9's division facts. Describe any patterns you see in your table.

Mixed Review and Test Prep

39. $8 \times 7 = $ ▦ (p. 152)

40. ▦ $= 6 \times 9$ (p. 148)

41. $24 \div 3 = $ ▦ (p. 202)

42. $28 \div 4 = $ ▦ (p. 202)

43. TEST PREP Which expression has the same product as $2 \times 3 \times 9$?
(p. 170)

A $2 \times 2 \times 8$ **C** $3 \times 3 \times 7$
B $5 \times 9 \times 1$ **D** $9 \times 6 \times 1$

Practice Division Facts Through 10

Quick Review

1. $7 \times \blacksquare = 56$

2. $\blacksquare \times 3 = 27$

3. $6 \times \blacksquare = 30$

4. $\blacksquare \times 4 = 16$

5. $5 \times \blacksquare = 40$

▶ **Learn**

BOXED CARS Bobby has 36 toy cars that he wants to put in display boxes. Each display box holds 9 cars. How many display boxes will Bobby need?

$36 \div 9 = \blacksquare$

There are many ways to find the quotient.

A. Use counters.

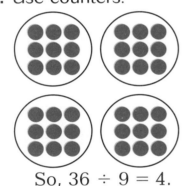

So, $36 \div 9 = 4$.

B. Use an array.

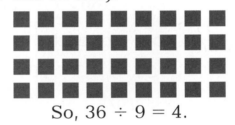

So, $36 \div 9 = 4$.

C. Use repeated subtraction.

$$
\begin{array}{cccc}
36 & 27 & 18 & 9 \\
-9 & -9 & -9 & -9 \\
\hline
27 & 18 & 9 & 0
\end{array}
$$

Number of times
you subtract 9:   1   2   3   4

So, $36 \div 9 = 4$.

D. Think about fact families.

Fact Family for 4, 9, and 36

factor	factor	product	dividend	divisor	quotient
4 × 9	= 36		36 ÷ 9	= 4	
9 × 4	= 36		36 ÷ 4	= 9	

So, $36 \div 9 = 4$.

CALIFORNIA STANDARDS ○┐NS 2.3 Use the inverse relationship of multiplication and division to compute and check results. **MR 2.3** Use a variety of methods, such as words, numbers, symbols, charts, graphs, tables, diagrams, and models, to explain mathematical reasoning. *also* NS 2.0, ○┐NS 2.2, NS 2.8, AF 1.0, AF 1.2, MR 1.0, MR 2.0, MR 2.4, MR 3.2

Find Missing Factors

E. Use a multiplication table.

Think: $9 \times \blacksquare = 36$

- Find the given factor 9 in the top row.

- Look down the column to find the product, 36.

- Look left across the row to find the missing factor, 4.

$9 \times 4 = 36 \qquad 36 \div 9 = 4$

So, Bobby needs 4 display boxes.

×	0	1	2	3	4	5	6	7	8	9	10
0	0	0	0	0	0	0	0	0	0	0	0
1	0	1	2	3	4	5	6	7	8	9	10
2	0	2	4	6	8	10	12	14	16	18	20
3	0	3	6	9	12	15	18	21	24	27	30
4	0	4	8	12	16	20	24	28	32	36	40
5	0	5	10	15	20	25	30	35	40	45	50
6	0	6	12	18	24	30	36	42	48	54	60
7	0	7	14	21	28	35	42	49	56	63	70
8	0	8	16	24	32	40	48	56	64	72	80
9	0	9	18	27	36	45	54	63	72	81	90
10	0	10	20	30	40	50	60	70	80	90	100

MATH IDEA Use equal groups, arrays, repeated subtraction, fact families, and multiplication tables to help you find quotients.

▶ Check

1. Explain how to find $56 \div 8$ in two different ways.

Write a division sentence for each.

2.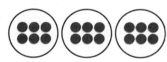

3.

4.
$$\begin{array}{r} 21 \\ -\ 7 \\ \hline 14 \end{array} \nearrow \begin{array}{r} 14 \\ -\ 7 \\ \hline 7 \end{array} \nearrow \begin{array}{r} 7 \\ -7 \\ \hline 0 \end{array}$$

Find the missing factor and quotient.

5. $3 \times \blacksquare = 15 \qquad 15 \div 3 = \blacksquare$

6. $8 \times \blacksquare = 32 \qquad 32 \div 8 = \blacksquare$

7. $4 \times \blacksquare = 40 \qquad 40 \div 4 = \blacksquare$

8. $7 \times \blacksquare = 56 \qquad 56 \div 7 = \blacksquare$

Find the quotient.

9. $10 \div 2 = \blacksquare$ **10.** $18 \div 9 = \blacksquare$ **11.** $\blacksquare = 49 \div 7$ **12.** $80 \div 10 = \blacksquare$

13. $6\overline{)0}$ **14.** $9\overline{)9}$ **15.** $8\overline{)40}$ **16.** $5\overline{)20}$

LESSON CONTINUES ▶

Write a division sentence for each.

17.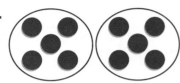

18.

19.
$$\begin{array}{r} 27 \\ -\ 9 \\ \hline 18 \end{array} \nearrow \begin{array}{r} 18 \\ -\ 9 \\ \hline 9 \end{array} \nearrow \begin{array}{r} 9 \\ -9 \\ \hline 0 \end{array}$$

Find the missing factor and quotient.

20. $10 \times \blacksquare = 90$ $90 \div 10 = \blacksquare$ **21.** $7 \times \blacksquare = 35$ $35 \div 7 = \blacksquare$

22. $4 \times \blacksquare = 16$ $16 \div 4 = \blacksquare$ **23.** $9 \times \blacksquare = 63$ $63 \div 9 = \blacksquare$

Find the quotient.

24. $10 \div 1 = \blacksquare$ **25.** $\blacksquare = 35 \div 5$ **26.** $50 \div 10 = \blacksquare$ **27.** $\blacksquare = 16 \div 2$

28. $\blacksquare = 81 \div 9$ **29.** $\blacksquare = 20 \div 10$ **30.** $60 \div 6 = \blacksquare$ **31.** $24 \div 3 = \blacksquare$

32. $9\overline{)54}$ **33.** $7\overline{)28}$ **34.** $8\overline{)72}$ **35.** $10\overline{)100}$

36. $8\overline{)24}$ **37.** $2\overline{)0}$ **38.** $4\overline{)4}$ **39.** $3\overline{)21}$

40. Divide 63 by 7.   **41.** Divide 30 by 5.   **42.** Divide 0 by 9.

Choose the letter of the division sentence that matches each.

   a. $42 \div 6 = 7$ **b.** $36 \div 9 = 4$ **c.** $56 \div 7 = 8$ **d.** $27 \div 9 = 3$

43.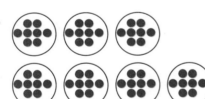

44.

Compare. Write <, >, or = for each ●.

45. 9×6 ● 9×5 **46.** $24 \div 6$ ● $16 \div 4$ **47.** $4 + 4$ ● $72 \div 8$

48. 8×5 ● 10×4 **49.** $23 - 18$ ● $45 \div 5$ **50.** 3×3 ● $70 \div 10$

51. REASONING Roberta has some boxes with 8 cars in each box. Could Roberta have 20 cars in all? Explain.

52. 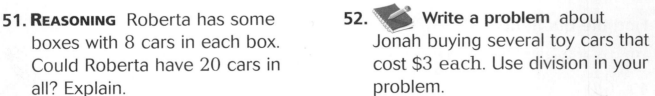 **Write a problem** about Jonah buying several toy cars that cost $3 each. Use division in your problem.

53. Chi has 2 sheets of stickers. Each sheet has 9 rows of 7 stickers. If Chi uses 2 rows of stickers, how many stickers will Chi have left?

54. Carla put 24 toy animals in 4 boxes. Each box has the same number of animals. How many animals are in 3 boxes?

Mixed Review and Test Prep

55.   35 (p. 36)
    22
 +15

56.   137 (p. 42)
  353
+229

Compare. Write <, >, or = for each ⬤. (p. 20)

57. 458 ⬤ 584

58. 1,602 ⬤ 1,062

59. 3,459 ⬤ 3,459

60. 5,891 ⬤ 5,981

61. **TEST PREP** Maria buys pencils for $1.39, a notebook for $1.79, and a marker for $0.85. If she pays with a $5 bill, how much change will she get? (p. 88)

A $4.03   **C** $1.82
B $1.97   **D** $0.97

62. **TEST PREP** Andres had 107 rocks in his collection. He gave 18 rocks to his sister. How many rocks does Andres have left? (p. 58)

F 89   **G** 91   **H** 99   **J** 125

LINKUP

to Social Studies

Hundreds of years ago, the Miwok Indians lived in the central part of California. Their villages spread from the Pacific Ocean to Yosemite Valley. The Miwoks' main source of food was acorns. Acorns were dried and stored in shelters made from bark, grass, and mud.

1. Acorns were soaked in running streams for 2 to 3 weeks to get rid of the bitter taste. If you began soaking acorns on October 1, on what dates might you check on them?

2. A good acorn harvest might come once every 4 to 5 years. If there was a good harvest of acorns in 1787, in what year might the next good harvest have occurred?

Extra Practice page H44, Set C

Algebra: Find the Cost

Quick Review

1. $2 \times 9 = \blacksquare$

2. $4 \times \blacksquare = 32$ 3. $6 \times 2 = \blacksquare$

4. $\blacksquare \times 5 = 45$ 5. $10 \times 2 = \blacksquare$

▶ Learn

WHAT'S FOR LUNCH? Mrs. Hugo buys 3 pizzas for her family. How much does Mrs. Hugo spend?

To find the total amount spent, multiply the number of pizzas by the cost of one pizza.

The Lunch Box

Pizza	$9
Box of 6 Tacos	$12
Salad	$3
Bag of 4 Cookies	$4

$$3 \qquad \times \qquad \$9 \qquad = \qquad \$27$$

number of pizzas — cost of one — total spent

So, Mrs. Hugo spends $27 for 3 pizzas.

Nicolas buys a box of 6 tacos. How much does each taco cost?

To find the cost of one taco, divide the total amount spent by the number of tacos bought.

$$\$12 \qquad \div \qquad 6 \qquad = \qquad \$2$$

total spent — number of tacos — cost of one

So, each taco costs $2.

MATH IDEA Multiply to find the cost of multiple items or divide to find the cost of one item.

▶ Check

1. **Name** the operation you can use to find the cost of one cookie. Explain.

For 2–3, write a number sentence. Then solve.

2. Alem bought 4 salads. Each salad cost $3. How much did Alem spend?

3. Kim spends $24 for a box of 4 burritos. How much does each burrito cost?

CALIFORNIA STANDARDS NS 2.7 Determine the unit cost when given the total cost and number of units.
O—n AF 2.1 Solve simple problems involving a functional relationship between two quantities. also NS 2.0,
**O—n NS 2.2, O—n NS 2.3, NS 2.8, AF 1.0, O—n AF 1.1, AF 1.2, AF 2.0, MR 1.0, MR 1.1 MR 2.0, MR 2.4,
MR 3.2, MR 3.3**

For 4–5, write a number sentence. Then solve.

4. Shakira bought 4 hot dogs. Each hot dog cost $4. How much did Shakira spend?

5. Mr. Hess spends $18 for an order of 6 sandwiches. How much does each sandwich cost?

USE DATA For 6–16, use the price list at the right to find the cost of each number of items.

6. 4 discs

7. 6 tapes

8. 8 tapes

9. 7 books

10. 2 tapes

11. 5 discs

12. 3 books

13. 9 discs

14. 5 books

PRICE LIST	
Books	$4 each
Tapes	$7 each
Discs	$9 each

15. 2 discs and 6 books

16. 4 books and 5 tapes

Find the cost of one of each item.

17. 9 markers cost $27.

18. 6 notepads cost $18.

19. 3 stamps cost $15.

20. 5 balls cost $30.

21. 8 pencils cost $8.

22. 4 games cost $32.

23. 7 toy cars cost $28.

24. 10 pens cost $20.

25. 2 T-shirts cost $12.

26. **REASONING** Ako has $20. She wants to buy rubber stamps that cost $6 each. How many rubber stamps can she buy? Explain.

27. Heidi buys 3 videotapes for $24. She gives the clerk $30. How much does each tape cost? How much change does she get?

Mixed Review and Test Prep

Continue each pattern. (p. 136)

28. 4, 9, 14, 19, ▪, ▪

29. 44, 40, 36, 32, ▪, ▪

30.   2,391 (p. 46)
  +1,236

31.   3,529 (p. 46)
  +9,382

32. **TEST PREP** Latasha makes 2 sandwiches for each of her 4 friends. She puts 2 slices of ham in each sandwich. How many slices of ham does she use? (p. 172)

A 2   C 8

B 4   D 16

LESSON

5

Problem Solving Strategy
Work Backward

Quick Review

1. $18 \div 3 = \blacksquare$
2. $12 \div 4 = \blacksquare$
3. $24 \div 6 = \blacksquare$
4. $20 \div 2 = \blacksquare$
5. $10 \div 5 = \blacksquare$

PROBLEM Henry used a popover tin to bake 3 batches of popovers. The extra batter made 4 more popovers. He made 31 popovers in all. How many popovers does the popover tin hold?

 • What are you asked to find?

• What information will you use?

 • What strategy can you use to solve the problem?

You can *work backward* to find how many popovers the popover tin holds.

 • How can you use the strategy to solve the problem?

Look at the diagram. Begin with the total number of popovers. Subtract the number of extra popovers from the total.

31	−	4	=	27
↑		↑		↑
total popovers		extra popovers		popovers in 3 batches

Divide to find the number of popovers in one batch.

27	÷	3	=	9
↑		↑		↑
popovers in 3 batches		number of batches		number in each batch

So, Henry's popover tin holds 9 popovers.

 • Look back. Does your answer make sense?

226

CALIFORNIA STANDARDS MR 1.1 Analyze problems by identifying relationships, distinguishing relevant from irrelevant information, sequencing and prioritizing information, and observing patterns. **MR 2.3** Use a variety of methods, such as words, numbers, symbols, charts, graphs, tables, diagrams, and models, to explain mathematical reasoning. *also* **NS 2.0,** **NS 2.2, NS 2.3, NS 2.8, AF 1.0, AF 1.1, AF 1.2, MR 1.0, MR 2.0, MR 2.4, MR 2.6, MR 3.1**

Problem Solving Practice

Work backward to solve.

1. **What if** Henry used a different tin to bake 4 batches of popovers? Then he used the extra batter to make 3 more popovers. He made 27 popovers in all. How many popovers does this tin hold?

2. Mr. Jones spent $30 at the sports shop. He bought a mitt for $10 and 4 balls. How much did each ball cost?

Draw a Diagram or Picture
Make a Model or Act It Out
Make an Organized List
Find a Pattern
Make a Table or Graph
Predict and Test
▶ **Work Backward**
Solve a Simpler Problem
Write an Equation
Use Logical Reasoning

Mr. Lo bought 2 books that cost the same amount. He gave the cashier $20 and received $6 in change. How much money did each book cost?

3. Which number sentence shows how to find the total cost of the 2 books?

 A $20 + $6 = $26
 B $2 \times \$6 = \$12$
 C $20 - $6 = $14
 D $20 \div 2 = \$10$

4. How much money did each book cost?

 F $6
 G $7
 H $10
 J $14

Mixed Strategy Practice

USE DATA For 5–7, use the price list.

5. Zach pays for an apple pie and a bag of cookies with a $10 bill. He gets his change in quarters and dimes. There are 13 coins in all. How many of each coin does Zach get?

6. **? What's the Error?** Lara says a lemon tart and 2 boxes of muffins cost $11.65. Describe Lara's error and give the correct cost of the items.

Bake Sale Price List
- Bag of 3 Cookies $0.75
- Box of 6 Muffins $3.45
- Apple Pie $6.75
- Lemon Tart $8.20

7. Kenneth has 3 quarters, 2 dimes, 2 nickels, and 5 pennies. List all the ways he can pay for a bag of cookies.

Review/Test

✓ CHECK CONCEPTS

Write a division sentence for each. (pp. 220–223)

1.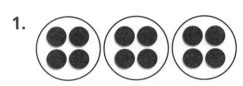

2. ■ ■ ■ ■
   ■ ■ ■ ■
   ■ ■ ■ ■
   ■ ■ ■ ■

3. $\begin{array}{r} 24 \\ -\ 8 \\ \hline 16 \end{array}$ ⟋ $\begin{array}{r} 16 \\ -\ 8 \\ \hline 8 \end{array}$ ⟋ $\begin{array}{r} 8 \\ -8 \\ \hline 0 \end{array}$

✓ CHECK SKILLS

Find the missing factor and quotient. (pp. 214–217)

4. $8 \times \blacksquare = 40$ $40 \div 8 = \blacksquare$

5. $6 \times \blacksquare = 42$ $42 \div 6 = \blacksquare$

6. $7 \times \blacksquare = 56$ $56 \div 7 = \blacksquare$

7. $8 \times \blacksquare = 32$ $32 \div 8 = \blacksquare$

Find the quotient. (pp. 214–223)

8. $14 \div 7 = \blacksquare$ 9. $\blacksquare = 30 \div 5$ 10. $40 \div 4 = \blacksquare$ 11. $\blacksquare = 63 \div 9$

12. $8\overline{)24}$ 13. $10\overline{)90}$ 14. $2\overline{)14}$ 15. $9\overline{)18}$ 16. $3\overline{)27}$

For 17–20, use the price list at the right to find the cost of each number of items. (pp. 224–225)

17. 4 balloons

18. 7 noise makers

19. 3 noise makers

20. 9 balloons

PARTY SUPPLIES PRICE LIST	
Balloons	$3
Noise Makers	$2

Find the cost of one of each item. (pp. 224–225)

21. 5 notebooks cost $10.   22. 6 markers cost $18.   23. 5 caps cost $35.

✓ CHECK PROBLEM SOLVING

Work backward to solve. (pp. 226–227)

24. Nikki used a tin to make 3 batches of popovers. Then she made 2 extra popovers. She made 26 popovers in all. How many popovers does the tin hold?

25. Roger earned $35. He made $15 from a paper route and walked 4 dogs after school. How much did Roger charge to walk each dog?

Cumulative Review

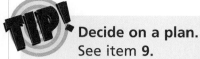

Decide on a plan.
See item **9.**

Find how many pages Jeremy did not read on the first night. Then you can find the number of pages he read on each of the next 6 nights.

Also see problem **4**, p. H64.

For 1–10, choose the best answer.

1. Mrs. Jenkins made 64 pretzels. She put 8 pretzels on each plate. How many plates did she use?

   A 4 **C** 8
   B 6 **D** 9

2. $42 \div 7 = \blacksquare$

   F 9 **H** 7
   G 8 **J** NOT HERE

3. $54 \div 9 = \blacksquare$

   A 5 **C** 8
   B 6 **D** NOT HERE

4. Mr. Murphy has 80 flowers to plant. He will plant 10 flowers in each flower box. How many flower boxes will he need?

   F 90 **H** 10
   G 70 **J** 8

5. The pep club members held a car wash. They charged $4 to wash each car. They washed 8 cars in the first 2 hours. How much did they make?

   A $32 **C** $12
   B $24 **D** $8

6. Which number makes this number sentence true?

   $$9 \times \blacksquare = 72$$

   F 6 **H** 8
   G 7 **J** 9

7. Carrie bought 7 slices of pizza for her family. Each slice cost $3. How much did Carrie spend on pizza?

   A $14 **C** $24
   B $21 **D** $28

8. A box of 8 beanbag animals costs $24. How much does 1 beanbag animal cost?

   F $6 **H** $3
   G $4 **J** $2

9. Jeremy read all 47 pages of his new book. The first night he read 17 pages. On each of the next 6 nights he read an equal number of pages. How many pages did he read on each of these 6 nights?

   A 5 **C** 30
   B 8 **D** 70

10. Leticia put 5 toys on each of 4 shelves. Which number sentence tells how many toys there were?

    F $5 + 4 = 9$ **H** $5 - 4 = 1$
    G $5 \times 4 = 20$ **J** $20 \div 5 = 4$

MATH DETECTIVE

Missing Parts

Put on your thinking cap and use what you know about the relationship between multiplication and division to find the missing factors and products.

Copy and complete each table. Use grid paper to help.

Table 1

×	■	■	4	■
5	15	■	■	■
7	■	56	■	■
■	24	■	■	48
■	■	■	36	■

Table 2

×	■	■	■	■
■	28	■	■	■
■	■	48	24	■
■	63	■	■	18
■	■	■	15	■

STRETCH YOUR THINKING

• Explain how you found the missing factors. Explain how you found the missing products.

• List as many multiplication and division sentences as you can that have a product or a quotient of 6.

• List as many multiplication and division sentences as you can that have a product or a quotient of 24.

230

Challenge

Divide by 11 and 12

Mr. Samson collected 132 eggs. He wants to put them into egg cartons that hold 12 eggs each. How many egg cartons does Mr. Samson need?

$$132 \div 12 = \blacksquare$$

↑   ↑   ↑

number of eggs   number in each carton   number of cartons

×	0	1	2	3	4	5	6	7	8	9	10	11	12
0	0	0	0	0	0	0	0	0	0	0	0	0	0
1	0	1	2	3	4	5	6	7	8	9	10	11	12
2	0	2	4	6	8	10	12	14	16	18	20	22	24
3	0	3	6	9	12	15	18	21	24	27	30	33	36
4	0	4	8	12	16	20	24	28	32	36	40	44	48
5	0	5	10	15	20	25	30	35	40	45	50	55	60
6	0	6	12	18	24	30	36	42	48	54	60	66	72
7	0	7	14	21	28	35	42	49	56	63	70	77	84
8	0	8	16	24	32	40	48	56	64	72	80	88	96
9	0	9	18	27	36	45	54	63	72	81	90	99	108
10	0	10	20	30	40	50	60	70	80	90	100	110	120
11	0	11	22	33	44	55	66	77	88	99	110	121	132
12	0	12	24	36	48	60	72	84	96	108	120	132	144

Use a multiplication table to find the quotient.

Think: $12 \times \blacksquare = 132$

Find the factor 12 in the top row. Look down the column to find the product, 132. Look left along the row to find the missing factor, 11.

$$12 \times 11 = 132$$

$$132 \div 12 = 11$$

So, Mr. Samson needs 11 egg cartons.

Talk About It

- Explain how to use repeated subtraction to find $48 \div 12$.

- What patterns do you notice in the elevens column of the multiplication table?

Try It

Use the multiplication table to solve.

1. $99 \div 11 = \blacksquare$ **2.** $\blacksquare = 108 \div 12$ **3.** $110 \div 11 = \blacksquare$ **4.** $\blacksquare = 84 \div 12$

5. $72 \div 12 = \blacksquare$ **6.** $66 \div 11 = \blacksquare$ **7.** $\blacksquare = 144 \div 12$ **8.** $0 \div 11 = \blacksquare$

9. $12\overline{)120}$ **10.** $11\overline{)121}$ **11.** $11\overline{)77}$ **12.** $12\overline{)36}$

Study Guide and Review

VOCABULARY

Choose the best term from the box.

1. A set of related multiplication and division sentences is a __?__ . (p. 190)

2. Any number divided by __?__ is that number. (p. 204)

> one
> fact family
> quotient
> zero

STUDY AND SOLVE

Chapter 11

Use repeated subtraction to divide.

$28 \div 7 = \blacksquare$

$$
\begin{array}{ccccc}
28 & 21 & 14 & 7 \\
-7 & -7 & -7 & -7 \\
\hline
21 & 14 & 7 & 0
\end{array}
$$

You subtracted 7 from 28 four times.
So, $28 \div 7 = 4$.

Write the division sentence shown by the repeated subtraction. (pp. 186–187)

3.
$$
\begin{array}{ccc}
15 & 10 & 5 \\
-5 & -5 & -5 \\
\hline
10 & 5 & 0
\end{array}
$$

4.
$$
\begin{array}{cccc}
32 & 24 & 16 & 8 \\
-8 & -8 & -8 & -8 \\
\hline
24 & 16 & 8 & 0
\end{array}
$$

Use arrays to divide.

$20 \div 4 = \blacksquare$

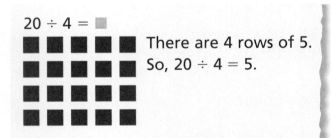

There are 4 rows of 5.
So, $20 \div 4 = 5$.

Use the array to find the quotient. (pp. 188–189)

5.

2 rows of $\blacksquare$ = 8
$8 \div 2 = \blacksquare$

6.

3 rows of $\blacksquare$ = 9
$9 \div 3 = \blacksquare$

Write fact families.

This is the fact family for 3, 4, and 12.
$3 \times 4 = 12 \quad 12 \div 4 = 3$
$4 \times 3 = 12 \quad 12 \div 3 = 4$

Write the fact family for each set of numbers. (pp. 190–193)

7. 3, 9, 27

8. 5, 7, 42

9. 5, 8, 40

10. 4, 9, 36

Chapter 12

Use related multiplication facts to find quotients.

$45 \div 5 = \blacksquare$
Think: $5 \times \blacksquare = 45$
$ 5 \times 9 = 45$
So, $45 \div 5 = 9$, or $5\overline{)45}^{\,9}$.

Find each quotient. (pp. 200–203)

11. $18 \div 3 = \blacksquare$ **12.** $30 \div 5 = \blacksquare$

13. $20 \div 4 = \blacksquare$ **14.** $16 \div 2 = \blacksquare$

15. $40 \div 5 = \blacksquare$ **16.** $24 \div 4 = \blacksquare$

17. $4\overline{)32}$ **18.** $3\overline{)9}$ **19.** $2\overline{)10}$

Chapter 13

Multiply to find the cost of multiple items.

Pens cost $4 each. Find the cost of 7 pens.
$7 \times \$4 = \$28$
So, 7 pens cost $28.

Beach balls cost $3 each. Find the cost of each number of items.
(pp. 224–225)

20. 6 beach balls

21. 2 beach balls

22. 5 beach balls

Divide to find the cost of one item.

8 erasers cost $16.
$\$16 \div 8 = \$2$
So, each eraser costs $2.

Find the cost of one of each item.
(pp. 224–225)

23. 9 tennis balls cost $18.

24. 7 baskets cost $42.

25. 10 notepads cost $10.

PROBLEM SOLVING PRACTICE

Solve. (pp. 194–195, 208–209, 226–227)

26. Marcie bought 5 packs of juice. There are 3 juice boxes in each pack. How many juice boxes did Marcie buy? Write a number sentence and solve.

27. Noah spent 15 minutes eating lunch and then played kickball for 25 minutes. Now it is 12:45 P.M. What time did Noah start eating lunch?

28. Janet had $7.35 and spent $2.50 on a snack. How much money does Janet have left?

California Connections

SACRAMENTO

Sacramento, the capital of California, has important highways, railroads, and rivers going through it.

◀ The State Capitol in Sacramento is noted for its class dome, which rises 210 feet above the street.

1. There are 56 tours of the capitol building given each week. If the same number of tours is given each day, how many tours are given each day? Explain how you know.

2. There are 21 locomotives and train cars at the California State Railroad Museum. Suppose you look at 7 each hour. How many hours will it take to see all 21?

3. Sightseeing trips on the *Spirit of Sacramento* riverboat cost $10 for adults and $5 for children. If you pay $35 for one adult and some children, how many children have you paid for? Explain how you know.

SUTTER'S FORT

Sutter's Fort was Sacramento's earliest settlement. Now visitors can see what pioneer life was like, learn about California history, and see the daily cannon firings at the fort.

The main building near the center of the fort is the only part of the original fort still standing. ▼

▲ Sutter's Fort is located in midtown Sacramento.

1. Suppose a blacksmith made 12 horseshoes in 3 hours. If he made the same number each hour, how many horseshoes did he make each hour? Write a division sentence you could use to solve the problem.

2. **What if** 36 barrels were made and an equal number of them were stored along each of 4 walls? How many barrels would be along each wall? Write an expression to describe the problem.

3. Suppose 28 workers and 26 guests ate a dinner of bread and beef at the fort. If 1 loaf of bread served about 6 people, how many loaves did the bakers prepare for dinner? Explain how you found the answer.

▲ The cooper, or barrelmaker, made wooden buckets, barrels, butter churns, and tubs.

▲ Bread was prepared in the bakery and baked in the brick oven in the courtyard.

Collect and Record Data

Dinosaurs of all shapes and sizes lived on Earth many years ago. This Tyrannosaurus Rex skeleton, named Sue, was found in South Dakota in 1990 by Sue Hendrickson. It is on display at the Field Museum of Natural History in Chicago. This skeleton is 41 feet long. Look at the chart. Make a table that shows the dinosaurs in order from smallest to largest.

Sue is the largest, most complete, Tyrannosaurus Rex fossil skeleton ever found.

DINOSAURS AND THEIR LENGTHS

Tyrannosauras Rex

Compsognathus

Diplodocus

Pentaceratops

Apatosaurus

0 10 20 30 40 50 60 70 80 90 100

CHECK WHAT YOU KNOW ✓

Use this page to help you review and remember
important skills needed for Chapter 14.

✓ COLUMN ADDITION (See p. H10.)

Find the sum.

1.	2.	3.	4.	5.
8	9	8	5	3
3	3	6	4	4
4	1	4	7	7
+7	+6	+2	+3	+1

6.	7.	8.	9.	10.
4	2	5	7	9
3	4	5	3	5
8	7	5	6	9
+5	+4	+6	+1	+3

11. $5 + 4 + 7 + 2 + 8$ **12.** $1 + 7 + 5 + 9 + 3$ **13.** $6 + 7 + 3 + 4 + 7$

✓ READ A CHART (See p. H17.)

For 14–17, use the information in this chart.

14. What is the title of the chart?

15. How many students like snowy weather best?

16. What kind of weather had the fewest votes?

17. How many students were asked?

OUR FAVORITE WEATHER	
Type	**Students**
☀ Sunny	7
💧 Rainy	4
❄ Snowy	8
☁ Cloudy	3

For 18–20, use the information in this chart.

18. How many students were asked?

19. How many students eat dinner at 6:00?

20. At what time do the greatest number of students eat dinner?

TIME FOR DINNER	
Time	**Students**
5:00	8
5:30	9
6:00	5
6:30	2

HANDS ON
Collect and Organize Data

▶ Explore

Information collected about people or things is called **data**.

Kelly's class voted for their favorite dinosaurs and made tables to show the results.

A **tally table** uses tally marks to record data.

A **frequency table** uses numbers to record data.

FAVORITE DINOSAUR

Name	Tally
Anatosaurus	ⅢⅡ|
Brachiosaurus	ⅢⅡ||
Tyrannosaurus	ⅢⅡ ⅢⅡ||
Stegosaurus	|||

FAVORITE DINOSAUR

Name	Number
Anatosaurus	6
Brachiosaurus	7
Tyrannosaurus	12
Stegosaurus	3

Collect data about your classmates' favorite dinosaurs. Organize the data in a tally table.

STEP 1

Make a tally table. List four answer choices.

STEP 2

Ask classmates *What's your favorite dinosaur?* Make a tally mark for each answer.

Favorite Dinosaur

Name	Tally

• Why is a tally table good for recording data?

Try It

Decide on a question to ask your classmates.

a. Write four answer choices in a tally table.

b. Ask your classmates the question. Complete the tally table.

CALIFORNIA STANDARDS MR 2.3 Use a variety of methods, such as words, numbers, symbols, charts, graphs, tables, diagrams, and models, to explain mathematical reasoning. **MR 2.0** Students use strategies, skills, and concepts in finding solutions. *also* **MR 3.0, MR 3.2**

▶ Connect

Use the data you collected in the tally table on page 238 to make a frequency table of your classmates' choices for their favorite dinosaur.

STEP 1	STEP 2
Write the title and headings. List the four answer choices.	Count the number of tallies in each row. Write each number in the frequency table.

Favorite Dinosaur

Name	Number

• Why is a frequency table a good way to show data?

▶ Practice

1. Make a tally table of three after-school activities. Ask your classmates which activity they like best. Make a tally mark beside the activity for each answer.

2. Use the data from your tally table to make a frequency table. Which after-school activity did the greatest number of classmates choose? the fewest?

3. Which table is better for reading results? Which table is better for collecting choices?

For 4–5, use the tally table.

4. Make a frequency table of the data in the tally table. How many students voted for their favorite fruit?

5. ✎ **Write a problem** using the information in the Favorite Fruit tally table.

FAVORITE FRUIT	
Name	**Tally**
Grapes	IIII IIII IIII I
Oranges	IIII
Apples	IIII
Bananas	IIII III

Mixed Review and Test Prep

6. $2 \times 10 = $ ▨ (p. 164)

7. $7 \times 10 = $ ▨ (p. 164)

8. Which number is greater: 4,545 or 4,454? (p. 20)

9. Round 3,495 to the nearest thousand. (p. 30)

10. **TEST PREP** Ebony buys a sandwich for $3.49. She pays with a $5 bill. How much change will she receive? (p. 86)

   A $1.31   C $1.51
   B $1.41   D $1.61

Understand Data

Quick Review

1. $17 + 21 + 12$

2. $13 + 9 + 6$

3. $24 + 5 + 7$

4. $8 + 14 + 7$

5. $19 + 13 + 21 + 5$

VOCABULARY
survey
results

▶ **Learn**

SURVEY SAYS . . . A **survey** is a question or set of questions that a group of people are asked. The answers from a survey are called the **results** of the survey.

Jillian and Ted took a survey to find their classmates' favorite snack. The tally table shows the choices and votes of their classmates.

What is the favorite snack of their classmates?

Since cookies got the greatest number of votes, 12, cookies are the favorite snack.

• How many students answered Jillian's and Ted's survey? How do you know?

▶ **Check**

1. **List** the snacks in order from the most votes to the fewest votes.

FAVORITE SNACK	
Snack	**Tally**
Popcorn	IIII
Cookies	₩₩ ₩₩ II
Granola bars	₩₩
Apples	₩₩ II
Pretzels	II

For 2–3, use the tally table.

DO YOU HAVE AN OLDER BROTHER OR SISTER?	
Answer	**Tally**
Yes	₩₩ ₩₩ II
No	₩₩ ₩₩ ₩₩ II

TECHNOLOGY LINK

More Practice:
Use E-Lab, *Collecting and Organizing Data.*

www.harcourtschool.com/elab2002

2. How many people were surveyed?

3. Write a statement that describes the survey results.

CALIFORNIA STANDARDS MR 2.3 Use a variety of methods, such as words, numbers, symbols, charts, graphs, tables, diagrams, and models, to explain mathematical reasoning. **NS 2.0** Students calculate and solve problems involving addition, subtraction, multiplication, and division. *also* **NS 1.2,** **NS 2.1, MR 2.0, MR 2.4, MR 3.0**

Practice and Problem Solving

For 4–7, use the tally table.

4. List the subjects in order from the most votes to the fewest votes.

5. How many students answered the survey?

6. How many more students chose math than chose social studies?

7. How many fewer students chose art than chose reading?

FAVORITE SCHOOL SUBJECT

Subject	Tally
Math |卌 卌 |||
Science	卌 ||||
Reading	卌 卌 卌
Social studies	卌	
Art	卌 ||

For 8–11, use the frequency table.

8. How many more students chose basketball than chose baseball?

9. Which sport did the greatest number of students choose?

10. How many students answered this survey?

11. **What if** 5 more students chose basketball? How would that change the results of the survey?

FAVORITE SPORTS

Sport	Number
Basketball	26
Football	15
Baseball	17
Hockey	21
Swimming	30

12. **Write About It** Take a survey to find your classmates' favorite pizza topping. Make a tally table and a frequency table of the data. Explain the results.

13. **Write a problem** using the information from any of the tables in this lesson.

14. **? What's the Error?** Lily wrote the following number sentence: $24 \times 0 = 24$. What's her error?

Mixed Review and Test Prep

15. (p. 42) $\begin{array}{r} 134 \\ +159 \\ \hline \end{array}$

16. (p. 42) $\begin{array}{r} 365 \\ +256 \\ \hline \end{array}$

17. (p. 42) $\begin{array}{r} 166 \\ +384 \\ \hline \end{array}$

18. Three friends shared 27 baseball cards equally. How many cards did each friend receive? (p. 202)

19. **TEST PREP** Toshio has 4 packages of 6 bagels. How many bagels does he have in all? (p. 134)

   A 10   C 28
   B 24   D 30

Extra Practice page H45, Set A

Classify Data

Quick Review
Tell if each number is *less than* or *greater than* 345.

1. 298   **2.** 404

3. 344   **4.** 354

5. 346

▶ **Learn**

MARBLES IN MOTION You can group, or classify, data in many different ways such as by size, color, or shape.

Ms. Vernon gave each pair of students in her class a bag of marbles. She asked students to think about ways that they could group the marbles.

Luis and Joey made a table to show what they did.

VOCABULARY
classify

Bag of Marbles

	Small	Medium	Large
Blue	2	2	5
Red	3	1	5
Multicolor	3	3	2

• How did Luis and Joey classify their data?

▶ **Check**

1. **Tell** 3 things that the groupings in the table helped you know about the marbles.

2. How many small blue marbles are there?

3. How many large red marbles are there?

4. How many multicolored marbles are there?

5. How many medium-size marbles are there?

242

CALIFORNIA STANDARDS MR 2.3 Use a variety of methods, such as words, numbers, symbols, charts, graphs, tables, diagrams, and models, to explain mathematical reasoning. **MR 1.1** Analyze problems by identifying relationships, distinguishing relevant from irrelevant information, sequencing and prioritizing information, and observing patterns. *also* **NS 2.0, ⊶NS 2.1, MR 1.0, MR 2.0**

Practice and Problem Solving

For 6–9, use the table.

6. How many girls are wearing red shirts?

7. How many students are wearing blue shirts?

8. How many of the students wearing green shirts are boys?

COLOR OF SHIRTS IN OUR CLASS				
	Green	Blue	Red	Yellow
Girls	1	6	2	5
Boys	3	4	4	0

9. How many students are in the class?

For 10–13, use the table.

10. How many pictures are in the art show in all?

11. What is the subject of the largest number of pictures in the art show: people, animals, or plants?

PICTURES IN THE ART SHOW				
	Chalk	Crayon	Paint	Pencil
People	3	5	8	1
Animals	4	5	4	3
Plants	6	2	9	5

12. What are the fewest pictures made with: chalk, crayon, paint, or pencil?

13. There are 12 possible categories of pictures in the art show. Which one was represented the least?

14. Look at the figures at the right. Make a table to classify, or group, the figures.

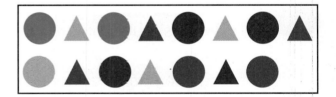

15. There are 11 girls and 8 boys in a music club. Of the girls, 6 are dancers and the rest are singers. Of the boys, 3 are dancers and the rest are singers. Make a table to classify, or group, the students.

Mixed Review and Test Prep

16. $700 - 238 =$ ■ (p. 58)

17. $306 - 67 =$ ■ (p. 58)

18. $5 \times 2 \times 5 =$ ■ (p. 170)

19. If 4 tapes cost $32, how much does 1 tape cost? (p. 224)

20. **TEST PREP** Caroline left her house at 7:15 A.M. She arrived at school 20 minutes later. At what time did she arrive at school? (p. 98)

   A 6:55 A.M.   C 7:30 A.M.

   B 7:20 A.M.   D 7:35 A.M.

LESSON

4

Problem Solving Strategy
Make a Table

Understand ➡ Plan ➡ Solve ➡ Check

$6 + 4$ **2.** $3 + 5$

3. $8 + 7$ **4.** $9 + 6$

5. $6 + 7$

PROBLEM Leo and Sally did an experiment with the two spinners shown at the right. They spun the pointers and recorded the sum of the two numbers. They spun 20 times. Their results were 2, 4, 6, 5, 4, 5, 4, 4, 2, 4, 4, 3, 4, 5, 6, 4, 6, 4, 3, and 4. Which sum occurred most often?

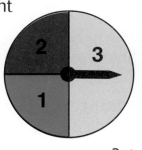
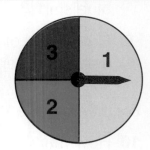

$3 + 1 = 4$

Understand

• What are you asked to find?

• What information will you use?

Plan

• What strategy can you use to solve the problem?

You can *make a table* to organize the data.

Solve

• What should you put in the table?

Leo and Sally recorded the sums in their experiment. So, label one column *Sums*. In this column, list all the sums that are possible. Label another column *Number of Spins*. Use a tally mark to record each spin.

SPINNER EXPERIMENT	
Sums	Number of Spins
2			
3			
4											
5				
6				

So, the sum 4 occurred most often.

Check

• What other strategy could you use to solve the problem?

CALIFORNIA STANDARDS MR 2.3 Use a variety of methods, such as words, numbers, symbols, charts, graphs, tables, diagrams, and models, to explain mathematical reasoning. **MR 1.0** Students make decisions about how to approach problems. *also* **MR 2.0, MR 3.0**

244

Problem Solving Practice

PROBLEM SOLVING STRATEGIES

Draw a Diagram or Picture
Make a Model or Act It Out
Make an Organized List
Find a Pattern
► **Make a Table or Graph**
Predict and Test
Work Backward
Solve a Simpler Problem
Write an Equation
Use Logical Reasoning

Solve.

1. **What if** Leo and Sally spin the pointers 25 times and record the sums in a table? How many tallies should there be?

2. Marta and Dan rolled two number cubes 15 times and recorded the sums. Their results were 9, 9, 5, 6, 9, 3, 7, 7, 3, 5, 12, 12, 10, 8, and 5. Make a table and find the sum rolled most often.

Your experiment is to roll two number cubes, numbered 1–6, twenty times to find out what difference you will roll most often.

3. What are all the differences you will list on your table?

   A 0, 2, 4, 6
   B 0, 1, 2, 3, 4, 5, 6
   C 0, 1, 2, 3, 4, 5
   D 1, 2, 3, 4, 5, 6

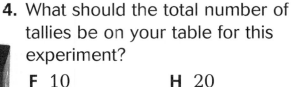

4. What should the total number of tallies be on your table for this experiment?

   F 10   **H** 20
   G 15   **J** 25

Mixed Strategy Practice

5. **REASONING** Louise is older than Marsha and Luke. Marsha is younger than Jim. Jim is older than Luke. Al is the youngest of the group. What is the order of the group from youngest to oldest?

6. This was the time when the game ended. If the game lasted 2 hours and 15 minutes, when did the game start?

7. In the library, there are 3 shelves with 9 new books on each shelf. How many new books are on display?

8. Lonnie has 48 dinosaur models. He puts them into 8 bags so that there are the same number in each bag. How many models do 2 bags contain?

9. **? What's the Question?** Sylvia spent $7 at the movies. She still had $8 when she got home. The answer is $15.

Review/Test

✔ CHECK VOCABULARY

Choose the best term from the box.

Box: data, survey, classify, tally table, frequency table

1. A question or set of questions that a group of people are asked is a __?__ . (p. 240)

2. Information collected about people or things is called __?__ . (p. 238)

3. A table that uses tally marks to record data is a __?__ . (p. 238)

4. A table that uses numbers to record data is a __?__ . (p. 238)

✔ CHECK SKILLS

For 5–6, use the tally table. (pp. 240–241)

5. List the instruments in order from the most votes to the fewest votes.

6. How many people answered the survey?

FAVORITE MUSICAL INSTRUMENT	
Musical Instrument	Tally
Guitar	ЖЖЖ
Flute	ЖIIII
Drums	ЖЖЖIII
Piano	ЖЖIII

For 7–8, use the frequency table. (pp. 242–243)

7. How many students have brown eyes?

8. How many girls are in the class?

EYE COLOR OF STUDENTS			
	Blue	Brown	Green
Girls	4	3	1
Boys	5	7	2

✔ CHECK PROBLEM SOLVING

Solve. (pp. 244–245)

9. There are 12 fourth graders and 13 fifth graders in the Science Club. Five of the fourth graders and 9 of the fifth graders are girls. Make a table to group the students in the Science Club.

10. Sancho rolled two number cubes 10 times and recorded the sums. His results were 7, 6 11, 2, 8, 9, 6, 8, 11, and 8. Make a table and find the sum rolled most often.

Cumulative Review

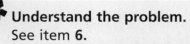

Understand the problem.
See item **6**.

How many people voted means the same as the Number of Votes in the frequency table. Decide which operation to use to answer the question.

Also see problem **1**, p. H62.

For 1–9, choose the best answer.

For 1–3, use the tally table.

SHIRT COLOR	
Color	**Tally**
Red	⅃Ⅱ
Yellow	Ⅲ
Green	⅃
Purple	⅃ⅢⅠ

1. How many shirts were counted?

A 4   **C** 21
B 14   **D** 24

2. Which color did the fewest children wear?

F red   **H** yellow
G green   **J** purple

3. How many more children wore purple shirts than yellow shirts?

A 4   **C** 6
B 5   **D** 7

4. Bill made a tally table to record the results of 25 spins on a spinner. How many tallies are on his table?

F 25   **H** 15
G 20   **J** 5

5. Mrs. Martin has 8 vases in her shop. She wants to put 9 tulips in each vase. She has 55 tulips. How many more tulips does she need?

A 17   **C** 46
B 27   **D** 72

For 6–8, use the frequency table.

FAVORITE ICE-CREAM TOPPING	
Topping	**Number of Votes**
Marshmallow	16
Chocolate	32
Nuts	19
Sprinkles	11

6. How many people voted for their favorite topping?

F 32   **H** 76
G 68   **J** 78

7. How many more people voted for chocolate than nuts?

A 27   **C** 13
B 21   **D** 11

8. How many votes were there in all for nuts and sprinkles?

F 11   **H** 36
G 30   **J** NOT HERE

9. $32 \div 4 = \blacksquare$

A 9   **C** 6
B 8   **D** NOT HERE

CHAPTER 15

Analyze and Graph Data

There are about 1,000 giant pandas in the world today. Giant pandas can be 5 feet tall and can weigh more than 300 pounds! Two of the zoos in the United States that have giant pandas are the San Diego Zoo and Zoo Atlanta. Zoos also have many other types of animals. Look at the pictograph and make a bar graph of some of the animals at Zoo Atlanta.

ZOO ATLANTA	
Giant panda	🐾 🐾
Red panda	🐾 🐾
African elephant	🐾 🐾 🐾
Gorilla	🐾 🐾
Black rhinoceros	🐾 🐾
Orangutan	🐾 🐾 🐾 🐾 🐾 🐾 🐾 🐾 🐾 🐾

Key: Each 🐾 = 1 animal.

Giant panda eating bamboo

CHECK WHAT YOU KNOW ✓

Use this page to help you review and remember
important skills needed for Chapter 15.

✓ TALLY DATA (See p. H17.)

For 1–4, use the tally table.

HOW DO YOU GET TO SCHOOL?	
Car	卌 |||
Walk	卌 卌 卌
Bus	卌 卌 ||
Bike	卌 卌	

1. How many children ride in a car to school?

2. How many children ride their bikes?

3. How many children answered the survey question?

4. How many more children walk than ride the bus?

✓ SKIP-COUNT (See p. H9.)

Find the missing numbers in the pattern.

5. 2, 4, 6, 8, 10, ■, ■, ■

6. 4, 8, 12, 16, 20, ■, ■, ■

7. 45, 40, 35, 30, ■, ■, ■

8. 3, 6, 9, 12, 15, ■, ■, ■

9. 6, 12, 18, 24, 30, ■, ■, ■

10. 80, 70, 60, 50, ■, ■, ■

✓ READ PICTOGRAPHS (See p. H18.)

Use the value of the symbol to find the missing number.

11. If $\Delta = 2$, then

$\Delta + \Delta + \Delta = $ ■.

12. If $\square = 3$, then

$\square + \square + \square + \square = $ ■.

13. If ✿ = 4, then

✿ + ✿ + ✿ + ✿ + ✿ + ✿ = ■.

14. If ♥ = 5, then

♥ + ♥ + ♥ + ♥ + ♥ + ♥ = ■.

15. If ✪ = 2, then

✪ + ✪ + ✪ + ✪ + ✪ = ■.

16. If ♦ = 4, then

♦ + ♦ + ♦ + ♦ + ♦ = ■.

17. If ☺ = 3, then

☺ + ☺ + ☺ + ☺ + ☺ + ☺ = ■.

18. If ☆ = 10, then

☆ + ☆ + ☆ + ☆ + ☆ = ■.

Problem Solving Strategy
Make a Graph

 Understand ➤ **Plan** ➤ **Solve** ➤ **Check**

Quick Review

1. $4 + 4 + 4 = $ ■

2. $3 + 3 + 3 + 3 = $ ■

3. $2 + 2 + 2 + 2 = $ ■

4. $5 + 5 + 5 = $ ■

5. $2 + 2 + 2 = $ ■

PROBLEM The soccer team sold boxes of greeting cards to raise money. Rafael sold 14 boxes, Joselyn sold 7, Phil sold 24, Ken sold 12, and Felicia sold 10. What is one way the sales could be shown in a graph?

 Understand • What are you asked to do?

 Plan • What strategy can you use to solve the problem?

You can *make a pictograph*.

Solve • How can you show the data in a pictograph?

    a. Choose a **title** that tells about the graph.

    b. Write a **label** for each row.

    c. Look at the numbers. Choose a **key** to tell how many each picture stands for.

    d. Decide how many **pictures** should be next to each person's name.

GREETING CARDS SOLD

Rafael	✉✉✉✉✉✉✉
Joselyn	✉✉✉✉
Phil	✉✉✉✉✉✉✉✉ ✉✉✉
Ken	✉✉✉✉✉✉
Felicia	✉✉✉✉✉

Key: Each ✉ = 2 boxes.

 Check • How can you know if the number of pictures in your graph is correct?

CALIFORNIA STANDARDS MR 2.3 Use a variety of methods, such as words, numbers, symbols, charts, graphs, tables, diagrams, and models, to explain mathematical reasoning. **MR 2.0** Students use strategies, skills, and concepts in finding solutions. *also* NS 2.0, MR 1.0, MR 1.1, MR 2.4, MR 2.6

▶ Problem Solving Practice

PROBLEM SOLVING STRATEGIES

Draw a Diagram or Picture
Make a Model or Act It Out
Make an Organized List
Find a Pattern
▶ **Make a Table or Graph**
Predict and Test
Work Backward
Solve a Simpler Problem
Write an Equation
Use Logical Reasoning

1. **What if** Derek sold 30 boxes of cards, Andy sold 25 boxes, and Kay sold 15 boxes? Make a pictograph to show the information.

2. Barry sold 11 boxes of cards. Using a key of 2, explain how you would show his sales in a pictograph.

Torrie and Jeremy made pictographs using the data in the table.

3. Torrie used a key of 3. How many symbols should she draw to show the votes for soccer?

   A 2   **C** 5
   B 3   **D** 6

4. What key did Jeremy use if he drew $7\frac{1}{2}$ symbols to show the votes for football?

   F key of 1   **H** key of 3
   G key of 2   **J** key of 4

FAVORITE SPORTS	
Sport	Number of Votes
Soccer	18
Softball	12
Basketball	21
Football	15

┌ *Mixed Strategy Practice* ┐

5. It was 12:05 P.M. when Liza and Nick began eating lunch. Nick finished in 15 minutes. Liza finished 8 minutes later than Nick. At what time did Liza finish lunch?

6. Regina bought a book for $1.80. She gave the cashier a $1 bill and 4 coins. She did not get any change. What coins could she have used?

7. Lloyd spent $3.59, $4.50, and $9.75 for games. He gave the clerk $20.00. How much change should he receive?

8. **REASONING** If December 1 is on a Wednesday, on which day of the week is December 15?

9. Sydney's team scored 89 points, 96 points, 98 points, and 107 points. How many points did the team score in all?

10. **? What's the Question?** Marty bought 2 packs of trading cards. He gave the clerk $10. His change was $4. The answer is $3.

2 Read Bar Graphs

► Learn

YEARS AND YEARS A **bar graph** uses bars to show data. A **scale** of numbers helps you read the number each bar shows.

These bar graphs show the same data.

VOCABULARY
bar graph   scale
horizontal bar graph
vertical bar graph

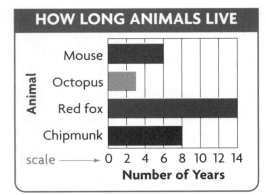

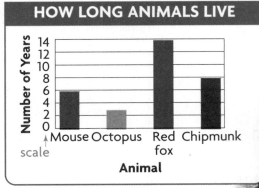

In a **horizontal bar graph**, the bars go across from left to right.

In a **vertical bar graph**, the bars go up.

- What scale is used on the bar graphs? Why is this a good scale?

- How do you read the bar for the octopus, which ends halfway between two lines?

Examples

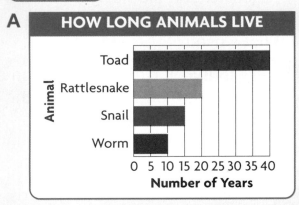

- How are these graphs alike? How are they different?

CALIFORNIA STANDARDS MR 1.0 Students make decisions about how to approach problems. **MR 1.1** Analyze problems by identifying relationships, distinguishing relevant from irrelevant information, sequencing and prioritizing information, and observing patterns. *also* **MR 2.0, MR 2.3, MR 3.2**

1. **Explain** how you would use the graph in Example B to tell how long a snail lives.

For 2–3, use the bar graphs in Examples A and B.

2. How long can a toad live?

3. How long can a worm live?

▶ **Practice and Problem Solving**

For 4–6, use the Sea Animals bar graph.

4. Is this a vertical or horizontal bar graph?

5. How long is a giant squid?

6. How much longer is a gray whale than a bottlenose dolphin?

For 7–9, use the Favorite Wild Animals bar graph.

7. Which animal received the most votes?

8. How many students in all voted for their favorite wild animal?

9. **? What's the Question?** The answer is 15.

10. **REASONING** Namiko saw a bar graph showing favorite foods. How could she tell which food got the most votes without looking at the scale?

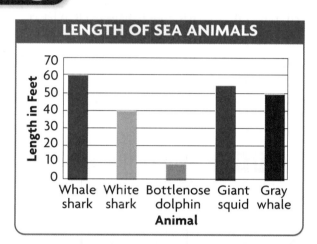

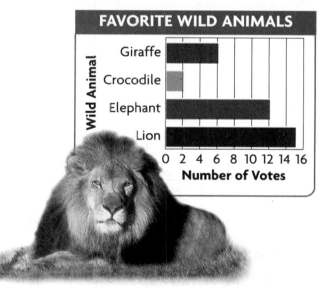

Mixed Review and Test Prep

11. $6 \div 2 = $ ■ (p. 200)

12. $14 \div 2 = $ ■ (p. 200)

13. $1,000 + 300 + 5 = $ ■ (p. 6)

14. Find the product of 7 and 8. (p. 150)

15. **TEST PREP** Meg put 24 counters in 4 equal piles. How many counters are in each pile? (p. 184)

A 4   C 7

B 6   D 20

4 Line Plots

Quick Review

1. $8 \times \blacksquare = 24$
2. $5 \times 4 = \blacksquare$
3. $\blacksquare \times 6 = 48$
4. $\blacksquare \times 5 = 40$
5. $7 \times \blacksquare = 49$

VOCABULARY

line plot
mode
range

▶ **Learn**

LOTS OF PLOTS All 24 third-grade students measured to find their heights in inches.

HEIGHTS OF THIRD GRADERS	
Height in Inches	Number of Students
49		
50				
51			
52						
53							
54					
55		

You can make a **line plot** to record each piece of data on a number line.

The **mode** is the number found most often in a set of data.

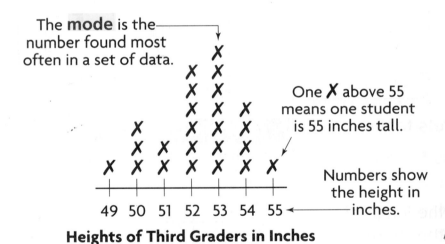

One **X** above 55 means one student is 55 inches tall.

Numbers show the height in inches.

Heights of Third Graders in Inches

The **range** is the difference between the greatest number and the least number.

greatest number → 55
least number → 49
range → 6

55 − 49 = 6

- What are the range and mode for this data?
- How many students are 50 inches tall?

CALIFORNIA STANDARDS NS 1.2 Compare and order whole numbers to 10,000. MR 2.1 Analyze problems by identifying relationships, distinguishing relevant from irrelevant information, sequencing and prioritizing information, and observing patterns. *also* NS 2.0, O—n SDAP 1.3, SDAP 1.4, MR 1.0, MR 2.0, MR 2.3

Make a Line Plot

Debbie took a survey in her third-grade class to find out how many peanut butter and jelly sandwiches the students ate last week.

She put the survey data in a tally table.

PEANUT BUTTER AND JELLY SANDWICHES	
Number of Sandwiches	Tallies
0		
1						
2				
3							
4					
5			

Activity

Make a line plot of the data in the table.

MATERIALS: number line

STEP 1 Write a title for the line plot. Label the numbers from 0 to 5.

```
 +---+---+---+---+---+
 0   1   2   3   4   5
```

Peanut Butter and Jelly Sandwiches Eaten Last Week

STEP 2 Draw **X**'s above the number line to show how many students ate each number of sandwiches.

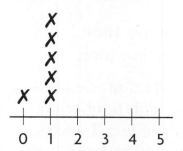

Peanut Butter and Jelly Sandwiches Eaten Last Week

• How are the tally table and the line plot alike? How are they different?

▶ Check

1. **Tell** what the mode is for the data above. Explain.

For 2–3, use your line plot.

2. How many students ate exactly 3 peanut butter and jelly sandwiches last week?

3. How many more students ate 1 or 2 sandwiches than ate 4 or 5 sandwiches?

LESSON CONTINUES

▶ Practice and Problem Solving

For 4–7, use the line plot at the right.

4. The **X**'s on the line plot stand for the band members. What do the numbers stand for?

5. How many band members practiced for just 6 hours?

6. What is the greatest number of hours any student practiced? What is the least number? What is the range for this set of data?

7. For how many hours did the greatest number of band members practice?

Hours Band Members Practiced

For 8–11, use the High Temperatures line plot.

Mrs. Brown's third-grade class recorded the high temperature each day for the last 3 weeks.

8. What high temperature occurred most often in the last 3 weeks? On how many days?

9. On how many days was the high temperature below 70 degrees?

10. What was the range of high temperatures the last 3 weeks?

11. Predict what you think the high temperature will be this week. Explain.

12. Take a survey to find out how many pets each student has. Show the results in a tally table and a line plot.

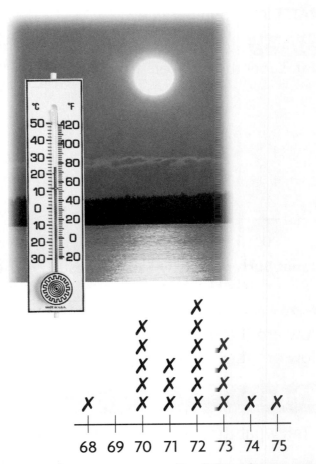

High Temperatures Each Day (in Degrees)

13. **Write About It** Explain how you decided which numbers to use in your line plot.

Mixed Review and Test Prep

Find each product. (p. 156)

14. $4 \times 7 = \blacksquare$ **15.** $3 \times 8 = \blacksquare$

16. $6 \times 6 = \blacksquare$ **17.** $7 \times 8 = \blacksquare$

Find each quotient. (p. 220)

18. $35 \div 7 = \blacksquare$ **19.** $27 \div 9 = \blacksquare$

20. $48 \div 8 = \blacksquare$ **21.** $49 \div 7 = \blacksquare$

22. **TEST PREP** What multiplication fact could you use to solve $8 \div 2$? (p. 188)

A $2 \times 2 = 4$ **C** $8 \times 2 = 16$
B $2 \times 4 = 8$ **D** $4 \times 8 = 32$

23. **TEST PREP** Sonya bought juice for $1.09 and a sandwich for $3.45. She paid the cashier with a $10 bill. How much change should she receive? (p. 88)

F $5.46 **H** $6.46
G $5.56 **J** $6.56

--LINKUP---
to Reading

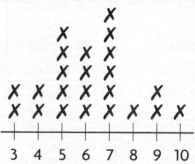

STRATEGY • USE GRAPHIC AIDS Graphic aids can be compared and analyzed. When you use a graphic aid, you "read" a picture rather than just words.

Brandi took a survey to find how many books her classmates read during the summer. She made a line plot and a bar graph of the data.

1. How many classmates did Brandi survey?

2. What are the mode and range of the data?

3. **Write About It** Tell whether you think the bar graph or the line plot is easier to read. Explain the reason for your choice.

Books Read During the Summer

Locate Points on a Grid

▶ Learn

GET TO THE POINT Jack is using a map to help him find different animals at the zoo. He wants to see the elephants first. How can this map help Jack find the elephants?

VOCABULARY

grid
ordered pair

The horizontal and vertical lines on the map make a **grid**.

Start at 0. Move 2 spaces to the right.
Then, move 3 spaces up.
The elephants are located at (2,3).

(2,3)

The first number tells how many spaces to move to the right.

The second number tells how many spaces to move up.

MATH IDEA An **ordered pair** of numbers within parentheses, like (2,3), names a point on a grid.

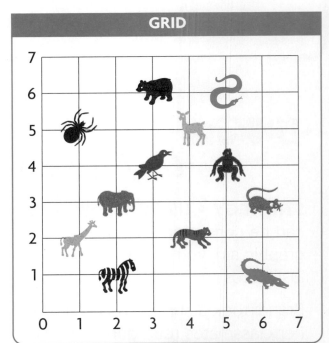

GRID

Example

Which animal is found at (5,4) on the grid?

Start at 0.
Move 5 spaces to the right.
Then move 4 spaces up.

So the monkey is found at (5,4).

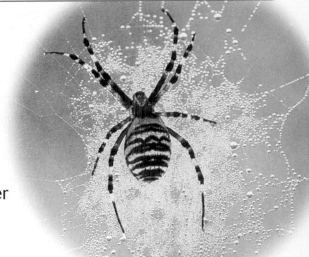

• Tell how to find the ordered pair for the spider on the grid.

▲ Wasp spider

 CALIFORNIA STANDARDS MR 1.1 Analyze problems by identifying relationships, distinguishing relevant from irrelevant information, sequencing and prioritizing information, and observing patterns. **MR 2.3** Use a variety of methods, such as words, numbers, symbols, charts, graphs, tables, diagrams, and models, to explain mathematical reasoning. *also* **MR 1.0, MR 2.0, MR 3.0**

1. **Explain** how you would find (11,12) on a 12-by-12 grid.

For 2–5, use the zoo grid on page 260.

2. Does (4,3) show the same point on the grid as (3,4)? Explain.

Write the ordered pair for each animal.

3. tiger   4. bear   5. giraffe

▶ **Practice and Problem Solving**

▲ Emerald tree boa

For 6–11, use the zoo grid on page 260. Write the ordered pair for each animal.

6. bird   7. snake   8. zebra

9. mouse   10. deer   11. alligator

For 12–17, use the grid at the right. Write the letter of the point named by the ordered pair.

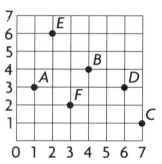

12. (6,3)   13. (1,3)
14. (4,4)   15. (2,6)
16. (7,1)   17. (3,2)

18. **REASONING** Rachel saw the zebra at (2,1) on the zoo grid. Mary saw the elephant at (2,3). What ordered pair names the point between the zebra and the elephant?

19. **? What's the Error?** Jerry said, "To find the point (2,3), start at 0, move 3 spaces to the right and 2 spaces up." What error did Jerry make?

20. Walt has 45 pennies. Arthur has 11 nickels. Which boy has more money? How much more does he have?

Mixed Review and Test Prep

21. $5 \times 9 = $ ■
(p. 118)

22. $7 \times 6 = $ ■
(p. 150)

23. $6 \times 8 = $ ■
(p. 148)

24. $4 \times 7 = $ ■
(p. 134)

25. **TEST PREP** Find the difference of 567 and 288. (p. 58)
   A 279   C 379
   B 289   D 855

Extra Practice page H46, Set C

Read Line Graphs

Quick Review

1. $5 \times \blacksquare = 0$

2. $\blacksquare \times 9 = 63$ 3. $30 = \blacksquare \times 6$

4. $9 \times 4 = \blacksquare$ 5. $7 \times \blacksquare = 7$

▶ Learn

LINE UP A line graph is a graph that uses a line to show how something changes over time. What was the normal, or average, temperature in Des Moines, Iowa, in March?

VOCABULARY
line graph

A line graph is like a grid.

a. From 0 find the vertical line for March. Move up to the point.

b. Follow the horizontal line left to the scale.

c. The point for March is at 40 degrees.

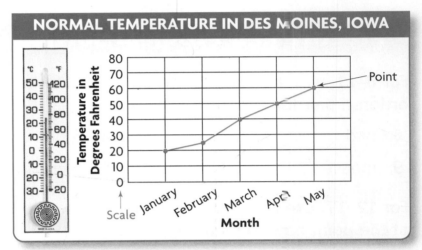

NORMAL TEMPERATURE IN DES MOINES, IOWA

So, the normal temperature in Des Moines in March is 40 degrees.

TECHNOLOGY LINK

To learn more about data and graphs, watch the **Harcourt Math Newsroom Video**, *Fish Census*.

▶ Check

1. **Tell** what you notice about the temperature in Des Moines from January to May.

For 2–3, use the line graph above.

2. What is the normal temperature in January?

3. In what month is the normal temperature 60 degrees?

CALIFORNIA STANDARDS NS 1.2 Compare and order whole numbers to 10,000. **MR 1.1** Analyze problems by identifying relationships, distinguishing relevant from irrelevant information, sequencing and prioritizing information, and observing patterns. *also* NS 2.0, NS 2.8, MR 1.0, MR 2.0, MR 2.3

► Practice and Problem Solving

For 4–6, use the Alaska line graph.

4. What is the normal temperature in December?

5. In what month is the normal temperature 50 degrees?

6. How many degrees higher is the normal temperature in August than in November?

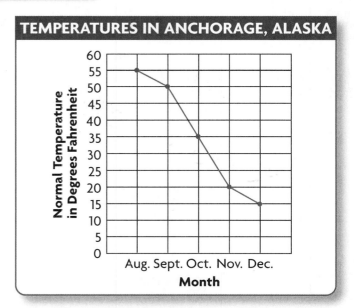

TEMPERATURES IN ANCHORAGE, ALASKA

For 7–9, use the Pumpkin line graph.

7. On what day were the most pumpkins sold? the fewest?

8. On which days were more than 8 pumpkins sold?

9. On which two days were the same number of pumpkins sold? How many pumpkins?

10. ✏️ **Write a problem** using one of the line graphs above.

PUMPKINS SOLD

11. Sachio decorated 17 cookies and Linda decorated 12 cookies. They gave 20 cookies to their neighbors. How many cookies are left?

12. The total cost of a watch and a calculator is $50. The watch costs $20 more than the calculator. Find the cost of the calculator.

Mixed Review and Test Prep

Find the quotient.

13. $3\overline{)12}$ (p. 202) 14. $4\overline{)24}$ (p. 202)

15. $5\overline{)15}$ (p. 200) 16. $2\overline{)18}$ (p. 200)

17. **TEST PREP** In a multiplication sentence, the product is 56. One factor is 8. What is the other factor? (p. 142)

   A 6   **B** 7   **C** 8   **D** 9

Review/Test

✓ CHECK VOCABULARY AND CONCEPTS

Choose the best term from the box.

> line plot
> line graph
> mode
> range

1. The number that occurs most often in a set of data is the __?__ . (p. 256)

2. A number line that is used to record each piece of data is a __?__ . (p. 256)

3. A graph that shows change over time is a __?__ . (p. 262)

4. How would you find (4,5) on a grid? (p. 260)

For 5–7, choose the letter that names each. (pp. 250–253; 256–259; 262–263)

A. bar graph   **B.** pictograph   **C.** line plot   **D.** line graph

5.
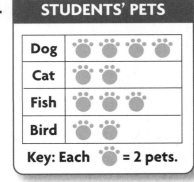

STUDENTS' PETS	
Dog	🐾 🐾 🐾 🐾
Cat	🐾 🐾
Fish	🐾 🐾 🐾
Bird	🐾 🐾

Key: Each 🐾 = 2 pets.

6.
```
X
X
X  X
X  X  X     X
X  X  X  X  X
X  X  X  X  X
+--+--+--+--+
0  1  2  3  4
```
Number of Pets

7.
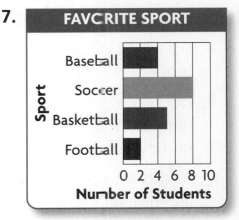

FAVORITE SPORT

Baseball, Soccer, Basketball, Football

Sport

0 2 4 6 8 10
Number of Students

✓ CHECK SKILLS

For 8–9, use the bar graph above. (pp. 252–255)

8. What scale is used on this graph?

9. How many students chose soccer?

For 10–11, use the line plot above. (pp. 256–259)

10. What is the mode for this data?

11. What is the range?

✓ CHECK PROBLEM SOLVING

For 12–15, use the pictograph above. (pp. 250–251)

12. What key is used on this graph?

13. How many pets does a 🐾 equal?

14. How many more dogs than cats do the students have?

15. How many pets do the students have in all?

Cumulative Review

Get the information you need.
See item **3.**

You know that each symbol means 3 votes. You need to find the number of votes for raisins in the data for Favorite Dried Fruit.

Also see problem **3**, p. H63.

For 1–7, choose the best answer.

For 1–2, use the bar graph.

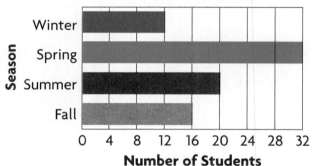

FAVORITE SEASON

1. How many students chose summer?

 A 32   **C** 12
 B 20   **D** 5

2. How many more students chose spring than chose fall?

 F 30   **H** 16
 G 26   **J** 4

Kelly made a pictograph using this data.

Favorite Dried Fruit

Raisins	24
Bananas	12
Cherries	18

3. If each symbol represents 3 votes, how many symbols will show the votes for raisins?

 A 3   **B** 6   **C** 8   **D** 12

4. 4,892 − 304

 F 4,498   **H** 4,592
 G 4,588   **J** NOT HERE

5. How many students did exactly 6 push-ups in one minute?

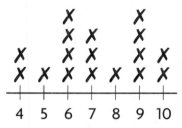

Number of Push-Ups in One Minute

 A 4   **C** 2
 B 3   **D** 1

For 6–7, use the grid.

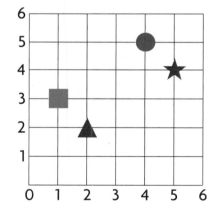

6. What is the ordered pair for the star?

 F (1,3)   **H** (2,2)
 G (4,5)   **J** (5,4)

7. What shape is found at (2,2) on the grid?

 A square   **C** circle
 B star   **D** triangle

Probability

Temperatures differ from place to place. In McMurdo, Antarctica, the temperature is very cold all year. In Nairobi, Kenya, it is always warm. In Aspen, Colorado, the temperature changes from season to season. Use the line graph to predict what the temperature will be in each place the next day.

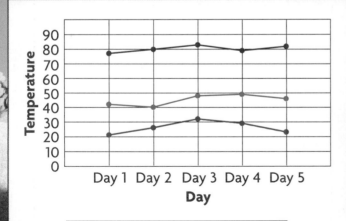

AVERAGE TEMPERATURES (°F) FOR FIVE DAYS IN JANUARY

Temperature: 0, 10, 20, 30, 40, 50, 60, 70, 80, 90

Day 1 Day 2 Day 3 Day 4 Day 5

Day

Key
Nairobi, Kenya ——
Aspen, Colorado ——
McMurdo, Antarctica ——

Joshua trees, like this one, grow in the desert.

CHECK WHAT YOU KNOW ✓

Use this page to help you review and remember
important skills needed for Chapter 16.

✓ IDENTIFY PARTS OF A WHOLE (See p. H18.)

For 1–3, write the fraction that names the red part
of the spinner.

1.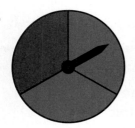

2.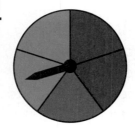

3.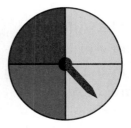

✓ COMPARE PARTS OF A WHOLE (See p. H19.)

For 4–7, write the color shown by the largest part of
each spinner.

4.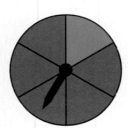

5.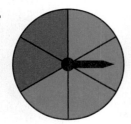

6.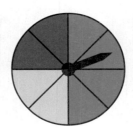

7.

✓ USE A TALLY TABLE (See p. H19.)

For 8–12, use the tally table.

8. What color was landed on the most times?

9. What color was landed on the fewest times?

10. How many more times was purple landed
on than blue?

11. How many fewer times was red landed on
than green?

12. How many times was the spinner used in all?

SPINNER RESULTS	
Color	**Tallies**
yellow	||
green	||||
red	||
blue		
purple	||||| |||

Certain and Impossible

Quick Review

Complete. Use <, >, or = for each ⬤.

1. $5 + 23$ ⬤ 28
2. 54 ⬤ $95 - 40$
3. $436 - 25$ ⬤ 409
4. $87 + 6$ ⬤ 95
5. $32 ÷ 4$ ⬤ $64 ÷ 8$

▶ Learn

WAY, NO WAY An **event** is something that happens.

An event is **certain** if it will always happen. An event is **impossible** if it will never happen.

CERTAIN	IMPOSSIBLE
A. You get wet if you jump into a swimming pool full of water.	A. A rock will turn into a piece of cheese.
B. Ice cubes feel cold.	B. A mouse will talk to you today.

VOCABULARY

event
certain
impossible

- Name some events that are certain. Name some that are impossible.

Alonso drew this spinner for a game he made. Is it *impossible* for him to spin orange on his spinner?

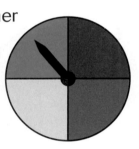

- How do you know?

Leslie put marbles in a bag for an experiment. Is it *certain* that she will pull a blue or a yellow marble?

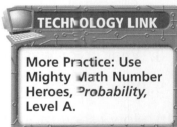

- How do you know?

▶ Check

1. **Decide** if it is certain or impossible that you will spin yellow or red on this spinner. Is it certain or impossible that you will spin green?

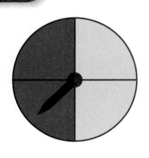

2. Is a spin of orange certain or impossible? Why?

TECHNOLOGY LINK

More Practice: Use Mighty Math Number Heroes, Probability, Level A.

CALIFORNIA STANDARDS SDAP 1.1 Identify whether common events are certain, likely, unlikely, or improbable. **MR 2.0** Students use strategies, skills, and concepts in finding solutions. *also* **NS 2.0, MR 2.2, MR 2.4, MR 3.0, MR 3.2, MR 3.3**

Tell whether each event is *certain* or *impossible*.

3. You could pull a red marble from this bag.

4. You could spin green or yellow on this spinner.

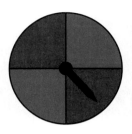

▶ Practice and Problem Solving

Tell whether each event is *certain* or *impossible*.

5. If you add any two 1-digit numbers the sum will be 19.

6. A box contains 100 pennies. You choose a coin from the box. It is a penny.

For 7–10, look at the bag of marbles at the right. Tell whether each pull is *certain* or *impossible*.

7. a marble

8. a yellow marble

9. two blue marbles

10. a red, blue, or green marble

11. Kaitlin rolled two cubes each numbered 1 through 6. If she adds the numbers, is it certain or impossible that she will get a sum less than 2?

12. Jeffrey has 21 marbles. If he gives 9 marbles away he will have twice as many marbles as Miguel. How many marbles does Miguel have?

13. REASONING Elena rolls a cube numbered 1 through 6. She rolls the cube 4 times and adds the numbers. Their sum is 18. Two of the numbers she rolls are sixes. What numbers could Elena have rolled?

─ *Mixed Review and Test Prep* ─

14. 246 (p. 42)
+323

15. 768 (p. 58)
−515

16. 4,819 (p. 46)
+5,073

17. 6,342 (p. 62)
−5,107

18. TEST PREP Verna's math class begins at 11:30 A.M. and lasts for 1 hour 15 minutes. At what time is the class over? (p. 98)

A 11:45 A.M.   **C** 12:30 P.M.

B 12:00 P.M.   **D** 12:45 P.M.

Extra Practice page H47, Set A

Likely and Unlikely

Quick Review

What time is 20 minutes earlier?

1. 11:25   2. 10:40

3. 3:05   4. 12:17

5. 9:00

VOCABULARY

likely   unlikely   outcome

▶ Learn

MAYBE, MAYBE NOT An event is **likely** to happen if it has a good chance of happening.

An event is **unlikely** if it does not have a good chance of happening.

An **outcome** is a possible result of an experiment.

Naomi and Ellis are playing a game. They use a spinner 25 times each.

• Look at the spinner. Which outcome is most likely? Which outcome is most unlikely? Explain.

MATH IDEA An event is sometimes likely or unlikely depending on its chance of happening.

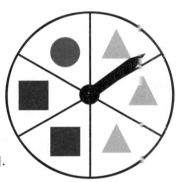

▶ Check

1. **Name** the color marble you are most likely to pull from this bag of marbles. Explain your choice.

2. **Name** the color marble you are most unlikely to pull. Explain.

For 3–4, look at the spinner.

3. Name the color you are most likely to spin.

4. Name the color you are most unlikely to spin.

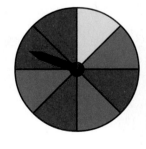

CALIFORNIA STANDARDS SDAP 1.1 Identify whether common events are certain, likely, unlikely, or improbable.
MR 2.0 Students use strategies, skills, and concepts in finding solutions. *also* **MR 2.4**

Name the color:

5. that is most likely to be pulled.

6. that you are most likely to spin.

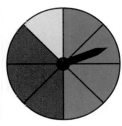

7. that is most unlikely to be pulled.

8. that you are most unlikely to spin.

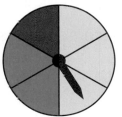

For 9–12, look at the bag of marbles at the right. Write *impossible, unlikely, likely,* **or** *certain* **to match each event.**

9. Pull a marble.

10. Pull a yellow marble.

11. Pull a black marble.

12. Pull a blue marble.

13. REASONING What is the difference between *likely* and *certain*?

14. REASONING What is the difference between *unlikely* and *impossible*?

15. Masami pulled a red marble from a bag 15 times and a yellow marble 5 times. He said it was likely there were more red marbles than yellow marbles. Do you agree? Explain.

16. ? What's the Question? Della bought five items at the store. She was given $5.27 change from a $10 bill. The answer is $4.73.

Mixed Review and Test Prep

17. $8)\overline{48}$ (p. 214)

18. $6)\overline{42}$ (p. 214)

19. $\begin{array}{r} 7 \\ \times 4 \end{array}$ (p. 150)

20. $\begin{array}{r} 5 \\ \times 6 \end{array}$ (p. 118)

21. TEST PREP Jiro's mother baked 18 cupcakes. She gave an equal number to each of 6 people. How many cupcakes did each person receive? (p. 214)

A 1   **B** 2   **C** 3   **D** 4

HANDS ON
Possible Outcomes

▶ Explore

A **possible outcome** is something that has a chance of happening.

Two outcomes are **equally likely** if they have the same chance of happening.

Record the outcomes of tossing a coin.

STEP 1 Look at a coin. Decide on the possible outcomes.

The possible outcomes are heads and tails.

STEP 2 Toss the coin 20 times. Record the results in a tally table.

COIN TOSS	
Outcome	Tallies
Heads	
Tails	

• What did you notice about your results?

Since they have an equal chance of happening, the outcomes *heads* and *tails* are equally likely.

When you do an experiment, you can **predict**, or tell what you think will happen.

We are tossing a coin 50 times. How many times do you think the coin will show heads?

Try It

a. **What if** you tossed the coin 50 times? Predict how many times you would expect to toss heads.

b. Now toss 2 coins. What are the possible outcomes? Toss the coins 25 times. Record the results.

CALIFORNIA STANDARDS O—n **SDAP 1.2** Record the possible outcomes for a simple event and systematically keep track of the outcomes when the event is repeated many times. **MR 2.0** Students use strategies, skills, and concepts in finding solutions. *also* **SDAP 1.0, SDAP 1.1, MR 2.4**

▶ Connect

Look at this spinner. There are 5 possible outcomes. The pointer can land on red, blue, yellow, green or white.

Each space on the spinner is the same size, so the chance is *equally likely* that you will spin any color.

The chance is *1 out of 5* that you will spin any color.

- What are the possible outcomes for this spinner? Which outcomes are equally likely? Explain.

- What is the chance that you will spin blue? that you will spin green? that you will spin red?

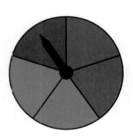

▶ Practice

For 1–4, list the possible outcomes of each event.

1. rolling a cube numbered 1–6

2. pulling blocks from this bag

3. using this spinner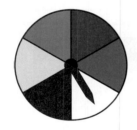

4. pulling blocks from this bag

5. Maggie used the spinner at the right. The pointer landed 6 times on red, 3 times on blue, and 1 time on yellow. Predict the color it will land on next. What is the chance that she will spin yellow?

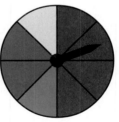

Mixed Review and Test Prep

6. 9×3 (p. 164)

7. 3×6 (p. 122)

8. $8 \times \blacksquare = 64$ (p. 142)

9. $12 \div 2$ (p. 200)

10. **TEST PREP** Gary has $2.78 left from the $5 bill his mother gave him. How much did he spend? (p. 88)

A $3.32   C $2.22

B $2.32   D $2.12

► Learn

TRY IT One way to find out how likely it is for outcomes to occur is to conduct experiments.

Activity 1

Materials: 4-part spinner pattern

STEP 1 Make a spinner that has four equal parts. Color the parts red, green, blue, and yellow.

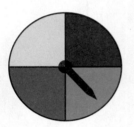

STEP 2 Make a tally table. List all the possible outcomes. Spin 20 times. Record the outcomes in the table.

OUTCOMES	
Color	Tallies
red	
blue	
green	
yellow	

STEP 3 Make a bar graph of your data to show the results of your experiment.

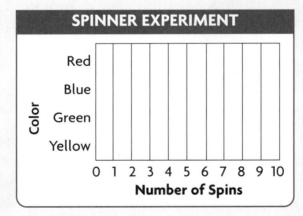

SPINNER EXPERIMENT

- Compare the graphs from all the experiments in the class. What do the graphs show about the outcomes?

- How would the outcome of your experiment change if 2 parts of the spinner were red?

CALIFORNIA STANDARDS SDAP 1.0 Students conduct simple probability experiments by determining the number of possible outcomes and make simple predictions. **O━┓SDAP 1.2** Record the possible outcomes for a simple event and systematically keep track of the outcomes when the event is repeated many times. *also* **NS 2.8, SDAP 1.1, O━┓SDAP 1.3, SDAP 1.4, MR 2.0, MR 2.4, MR 3.0, MR 3.2, MR 3.3**

Activity 2

Materials: color tiles; paper bag

STEP 1

Put 1 red tile, 3 green tiles, and 6 blue tiles in a paper bag.

STEP 2

Make a tally table. List all the possible outcomes.

STEP 3

Pull one tile, record the color, and put it back in the bag. Make 40 pulls.

STEP 4

Make a bar graph of the data to show the results of your experiment.

- Predict the color tile that is most likely to be pulled. Predict which tile is least likely to be pulled. Explain.

- What is the chance that you will pull either a red or a blue tile?

▶ Check

For 1–2, use the spinners.

1. **Choose** which spinner has equally likely outcomes. Explain.

2. On which spinner would spinning green be an unlikely outcome? Explain.

3. Draw a spinner on which all of the outcomes are equally likely.

4. Name all the possible outcomes for this bag of color tiles. Which are unlikely? Which are equally likely? Which is most likely?

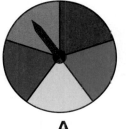

A

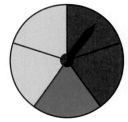

B

LESSON CONTINUES ▶

5. Name all the possible outcomes for this spinner. Which are unlikely? Which are equally likely? Which is most likely?

6. Keshawn pulled color tiles from this bag. What is the chance that he will pull a red tile?

7. USE DATA Pamela drew the bar graph at the right to show part of the results of an experiment. Predict which color marble she will pull most often.

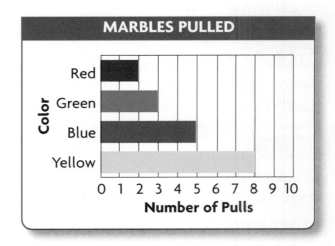

MARBLES PULLED

Color: Red, Green, Blue, Yellow

Number of Pulls: 0 1 2 3 4 5 6 7 8 9 10

8. Toss a coin 10 times. Record each toss in order. Then repeat the experiment. Is there a pattern in the results? What outcome do you notice?

9. Draw a spinner. Color it so that the chance of spinning red is 2 out of 4, of spinning green is 1 out of 4, and of spinning blue is 1 out of 4.

10. Spencer rolled a cube numbered 1, 2, 3, 4, 4, and 4. He rolled the cube 25 times. Which number do you think he rolled most often? Explain.

11. Andi started making a spinner at 11:45 A.M. She finished making it 25 minutes later. At what time did Andi finish making her spinner?

12. **Write About It** Explain what a spinner should look like if the three outcomes are all equally likely.

13. REASONING Gustavo had 15 cookies. He gave 3 cookies to each of his friends. He kept 3 cookies for himself. With how many friends did he share his cookies?

14. Heather went to the mall and bought 3 tapes. Each tape cost $9. Then she bought lunch for $6. She has $4 left. How much money did Heather start with?

Mixed Review and Test Prep

Add. (p. 42)

15.	127	16.	246	17.	485	18.	314	19.	380
	+418		+309		+438		+255		+163

Find the missing addend. (p. 68)

20. $15 + \blacksquare = 54$

21. $\blacksquare + 212 = 245$

22. $138 + \blacksquare = 250$

23. **TEST PREP** Hank bought a new comb for $1.59. He paid with a $5 bill. How much change should he get? (p. 88)

    A $2.41   **C** $3.41

    B $2.71   **D** $4.51

24. **TEST PREP** Katya bought a card for $1.85 and ribbon for $0.98. She paid with a $10 bill. How much change should she get? (p. 88)

    F $5.71   **H** $7.07

    G $6.77   **J** $7.17

--LINKUP--- to Science

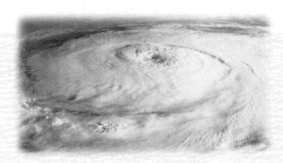

Scientists know that hurricanes most often form in September, October, and November. Hurricanes always have very high-speed winds that travel in circles. They form over warm, tropical ocean waters and then usually move west or northwest. Scientists use this information to predict hurricanes.

1. Which storm is most likely to become a hurricane? Explain.

2. Which storm is most unlikely to become a hurricane? Explain.

3. Which storms are equally likely to become hurricanes?

CHARACTERISTICS OF HURRICANES					
Storm	Circular winds	Warm ocean waters	High-speed winds	September to November	Moving west/ northwest
Anna	✓				
Ben	✓	✓		✓	
Chad	✓	✓	✓		
David	✓	✓	✓	✓	
Eli	✓			✓	✓

Extra Practice page H47, Set C

Predict Outcomes

Quick Review

1. 25¢ = ■ pennies
2. 9 nickels = ■¢
3. ■ dimes = $1.00
4. 4 quarters = ■ pennies
5. 50¢ = ■ nickels

▶ Learn

WEATHER OR NOT You can use the results of an experiment to predict what will happen in the future.

Meteorologists are scientists who predict weather. They study weather patterns from around the world to help them.

Activity

STEP 1	STEP 2	STEP 3
Take the outdoor temperature every day at the same time for two weeks.	Record each temperature on a line plot. Use a scale: 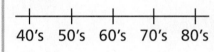 40's 50's 60's 70's 80's	Use the data to predict the temperature for the next day. Take the temperature the next day to check your prediction.

• How did you decide on your prediction? Explain.

• What was the temperature the next day? How did it compare with your prediction?

▶ Check

1. Look at the line plot. Tell what you would predict for the next day's temperature. Explain.

2. **What if** the temperatures for seven days in a row were: 46, 48, 52, 55, 46, 48, and 47? What temperature would you predict for the next day? Explain.

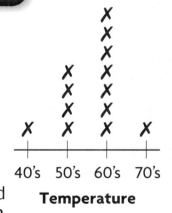

```
                    X
                    X
                    X
            X       X
            X       X
            X       X
    X       X   X X X   X
  --+-------+-------+-------+--
   40's   50's    60's    70's
         Temperature
```

CALIFORNIA STANDARDS O—n **SDAP 1.3** Summarize and display the results of probability experiments in a clear and organized way. **SDAP 1.4** Use the results of probability experiments to predict future events. also **NS 2.8,** **SDAP 1.0, SDAP 1.1,** O—n **SDAP 1.2, MR 1.2, MR 2.0, MR 2.4**

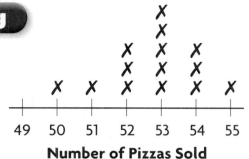

3. The line plot at the right shows the number of pizzas sold at Annie's Pizza Shop for the last two weeks. How many pizzas do you predict will be sold tomorrow?

Number of Pizzas Sold

4. This tally table shows the pulls from a bag of tiles. Predict which color is least likely to be pulled next.

OUTCOMES	
Color	Tallies
red	卌
blue	卌 卌 卌 ||
green	卌 ||
yellow	卌 ||||

5. The line plot below shows the results of rolling a number cube. Predict which number you would most likely roll.

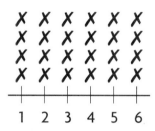

Number Cube Results

6. **REASONING** Martha rolls a number cube 30 times for an experiment. The number cube is labeled with these numbers: 1, 2, 3, 4, 5, 6. Predict about how many times she will roll an even number.

7. The library has 392 fiction books and 514 nonfiction books. Estimate how many of these books the library has in all.

8. Jared has a collection of 42 baseball cards. He trades 2 of his cards for 6 other cards. How many cards does he have now?

9. **REASONING** You toss a coin in the air one time. Predict how it will land. Explain.

10. **? What's the Error?** Todd says he will most likely pull a red cube from the bag. What's his error? Explain.

Mixed Review and Test Prep

Tell the value of the blue digit. (p. 6)

11. 7,593

12. 2,904

13. 8 (p. 152)
×7

14. 5)‾40 (p. 200)

15. **TEST PREP** Each of 4 students bought 8 packs of cards. How many packs of cards did they buy in all? (p. 134)

A 18 **B** 22 **C** 24 **D** 32

Problem Solving Skill

Draw Conclusions

Understand ➡ Plan ➡ Solve ➡ Check

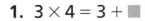

Quick Review

1. $3 \times 4 = 3 + \blacksquare$

2. $20 - 2 = 3 \times \blacksquare$

3. $3 \times 4 = \blacksquare \times 2$

4. $6 \times \blacksquare = 3 \times 8$

5. $6 \times 6 = \blacksquare \times 4$

VOCABULARY

fair

IT'S NOT FAIR All of the students in Omar's class made math games. Omar made a spinner game in which four cars move around a track. Each time a color is landed on, the car of that color moves forward a space. Which spinner would make Omar's game fair?

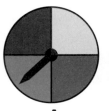 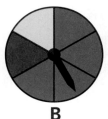

A   B

STEP 1 Find out what the problem asks.

Which spinner would make Omar's game fair?

STEP 2 Find out what a fair game is.

A game is **fair** if every player has an equal chance to win.

STEP 3 Look at the spinners to see if every player would have an equal chance to win.

Spinner A has 4 equal parts: 1 red, 1 green, 1 yellow, and 1 blue.

Spinner B has 3 parts, but the parts are not equal. The blue part is larger than the red or yellow parts.

STEP 4 Draw a conclusion. Decide which spinner would make Omar's game fair.

Every player has an equal chance to win with Spinner A.

So, Spinner A would make Omar's game fair.

Talk About It

- In Marcie's game, two players take turns pulling a marble from a bag to move their pieces. Which bag of marbles will make Marcie's game fair?

- **REASONING** In a fair game, will each player always win the same number of times? Explain.

A   B

CALIFORNIA STANDARDS SDAP 1.0 Students conduct simple probability experiments by determining the number of possible outcomes and make simple predictions. **MR 3.0** Students move beyond a particular problem by generalizing to other situations. *also* **SDAP 1.1, MR 1.0, MR 1.2, MR 2.0, MR 2.4, MR 3.2, MR 3.3**

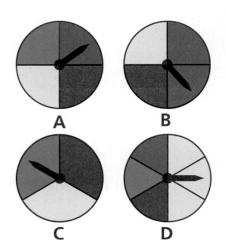

A   B

C   D

For 1–3, use the spinners.

1. Which of the Spinners A–D are fair. Explain.

2. Which of the Spinners A–D are unfair? Explain.

3. For each unfair spinner, what is the most likely outcome?

For 4–5, use these bags of marbles.

A   B   C   D

4. Which statement is true?

   A All bags have an equal number of marbles.

   B Every player has an equal chance to win with Bag D.

   C Every player has an equal chance to win with Bag C.

   D Bag A is fair.

5. Which of the bags of marbles are fair?

   F Bags A and B

   G Bags B, C, D

   H Bags B and C

   J Bags A and D

Mixed Applications

6. Pat is thinking of two numbers whose sum is 49 and whose difference is 1. What numbers is she thinking of?

7. Aaron has 8 nickels in one hand and 45¢ in nickels in the other hand. How many nickels does he have in all?

8. **REASONING** Anita and Cathy have $20 together. If Cathy gives Anita $3, they will each have the same amount of money. How much money does Anita have?

9. **REASONING** I am a number between 60 and 70. If you keep subtracting tens from me, you reach 2. What number am I?

10. **Write About It** Choose an unfair spinner or bag of marbles from above. Tell how to make it fair.

Review/Test

✔ CHECK VOCABULARY AND CONCEPTS

Choose the best term from the box.

1. An event that will always happen is __?__. (p. 268)

2. An outcome that probably won't happen is __?__. (p. 270)

> certain
> impossible
> likely
> unlikely

Tell whether each event is *certain* or *impossible*.

3. Next year, December 13 will come before December 12. (pp. 268–269)

4. Snow will melt when the temperature is above freezing. (pp. 268–269)

✔ CHECK SKILLS

5. What are the possible outcomes of using this spinner? (pp. 272–273)

6. What outcomes are equally likely when you use this spinner? (pp. 272–273)

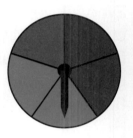

✔ CHECK PROBLEM SOLVING

Choose the spinner that is fair. Choose the bag of marbles that is fair. Write *A* or *B*. (pp. 280–281)

7.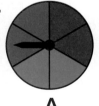

   A   B

8.

   A   B

For 9–10, use the spinner at the right. (pp. 280–281)

9. Is this spinner fair? Explain.

10. Are you likely or unlikely to spin yellow? Explain.

Cumulative Review

Eliminate choices.
See item **5**.
Relate each statement to the spinner until you find the one that is true.
Also see problem **5**, p. H64.

For 1–6, choose the best answer.

1. A spinner has four parts of equal size with the colors yellow, orange, purple, and green. Which event is **impossible**?

   A The spinner lands on green.
   B The spinner lands on orange.
   C The spinner lands on red.
   D The spinner lands on purple.

2. Jack will pull a marble out of the bag. Which color of marble is he most likely to get?

   F yellow
   G red
   H blue
   J green

3. $8 \times 7 = $ ▇

   A 48   **C** 64
   B 56   **D** NOT HERE

4. What is the chance that you will spin blue on this spinner?

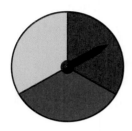

   F 1 out of 3   **H** 3 out of 4
   G 2 out of 3   **J** 3 out of 3

5. Which statement is true about the spinner?

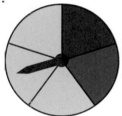

   A A player has an equal chance of spinning red, blue, or yellow.
   B A player is most likely to spin red.
   C A player is most likely to spin yellow.
   D A player is most likely to spin blue.

6. Which best describes a game that every player has an equal chance to win?

   F certain   **H** likely
   G fair   **J** impossible

MATH DETECTIVE

The Spinning Wheel

Put on your thinking cap. Use the clues to help you draw a spinner for each situation.

1. The chance of spinning red or blue is equally likely.

2. The chance of spinning blue is certain.

3. The chance of spinning red is impossible.

4. The chance of spinning red is slightly greater than the chance of spinning blue.

5. The chance of spinning red or yellow is equally likely, but the chance of spinning blue is greater than either red or yellow.

6. There is only a slight chance of spinning blue.

7. The chance of spinning either red, yellow, blue, or green is equally likely.

8. The chance of spinning blue or yellow is not very likely and the chance of spinning red is highly likely.

9. The chance of spinning red, blue, and yellow is equally likely.

STRETCH YOUR THINKING

If a spinner is divided into 12 equal sections, how would you label the spinner to have a slightly better chance of spinning red than blue?

Challenge

Find Mean and Median

Cari counted the number of robins she saw on her way to school. She recorded the data in a tally table.

Robins I Saw	
Monday	|||
Tuesday	卌 |||
Wednesday		
Thursday	卌	
Friday	||

Suppose Cari had seen the same total number of robins, but she had seen an equal number each day. How many robins would she have seen each day?

A **mean** is a number which can be used to represent all the numbers in a set of data.

MATERIALS: unit cubes
Make stacks of cubes to model the number of robins counted each day.

To find the mean, rearrange the stacks so that each stack has the same number of cubes. Count the number in each stack.

Mon. Tues. Wed. Thurs. Fri.   Mon. Tues. Wed. Thurs. Fri.

So, Cari would have seen 4 robins each day.
The mean is 4.

The **median** is the middle number in an ordered series of numbers.

Make stacks of cubes to model the number of robins counted each day. Place the stacks in order from least to greatest.

To find the median, count the number in the middle stack.

median →

So, the median of the set of data is 3 robins.

Try It

- Use cubes. Find the mean and median of 2, 8, and 5.

Study Guide and Review

VOCABULARY

Choose the best term from the box.

1. Something that has a chance of happening is a __?__ . (p. 272)

2. A graph that uses a line to show how something changes over time is a __?__ . (p. 262)

> data
> bar graph
> possible outcome
> line graph

STUDY AND SOLVE

Chapter 14

Interpret data from a survey.

FAVORITE JUICE

	Apple	Orange	Grape
Boys	6	8	3
Girls	5	4	6

How many boys voted in all?
Think: The boys are listed in the first row. Find the sum of these numbers.
6 + 8 + 3 = 17
So, 17 boys voted in all.

For 3–5, use the table. (pp. 240–243)

3. How many more boys than girls voted in the survey?

4. What is the most popular juice among the girls?

5. What is the most popular juice overall? Explain.

Chapter 15

Read a bar graph.

How many students play soccer?
Step 1: Find the bar for soccer.
Step 2: Compare the height of the bar with the scale to find the number of students who play soccer.
So, 8 of the students play soccer.

For 6–7, use the bar graph. (pp. 252–253)

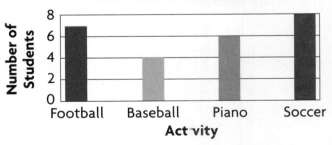
After-School Activities

6. How many more students play soccer than play baseball?

7. How many students play football?

Find points on a grid.

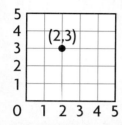

The **ordered pair** (2,3) tells you to move 2 spaces to the right of 0 and 3 spaces up.

Use the grid. (pp. 260–261)

8. Explain how to find (3,4) on the grid.

9. Explain how to find (4,3) on the grid.

Chapter 16

Predict outcomes of experiments.

Margo pulled a paper clip from a bag, recorded the color, and put it back. After 10 times she had pulled 1 orange, 3 red, and 6 blue paper clips.

What color paper clip do you predict she will pull next?

Think: Since she pulled a blue paper clip more than half the time, it is most likely that she will pull a blue paper clip next.

Use the tally table. (pp. 278–279)

This tally table shows the pulls from a bag of marbles.

OUTCOMES	
Color	Tallies
Yellow	IIII
Green	IЩ III
Purple	IЩ I

10. Name the color marble which is least likely to be pulled.

11. Name the color marble which is most likely to be pulled.

PROBLEM SOLVING PRACTICE

Solve. (pp. 250–251, 280–281)

12. Mr. Lind surveyed his customers to find out what to add to the menu. There were 20 votes for chili, 15 votes for ice cream, and 10 votes for pizza. Make a bar graph that shows the results of the survey.

13. Jill is making a game with a bag of marbles. She has 20 blue marbles and 12 orange marbles. How many blue marbles and orange marbles should she put in a bag to make the game fair?

California Connections

SAN FRANCISCO

San Francisco is one of the favorite tourist cities in the United States. Three of the many sights tourists enjoy are the cable cars, Fisherman's Wharf, and the Golden Gate Bridge.

Suppose you buy 3 postcards of cable cars, 2 postcards of Fisherman's Wharf, and 1 postcard of the Golden Gate Bridge. You put them all in a bag and, without looking, take out one card.

Fisherman's Wharf ▲

1. Is it *certain* or *impossible* that you will take out a postcard of Chinatown?

2. What are the possible postcard outcomes?

3. Is it more likely that you will take out a postcard of the Golden Gate Bridge or a cable car? Explain how you know.

cable car ▼

4. Is it *likely* or *unlikely* that you will take out a postcard of the Golden Gate Bridge?

5. What is the chance that you will take out a postcard of the Golden Gate Bridge? Explain how you know.

6. **REASONING** How could you change the postcards in the bag so it is equally likely that you will take out each kind of postcard?

◀ **Golden Gate Bridge**

SAN FRANCISCO DAY TRIPS

People visit the San Francisco area not only to see the sights of the city but also to explore some of the nearby places you can drive to in a day.

Here are four places you can visit on day trips from San Francisco.

▲ Stinson Beach

▲ Muir Woods National Monument

▲ Point Reyes National Seashore

▲ Año Nuevo State Reserve

1. Take a survey to find which of the four places your classmates would most like to visit. Make a tally table and a frequency table of the data.

2. What kind of graph would you use to display your data? Explain why you made that choice.

3. Make a graph to display your data.

4. **Write a problem** that your classmates could answer by looking at your graph. Exchange with a classmate and solve.

Multiply by 1-Digit Numbers

4-H is a club for young people. The four H's stand for Head, Heart, Hands, and Health. 4-H members living in rural areas often raise livestock, such as pigs, cows, and sheep. The animals are then shown at fairs. At the fairs, the pigs are kept in barns. Look at the diagram of a pig barn. If there is 1 pig in each stall and there are 4 barns like this one, how many pigs can be kept at the fair?

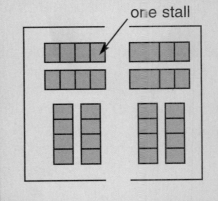

PIG BARN

one stall

CHECK WHAT YOU KNOW

Use this page to help you review and remember
important skills needed for Chapter 17.

✔ COLUMN ADDITION (See p. H10.)

Find the sum.

1. $\begin{array}{r} 47 \\ 47 \\ +47 \\ \hline \end{array}$	**2.** $\begin{array}{r} 36 \\ 36 \\ +36 \\ \hline \end{array}$	**3.** $\begin{array}{r} 54 \\ 54 \\ +54 \\ \hline \end{array}$	**4.** $\begin{array}{r} 18 \\ 18 \\ +18 \\ \hline \end{array}$
5. $\begin{array}{r} 29 \\ 29 \\ +29 \\ \hline \end{array}$	**6.** $\begin{array}{r} 65 \\ 65 \\ +65 \\ \hline \end{array}$	**7.** $\begin{array}{r} 72 \\ 72 \\ +72 \\ \hline \end{array}$	**8.** $\begin{array}{r} 83 \\ 83 \\ +83 \\ \hline \end{array}$

✔ REGROUP ONES AND TENS (See p. H20.)

Regroup. Write the missing number.

9. 6 tens 13 ones = ■ tens 3 ones

10. 2 tens 18 ones = 3 tens ■ ones

11. 4 tens ■ ones = 5 tens 8 ones

12. ■ tens 15 ones = 6 tens 5 ones

13. 6 tens 20 ones = 8 tens ■ ones

14. 7 tens 24 ones = ■ tens 4 ones

15. ■ tens 25 ones = 6 tens 5 ones

16. 5 tens ■ ones = 7 tens 3 ones

✔ MULTIPLICATION FACTS (See p. H20.)

Find the product.

17. $6 \times 7 = $ ■   **18.** $5 \times 5 = $ ■   **19.** ■ $= 9 \times 4$ **20.** $9 \times 1 = $ ■

21. $7 \times 10 = $ ■   **22.** ■ $= 8 \times 6$ **23.** $5 \times 9 = $ ■   **24.** ■ $= 2 \times 3$

25. ■ $= 4 \times 5$ **26.** $4 \times 8 = $ ■   **27.** ■ $= 0 \times 6$ **28.** $8 \times 7 = $ ■

29. $\begin{array}{r} 5 \\ \times 7 \\ \hline \end{array}$	**30.** $\begin{array}{r} 8 \\ \times 4 \\ \hline \end{array}$	**31.** $\begin{array}{r} 4 \\ \times 9 \\ \hline \end{array}$	**32.** $\begin{array}{r} 10 \\ \times 3 \\ \hline \end{array}$	**33.** $\begin{array}{r} 8 \\ \times 8 \\ \hline \end{array}$

HANDS ON
Multiply 2-Digit Numbers

Quick Review

1. $6 \times 5 = $ ■

2. ■ $= 4 \times 3$

3. $10 \times 6 = $ ■

4. ■ $= 6 \times 8$

5. $4 \times 10 = $ ■

MATERIALS
base-ten blocks

▶ Explore

Students in the school chorus stand in 4 rows of 16. How many students are in the chorus?

$$4 \times 16 = ■$$

Model the problem using an array of base-ten blocks. Find the product.

Remember
An array shows objects in rows and columns. An array with 3 rows of 5 shows 3×5.

STEP 1

Use 1 ten 6 ones to show 16. Make 4 rows of 16 to show 4×16.

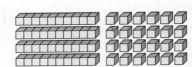

STEP 2

Combine the ones and the tens to find the product.

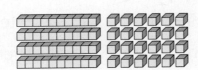

$4 \times 10 = 40 \quad 4 \times 6 = 24$

$40 + 24 = 64$

I have an array of 4 rows of 1 ten 3 ones. How do I find the product?

So, there are 64 students in the school chorus.

- How did you combine the ones and tens to find the product?

- **REASONING** Why is it helpful to show 16 by using 1 ten 6 ones instead of 16 ones?

Try It

Use base-ten blocks to find each product.

a. $4 \times 13 = ■$ b. $3 \times 18 = ■$

CALIFORNIA STANDARDS O—n NS 2.4 Solve simple problems involving multiplication of multidigit numbers by one-digit numbers. **MR 2.3** Use a variety of methods, such as words, numbers, symbols, charts, graphs, tables, diagrams, and models, to explain mathematical reasoning. *also* **NS 2.0,** O—n **AF 1.1, AF 1.2, MR 2.4, MR 3.2**

▶ Connect

You can use arrays on grid paper to multiply 2-digit numbers.

Find 3×17.

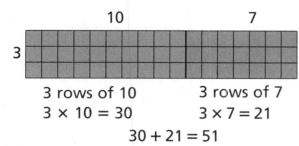

10   7

3

3 rows of 10   3 rows of 7
$3 \times 10 = 30$ $3 \times 7 = 21$
$30 + 21 = 51$

So, $3 \times 17 = 51$.

▶ Practice

Use the array to help find the product.

1.

$2 \times 10 = 20$ $2 \times 9 = 18$

$2 \times 19 = \blacksquare$

2.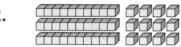

$3 \times 10 = 30$ $3 \times 4 = 12$

$3 \times 14 = \blacksquare$

3.

3 rows of 10   3 rows of 5

$3 \times 10 = 30$ $3 \times 5 = 15$

$3 \times 15 = \blacksquare$

4.

2 rows of 10   2 rows of 6

$2 \times 10 = 20$ $2 \times 6 = 12$

$2 \times 16 = \blacksquare$

Use base-ten blocks or grid paper to find the product.

5. $6 \times 14 = \blacksquare$ **6.** $5 \times 13 = \blacksquare$ **7.** $6 \times 16 = \blacksquare$

8. **REASONING** Draw an array on grid paper to find the missing factor in $\blacksquare \times 18 = 72$.

9. **Write About It** Describe how to model 3×21. Explain how to find the product.

Mixed Review and Test Prep

10. $30 \div 10$ (p. 218)   **11.** $90 \div 10$ (p. 218)

12. $\$1.14 + \$3.76 = \blacksquare$ (p. 88)

13. Round 24,515 to the nearest thousand. (p. 30)

14. **TEST PREP** Find the product.
$4 \times 2 \times 7 = \blacksquare$ (p. 170)

A 42   **C** 63
B 56   **D** 64

2 Record Multiplication

Quick Review

1. $2 \times 4 = \blacksquare$
2. $5 \times 7 = \blacksquare$
3. $\blacksquare = 2 \times 10$
4. $10 \times 7 = \blacksquare$
5. $\blacksquare = 8 \times 3$

▶ **Learn**

IN A HEARTBEAT The human heart exhibit at the Science Museum has 4 sections. Each section seats 30 students. How many students can sit in all 4 sections?

$$4 \times 30 = \blacksquare \quad \text{or} \quad \begin{array}{r} 30 \\ \times\ 4 \\ \hline \end{array}$$

Base-ten blocks can help you find the product.

STEP 1

Model 4 groups of 30.

STEP 2

Combine the tens to find the product. Regroup 12 tens as 1 hundred 2 tens.

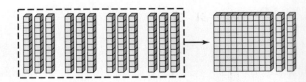

So, 120 students can sit in all 4 sections.

- What addition problem is the same as finding 4×30?

Examples

A $3 \times 20 = 60$

B $2 \times 40 = 80$

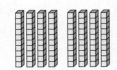

C $\begin{array}{r} 50 \\ \times\ 2 \\ \hline 100 \end{array}$

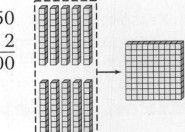

- What addition problem is the same as finding 2×50?

CALIFORNIA STANDARDS NS 2.4 Solve simple problems involving multiplication of multidigit numbers by one-digit numbers. **MR 2.3** Use a variety of methods, such as words, numbers, symbols, charts, graphs, tables, diagrams, and models, to explain mathematical reasoning. *also* NS 2.0, AF 1.1, MR 1.0, MR 1.1, MR 2.4

More Multiplication

For the field trip, there are 5 buses with 23 people on each bus. How many people are going on the field trip in all?

$$5 \times 23 = \blacksquare \quad \text{or} \quad \begin{array}{r} 23 \\ \times\ 5 \\ \hline \end{array}$$

STEP 1

Model 5 groups of 23. Multiply the ones.

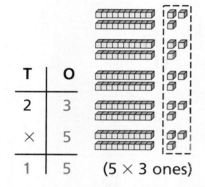

T	O
2	3
×		5
1	5

(5 × 3 ones)

STEP 2

Multiply the tens.

H	T	O
	2	3
×		5
1	0	0

(5 × 2 tens)

STEP 3

Add to find the product.

H	T	O
	2	3
×		5
	1	5
+ 1	0	0
1	1	5

So, 115 people are going on the field trip.

- How does knowing 5 × 2 help you multiply 5 × 20?

Examples

A
$$\begin{array}{r} 32 \\ \times\ 4 \\ \hline 8 \\ +120 \\ \hline 128 \end{array}$$
8 (4 × 2 ones)
+120 (4 × 3 tens)

B
$$\begin{array}{r} 51 \\ \times\ 3 \\ \hline 3 \\ +150 \\ \hline 153 \end{array}$$
3 (3 × 1 one)
+150 (3 × 5 tens)

C
$$\begin{array}{r} 73 \\ \times\ 5 \\ \hline 15 \\ +350 \\ \hline 365 \end{array}$$
15 (5 × 3 ones)
+350 (5 × 7 tens)

TECHNOLOGY LINK

To learn more about multiplication by 1-digit numbers, watch the **Harcourt Math Newsroom Video,** *Milk in a Bag.*

▶ **Check**

1. Model 5 × 28 with base-ten blocks.
Use paper and pencil to record what you did.

Find the product.

2.
$$\begin{array}{r} 30 \\ \times\ 2 \\ \hline \end{array}$$

3.
$$\begin{array}{r} 24 \\ \times\ 3 \\ \hline \end{array}$$

4.
$$\begin{array}{r} 38 \\ \times\ 2 \\ \hline \end{array}$$

▶ Practice and Problem Solving

Find the product.

5. $\begin{array}{r} 20 \\ \times\ 4 \\ \hline \end{array}$

6. $\begin{array}{r} 22 \\ \times\ 4 \\ \hline \end{array}$

7. $\begin{array}{r} 67 \\ \times\ 2 \\ \hline \end{array}$

Find the product. You may wish to use base-ten blocks.

8. $\begin{array}{r} 10 \\ \times\ 3 \\ \hline \end{array}$

9. $\begin{array}{r} 31 \\ \times\ 5 \\ \hline \end{array}$

10. $\begin{array}{r} 40 \\ \times\ 6 \\ \hline \end{array}$

11. $\begin{array}{r} 19 \\ \times\ 8 \\ \hline \end{array}$

12. $\begin{array}{r} 18 \\ \times\ 2 \\ \hline \end{array}$

13. $\begin{array}{r} 41 \\ \times\ 3 \\ \hline \end{array}$

14. $\begin{array}{r} 17 \\ \times\ 3 \\ \hline \end{array}$

15. $\begin{array}{r} 50 \\ \times\ 5 \\ \hline \end{array}$

16. $\begin{array}{r} 14 \\ \times\ 9 \\ \hline \end{array}$

17. $\begin{array}{r} 21 \\ \times\ 4 \\ \hline \end{array}$

18. $\begin{array}{r} 28 \\ \times\ 4 \\ \hline \end{array}$

19. $\begin{array}{r} 15 \\ \times\ 5 \\ \hline \end{array}$

20. $5 \times 14 =$ ▨

21. $6 \times 18 =$ ▨

22. $3 \times 27 =$ ▨

23. $5 \times 36 =$ ▨

Algebra For 24–27, use base-ten blocks to find the missing factor.

24. $6 \times$ ▨ $= 132$

25. $4 \times$ ▨ $= 56$

26. ▨ $\times 21 = 63$

27. ▨ $\times 25 = 100$

28. ✎ **Write a problem** There are 24 hours in a day. There are 7 days in a week. Write a problem using this information. Trade problems with a classmate and solve.

29. **? What's the Error?** Lori says that the product 3×16 is the same as the sum $16 + 16$. Describe Lori's error and then find the product.

30. REASONING Brandon says that 5×18 is the same as $40 + 50$. Do you agree or disagree? Explain.

31. The sum of two numbers is 20. The product of the two numbers is 75. What are the numbers?

32. Julio bought 8 dozen bagels. How many bagels did he buy? Write a number sentence and solve. (HINT: 1 dozen = 12)

33. Tanya bought 2 pairs of jeans for $18 each, a shirt for $12, and a sweatshirt for $8. How much did Tanya spend?

Mixed Review and Test Prep

Find each missing factor. (p. 142)

34. $2 \times \blacksquare = 8$ **35.** $\blacksquare \times 3 = 18$

36. $\blacksquare \times 4 = 16$ **37.** $8 \times \blacksquare = 56$

38. $5 \times \blacksquare = 35$ **39.** $\blacksquare \times 7 = 63$

40. $\blacksquare \times 6 = 42$ **41.** $\blacksquare \times 9 = 81$

42. **TEST PREP** The movie at the exhibit begins at 11:35 A.M. and ends at 12:15 P.M. How long is the movie? (p. 98)

A 25 min   **C** 40 min

B 35 min   **D** 45 min

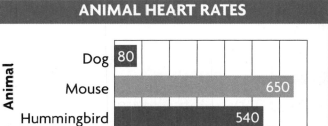

ANIMAL HEART RATES

Animal — Beats Per Minute

- Dog 80
- Mouse 650
- Hummingbird 540

(scale: 0, 100, 200, 300, 400, 500, 600, 700)

43. **TEST PREP** How many more times does a hummingbird's heart beat per minute than a dog's heart? (p. 252)

F 460   **G** 505   **H** 585   **J** 595

·· LINKUP ··· to Reading

STRATEGY • CHOOSE IMPORTANT INFORMATION

Some problems have information that you do not need.

Rock collections at the museum shop cost $5.98 each. Michael bought 2 boxes. There were 8 rocks in each box. How many rocks did Michael buy?

For 1–2, tell whether each fact gives information that you need to find how many rocks Michael bought. Write *yes* or *no*.

1. There were 8 rocks in a box.

2. Each box costs $5.98.

3. Use the information you need to find how many rocks Michael bought.

4. Use the information you need to find how much money he spent.

Practice Multiplication

► **Learn**

ROAD TRIP It took Mrs. Barrett 3 hours to drive from Portland to Seattle. She drove 56 miles each hour. How many miles did Mrs. Barrett drive?

$$3 \times 56 = \blacksquare \text{ or } \begin{array}{r} 56 \\ \times\ 3 \\ \hline \end{array}$$

STEP 1 Multiply the ones.

3×6 ones $= 18$ ones
Regroup 18 ones as 1 ten 8 ones.

Hundreds	Tens	Ones
	1	
	5	6
×		3
		8

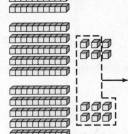

STEP 2 Multiply the tens. 3×5 tens $= 15$ tens
Add the 1 ten you regrouped. $(15 + 1)$ tens $= 16$ tens
Regroup 16 tens as 1 hundred 6 tens.

Hundreds	Tens	Ones
	1	
	5	6
×		3
1	6	8

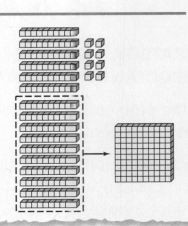

So, Mrs. Barrett drove 168 miles.

• The 8 ones were recorded in the ones column. How was the regrouped ten recorded? How was the regrouped hundred recorded?

► **Check**

1. **Explain** how to use regrouping to find 5×34. Then find the product.

CALIFORNIA STANDARDS O─┐ **NS 2.4** Solve simple problems involving multiplication of multidigit numbers by one-digit numbers. **MR 2.3** Use a variety of methods, such as words, numbers, symbols, charts, graphs, tables, diagrams, and models, to explain mathematical reasoning. *also* **NS 2.0, AF 1.0,** O─┐ **AF 1.1, AF 1.2, MR 1.0, MR 2.4**

Find the product. Tell whether you need to regroup.
Write *yes* or *no*.

2.	H	T	O		3.	H	T	O		4.	H	T	O		5.	H	T	O
		1	3				1	4				3	4				3	1
×			3		×			7		×			2		×			5

▶ **Practice and Problem Solving**

Find the product. Tell whether you need to regroup.
Write *yes* or *no*.

6.	19	7.	14	8.	74	9.	15	10.	43
	× 5		× 2		× 3		× 8		× 7

Find the product.

11.	42	12.	27	13.	95	14.	42	15.	39
	× 4		× 4		× 2		× 6		× 7

16.	54	17.	65	18.	37	19.	33	20.	74
	× 3		× 5		× 4		× 9		× 6

21. $2 \times 49 = \blacksquare$ **22.** $5 \times 23 = \blacksquare$ **23.** $8 \times 67 = \blacksquare$

$\frac{a+b}{c}$ **Algebra** Write <, >, or = for each ●.

24. 2×25 ● 7×15 **25.** 6×27 ● 9×13 **26.** 8×64 ● 64×8

27. REASONING Explain how you know without finding the product that 4×82 will have more than 2 digits.

28. **Write About It** Hiro said that 8×12 and 8×32 have the same ones digit. Explain how he knew this without finding each product.

Mixed Review and Test Prep

29. $\$5.00 - \$1.65 = \blacksquare$ (p. 88)

30. $48 + 484 = \blacksquare$ (p. 42)

31. $\$1.25 + \$2.55 = \blacksquare$ (p. 42)

32. $5,324 - 2,908 = \blacksquare$ (p. 62)

33. **TEST PREP** Taylor has 9 dimes, 8 nickels, and 20 pennies that he wants to exchange for quarters. How many quarters should he get? (p. 80)

A 6   **B** 7   **C** 8   **D** 9

Extra Practice page H48, Set B

Multiply Greater Numbers

More and more people are traveling on airplanes. Airlines must decide how many flights to schedule, and which types of planes to use. For example, could they carry more passengers on four 757s or on five 727s? Write a problem using the information in the table.

PLANES AND PASSENGERS	
Plane	Number of Passengers
727	149
737	119
757	190
767	254

O'Hare Airport,
Chicago, Illinois

CHECK WHAT YOU KNOW ✓

Use this page to help you review and remember
important skills needed for Chapter 18.

✓ REGROUP ONES AND TENS (See p. H20.)

Regroup. Write the missing number.

1. 5 tens 15 ones = ■ tens 5 ones

2. 4 tens 17 ones = 5 tens ■ ones

3. 3 tens ■ ones = 4 tens 4 ones

4. ■ tens 12 ones = 6 tens 2 ones

5. 6 tens 20 ones = 8 tens ■ ones

6. 7 tens 24 ones = ■ tens 4 ones

7. ■ tens 20 ones = 5 tens 0 ones

8. 5 tens ■ ones = 7 tens 1 one

✓ MULTIPLY 2-DIGIT NUMBERS (See p. H21.)

Find the product.

9. 14
× 3

10. 27
× 4

11. 56
× 5

12. 23
× 6

13. 71
× 9

14. 15
× 7

15. 43
× 3

16. 92
× 6

17. 85
× 2

18. 44
× 4

19. 58
× 8

20. 25
× 3

21. 62
× 5

22. 81
× 9

23. 64
× 6

24. 18
× 9

25. 53
× 6

26. 36
× 5

27. 75
× 2

28. 60
× 9

29. 45
× 3

30. 97
× 3

31. 33
× 8

32. 47
× 6

Mental Math: Patterns in Multiplication

▶ **Learn**

PRODUCT PATTERNS   Jena found different patterns in these multiplication exercises.

Jena

The product starts with the same digit as the first factor. The number of zeros in the second factor and the product increase by 1.

$3 \times 1 = 3$	$3 \times 1 = 3$
$3 \times 10 = 30$	$3 \times 10 = 30$
$3 \times 100 = 300$	$3 \times 100 = 300$
$3 \times 1,000 = 3,000$	$3 \times 1,000 = 3,000$

MATH IDEA Use place-value patterns to help you multiply by tens, hundreds, and thousands.

2×4	$=$	2×4 ones	$=$
2×40	$=$	2×4 tens	$=$
2×400	$=$	2×4 hundreds	$=$
$2 \times 4,000$	$=$	2×4 thousands	$=$

THOUSANDS	HUNDREDS	TENS	ONES
			8
		8	0
	8	0	0
8,	0	0	0

Examples

A $4 \times 1 = 4$
$4 \times 10 = 40$
$4 \times 100 = 400$
$4 \times 1,000 = 4,000$

B $3 \times 2 = 6$
$3 \times 20 = 60$
$3 \times 200 = 600$
$3 \times 2,000 = 6,000$

C $2 \times 5 = 10$
$2 \times 50 = 100$
$2 \times 500 = 1,000$
$2 \times 5,000 = 10,000$

• Why are there more zeros in the products in Example C than in Examples A and B?

CALIFORNIA STANDARDS O–¬NS 2.4 Solve simple problems involving multiplication of multidigit numbers by one-digit numbers. **MR 2.3** Use a variety of methods, such as words, numbers, symbols, charts, graphs, tables, diagrams, and models, to explain mathematical reasoning. *also* O–¬NS 1.3, NS 2.0, O–¬AF 1.1, MR 1.1, MR 3.2, MR 3.3

1. **Explain** how to use the basic fact $6 \times 4 = 24$ to help you find 6×40.

Copy and complete. Use patterns and mental math to help.

2. $6 \times 1 = \blacksquare$
$6 \times 10 = \blacksquare$
$6 \times 100 = \blacksquare$
$6 \times 1,000 = \blacksquare$

3. $3 \times 7 = \blacksquare$
$3 \times 70 = \blacksquare$
$3 \times 700 = \blacksquare$
$3 \times 7,000 = \blacksquare$

4. $4 \times 5 = \blacksquare$
$4 \times 50 = \blacksquare$
$4 \times 500 = \blacksquare$
$4 \times 5,000 = \blacksquare$

► **Practice and Problem Solving**

Copy and complete. Use patterns and mental math to help.

5. $5 \times 1 = \blacksquare$
$5 \times 10 = \blacksquare$
$5 \times 100 = \blacksquare$
$5 \times 1,000 = \blacksquare$

6. $7 \times 5 = \blacksquare$
$\blacksquare \times 50 = 350$
$7 \times \blacksquare = 3,500$
$7 \times 5,000 = \blacksquare$

7. $10 \times 1 = \blacksquare$
$\blacksquare \times 10 = 100$
$10 \times 100 = \blacksquare$
$10 \times \blacksquare = 10,000$

Use mental math and basic facts to find the product.

8. $2 \times 70 = \blacksquare$

9. $\blacksquare = 6 \times 400$

10. $4 \times 8,000 = \blacksquare$

11. $\blacksquare = 5 \times 300$

12. $5 \times 4,000 = \blacksquare$

13. $\blacksquare = 3 \times 90$

14. **REASONING** What basic multiplication fact could you use to help you find 5×600? Explain.

15. Emilio had 9 rolls of 50 pennies each. How many pennies did Emilio have?

16. **? What's the Question?** A case of staples contains 80 boxes. The answer is 480 boxes.

Mixed Review and Test Prep

Divide. (p. 220)

17. $4\overline{)32}$

18. $7\overline{)28}$

19. $5\overline{)40}$

20. Monica has 5 quarters. Her brother has 10 dimes. Who has more money? (p. 84)

21. **TEST PREP** A parking lot can fit 132 cars. There are 84 cars in the lot. How many more cars can park in the lot? (p. 58)

A 36 **B** 46 **C** 48 **D** 58

Problem Solving Strategy
Find a Pattern

Understand → Plan → Solve → Check

Quick Review

1. $4 \times 9 = \blacksquare$

2. $\blacksquare = 6 \times 7$

3. $3 \times 6 = \blacksquare$

4. $\blacksquare = 10 \times 8$

5. $2 \times 9 = \blacksquare$

PROBLEM A recycling container holds 200 pounds of newspaper. How many pounds of newspaper will 4 containers hold?

• What are you asked to find?

• What information will you use?

• What strategy can you use to solve the problem?

You can *find a pattern* to solve the problem.

• How can you use the strategy to solve the problem?

$$4 \quad \times \quad 200 \quad = \quad \blacksquare$$
↑   ↑   ↑

number of   pounds in   pounds in
containers   each container   all containers

Use a basic fact and place-value patterns to find the product.

$4 \times 2 = 8$
$4 \times 20 = 80$
$4 \times 200 = 800$

So, 4 containers will hold 800 pounds of newspaper.

• What other strategy could you use to solve the problem?

CALIFORNIA STANDARDS O┓NS 2.4 Solve simple problems involving multiplication cf multidigit numbers by one-digit numbers. **MR 1.1** Analyze problems by identifying relationships, distinguishing relevant from irrelevant information, sequencing and prioritizing information, and observing patterns. *also* **NS 2.0, MR 1.0, MR 2.0, MR 2.3, MR 2.4**

► Problem Solving Practice

PROBLEM SOLVING STRATEGIES

Draw a Diagram or Picture
Make a Model or Act It Out
Make an Organized List
► Find a Pattern
Make a Table or Graph
Predict and Test
Work Backward
Solve a Simpler Problem
Write an Equation
Use Logical Reasoning

Find a pattern to solve.

1. **What if** there are 6 containers to fill? How many pounds of newspaper will 6 containers hold?

2. A ferry boat carries 300 people on each trip. How many people can it carry on 6 trips?

A movie theater complex has 8 theaters. Each theater can seat 200 people. How many people can the theater complex seat?

3. Which number sentence shows how to solve this problem?

   A $8 \times 2 = 16$
   B $2 \times 80 = 160$
   C $2 + 800 = 802$
   D $8 \times 200 = 1,600$

4. **What if** the complex builds 2 more theaters that seat 500 people each? How many people could the theater complex seat in all?

   F 520   **H** 2,600
   G 1,000   **J** 4,000

Mixed Strategy Practice

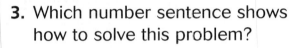

USE DATA For 5–6, use the table.

5. Make a bar graph of the data in the table. Be sure to write a title and labels.

6. How much longer is a blue whale than a whale shark? Write a number sentence to solve.

7. The difference between 2 numbers is 24. The product of the numbers is 112. What are the numbers?

8. Yuji wants to buy 5 boxes of crackers. Each box costs $3. How much will all of the boxes cost? Tell which operation you would use. Then solve the problem.

SEA ANIMAL LENGTHS	
Animal	**Length**
Blue whale	70 feet
Bottlenose dolphin	10 feet
Killer whale	25 feet
Whale shark	45 feet

9. Timothy wrote this number pattern.
   7, 14, 9, 18, 13, 26, 21, 42, 37
   What is the rule and the next 4 numbers in his pattern?

Estimate Products

Quick Review

1. $30 \times 9 = $ ■   2. ■ $= 6 \times 60$
3. $2 \times 70 = $ ■   4. ■ $= 100 \times 3$
5. $50 \times 4 = $ ■

▶ **Learn**

BUNCHES OF BISCUITS There are 28 dogs at the pet fair. If each dog gets 4 biscuits, about how many biscuits will be needed?

$4 \times 28 = $ ■ or $\begin{array}{r} 28 \\ \times\ 4 \\ \hline \end{array}$

MATH IDEA When you don't need an exact answer, you can estimate.

STEP 1 Round the first factor to the nearest ten.	**STEP 2** Find the estimated product.
$\begin{array}{r} 28 \\ \times\ 4 \\ \hline \end{array} \rightarrow \begin{array}{r} 30 \\ \times\ 4 \\ \hline \end{array}$	$\begin{array}{r} 30 \\ \times\ 4 \\ \hline 120 \end{array}$

> **Remember**
> To round a number:
> • Decide on the place to be rounded.
> • Look at the digit to its right.
> • If the digit is less than 5, the digit being rounded stays the same.
> • If the digit is 5 or more, the digit being rounded is increased by 1.

So, about 120 biscuits are needed.

• Is the actual answer less than or greater than 120? How do you know?

Examples

A Round to the nearest ten.

$\begin{array}{r} 52 \\ \times\ 7 \\ \hline \end{array} \rightarrow \begin{array}{r} 50 \\ \times\ 7 \\ \hline 350 \end{array}$

B Round to the nearest hundred.

$\begin{array}{r} 213 \\ \times\ 4 \\ \hline \end{array} \rightarrow \begin{array}{r} 200 \\ \times\ 4 \\ \hline 800 \end{array}$

▶ **Check**

1. **Explain** how to estimate 7×65.

CALIFORNIA STANDARDS O–π NS 2.4 Solve simple problems involving multiplication of multidigit numbers by one-digit numbers. **MR 2.3** Use a variety of methods, such as words, numbers, symbols, charts, graphs, tables, diagrams, and models, to explain mathematical reasoning. *also* O–π NS 1.3, NS 1.4, NS 2.0, O–π AF 2.1, **MR 2.0 MR 2.4, MR 2.5**

Estimate the product.

| **2.** 47 × 3 | **3.** 58 × 4 | **4.** 396 × 5 | **5.** 64 × 2 |

▶ Practice and Problem Solving

Estimate the product.

| **6.** 38 × 5 | **7.** 73 × 8 | **8.** 44 × 3 | **9.** 89 × 4 |

| **10.** 169 × 6 | **11.** 228 × 7 | **12.** 514 × 4 | **13.** 682 × 8 |

| **14.** 37 × 2 | **15.** 83 × 4 | **16.** 286 × 7 | **17.** 267 × 5 |

18. $3 \times 29 = \blacksquare$ **19.** $\blacksquare = 3 \times 78$ **20.** $5 \times 173 = \blacksquare$ **21.** $\blacksquare = 7 \times 731$

22. $\blacksquare = 9 \times 395$ **23.** $6 \times 419 = \blacksquare$ **24.** $4 \times 59 = \blacksquare$ **25.** $\blacksquare = 8 \times 612$

26. REASONING Gretchen says that $7 \times 43 = 3{,}001$. Estimate to decide if you agree or disagree. Explain.

27. REASONING Rodrigo said that 2×413 is the same as $800 + 20 + 6$. Do you agree or disagree? Explain.

28. Mr. Cory has 29 students and Miss Jan has 27 students. Each student needs 4 jars for an experiment. About how many jars are needed for all students?

29. ? What's the Error? Tui said that the product 7×488 is less than 2,800. Describe her error and give a more reasonable estimate.

Mixed Review and Test Prep

Subtract. (p. 58)

| **30.** 942 −258 | **31.** 604 −465 | **32.** 731 −283 |

33. Which product is greater: 4×1 or 13×0? (p. 132)

34. TEST PREP If Mr. Lewis paid $54 for 6 T-shirts, how much did 1 T-shirt cost? (p. 224)

A $6   **B** $7   **C** $8   **D** $9

Multiply 3-Digit Numbers

Quick Review

Find each product.

1. 32 × 8	**2.** 53 × 6	**3.** 36 × 4	
4. 24 × 7	**5.** 16 × 3		

▶ **Learn**

LUNCH TIME   The airline catering service is preparing lunches for 2 flights. There are 158 passengers on each flight. How many lunches will the service prepare?

$2 \times 158 = \blacksquare$ or $\begin{array}{r} 158 \\ \times\ \ 2 \\ \hline \end{array}$

Estimate. $2 \times 200 = 400$

STEP 1

Multiply the ones.
2×8 ones $= 16$ ones
Regroup 16 ones as 1 ten 6 ones.

Hundreds	Tens	Ones
	1	
1	5	8
×		2
		6

STEP 2

Multiply the tens.
2×5 tens $= 10$ tens
10 tens $+$ 1 ten $= 11$ tens
Regroup 11 tens as 1 hundred 1 ten.

Hundreds	Tens	Ones
1	1	
1	5	8
×		2
	1	6

STEP 3

Multiply the hundreds.
2×1 hundred $= 2$ hundreds
2 hundreds $+$ 1 hundred $= 3$ hundreds

Hundreds	Tens	Ones
1	1	
1	5	8
×		2
3	1	6

So, the catering service will prepare 316 lunches.
Since 316 is close to 400, the product is reasonable.

• How is multiplying 3-digit numbers like multiplying 2-digit numbers? How is it different?

CALIFORNIA STANDARDS O⊓NS 2.4 Solve simple problems involving multiplication of multidigit numbers by one-digit numbers. **MR 2.3** Use a variety of methods, such as words, numbers, symbols, charts, graphs, tables, diagrams, and models, to explain mathematical reasoning. *also* O⊓NS 1.3, NS 2.0, O⊓AF 2.1, MR 2.4, MR 2.6, MR 3.0, MR 3.1, MR 3.2, MR 3.3

More Multiplication

Examples

A	Hundreds	Tens	Ones
		1	
	1	2	5
×			3
	3	7	5

B	Hundreds	Tens	Ones
		1	
	3	0	6
×			2
	6	1	2

C	Hundreds	Tens	Ones
	2	4	
	1	4	8
×			6
	8	8	8

- Why is there only one regrouping in Example A and in Example B?

You may need to regroup 10 hundreds as 1 thousand when multiplying 3-digit numbers.

Find 3×457.

STEP 1

Multiply the ones. Regroup if needed.

$$\begin{array}{r} \overset{2}{4}57 \\ \times3 \\ \hline 1 \end{array}$$

STEP 2

Multiply the tens. Regroup if needed.

$$\begin{array}{r} \overset{12}{4}57 \\ \times3 \\ \hline 71 \end{array}$$

STEP 3

Multiply the hundreds. Regroup if needed.

$$\begin{array}{r} \overset{12}{4}57 \\ \times3 \\ \hline 1,371 \end{array}$$

So, $3 \times 457 = 1,371$.

You can use an estimate to check that your answer is reasonable.

Estimate. $3 \times 500 = 1,500$

Since 1,371 is close to 1,500, the answer is reasonable.

▶ Check

1. **Explain** how the 13 hundreds were regrouped and recorded in Step 3 above.

Multiply. Tell each place you need to regroup.

2. $\begin{array}{r} 148 \\ \times4 \\ \hline \end{array}$

3. $\begin{array}{r} 136 \\ \times5 \\ \hline \end{array}$

4. $\begin{array}{r} 208 \\ \times3 \\ \hline \end{array}$

5. $\begin{array}{r} 431 \\ \times2 \\ \hline \end{array}$

LESSON CONTINUES ▶

► Practice and Problem Solving

Multiply. Tell each place you need to regroup.

6. 203
    × 7

7. 411
    × 9

8. 381
    × 3

9. 428
    × 5

Find the product. Estimate to check.

10. 326
     × 4

11. 187
     × 3

12. 202
     × 7

13 423
    × 3

Find the product.

14. 294
     × 3

15. 419
     × 3

16. 120
     × 6

17. 168
     × 4

18. 121
     × 8

19. 177
     × 3

20. 395
     × 4

21. 241
     × 5

22. $4 \times 345 =$ ■

23. $2 \times 499 =$ ■

24. ■ $= 6 \times 322$

25. $3 \times 279 =$ ■

26. $5 \times 118 =$ ■

27. ■ $= 4 \times 421$

28. $8 \times 123 =$ ■

29. ■ $= 2 \times 391$

USE DATA For 30–34, use the table.

30. Diesta and her family took a train to Washington, D.C. How many seats were on the train in all?

31. How many coach seats are there on 4 trains?

32. **? What's the Question?** The answer is 192 seats.

33. Claudia needs to order pillows for the sleeper cars for three trains. How many pillows should she order?

34. **? What's the Error?** George says there are 620 coach seats on two trains. Describe George's error.

TRAIN	
Type of Car	Total Seats
Coach	360
Sleeper	168
Deluxe	90

35. ✏️ **Write a problem** in which regrouping is needed in both the ones and the tens places.

36. REASONING Is 3×672 greater than or less than 4×415? Explain.

Mixed Review and Test Prep

Find the sum or difference.

37. $\quad 275$ (p. 42)
$\quad +376$

38. $\quad 382$ (p. 58)
$\quad -158$

39. $\quad 822$ (p. 58)
$\quad -568$

40. $\quad 936$ (p. 42)
$\quad +826$

41. $\quad \$7.94$ (p. 88)
$\quad +\$3.54$

42. $\quad \$4.73$ (p. 88)
$\quad -\$1.89$

43. **TEST PREP** Eight friends picked 56 apples. If they share the apples equally, how many apples will each friend get? (p. 214)

A 4 **B** 5 **C** 6 **D** 7

44. **TEST PREP** What is 5,976 rounded to the nearest thousand? (p. 30)

F 5,000 **H** 5,980
G 5,900 **J** 6,000

··LINKUP··· to Math History

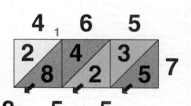

Lattice multiplication was used in Europe during the fourteenth and fifteenth centuries. You can use basic multiplication facts and a grid to find 7×465.

A Draw 1 row of 3 squares with a diagonal line in each square. Then write the factors as shown.

B Multiply each pair of digits.

$7 \times 5 = 35$
$7 \times 6 = 42$
$7 \times 4 = 28$

Write each product as shown.

C Start at the right. Add the digits in each diagonal. Regroup if needed.

So, $7 \times 465 = 3,255$.

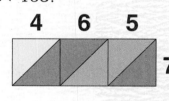

Use a grid to find the product.

1. $6 \times 722 = $ ■

2. $8 \times 346 = $ ■

3. $4 \times 296 = $ ■

4. $3 \times 435 = $ ■

Find Products Using Money

Quick Review

Find each product.

1.	57	2.	35	3.	27
	× 6		× 4		× 5

4.	59	5.	27
	× 2		× 7

▶ **Learn**

PLANE FUN   Mr. Haines and his son bought 3 puzzle books for their flight. Each book cost $3.85. How much did the books cost in all?

Estimate. $3.85 rounded to the nearest dollar is $4.00. 3 × $4.00 = $12.00

STEP 1   Write the problem in cents.

Think: $3.85 = 385 cents

$3.85 → 385
× 3   × 3

STEP 2   Multiply to find the product in cents.

$$\begin{array}{r} {\scriptstyle 2\,1} \\ 385 \\ \times\quad 3 \\ \hline 1,155 \end{array}$$

Remember
When you write money amounts, separate the dollars and cents with a decimal point and write a dollar sign on the left.

decimal point ⌐
dollar sign → $4.78

STEP 3   Write the product in dollars and cents.
1,155 cents = $11.55
Since $11.55 is close to $12.00, the answer is reasonable.

So, the puzzle books cost $11.55.

• How can estimation help you decide if your answer is reasonable?

MATH IDEA Multiply money amounts the same way you multiply whole numbers. Then write the product in dollars and cents.

Examples

A	B	C
$\scriptstyle 2$	$\scriptstyle 4\,4$	$\scriptstyle 3$
$2.60	$2.78	$4.16
× 4	× 6	× 5
$10.40	$16.68	$20.80

CALIFORNIA STANDARDS O⌐n NS 3.3 Solve problems involving addition, subtraction, multiplication, and division of money amounts in decimal notation and multiply and divide money amounts in decimal notation by using whole-number multipliers and divisors. *also* NS 2.0, O⌐n AF 2.1, MR 2.1, MR 2.3 MR 2.4, MR 2.6, MR 3.0, MR 3.1, MR 3.2, MR 3.3

1. **Explain** why you do not need to regroup when you multiply 5×1 in Example C.

Find the product in dollars and cents. Estimate to check.

2. $2.95
   $\times$ 6

3. $1.38
   $\times$ 8

4. $4.76
   $\times$ 2

5. $3.75
   $\times$ 3

► **Practice and Problem Solving**

Find the product in dollars and cents. Estimate to check.

6. $5.09
   $\times$ 9

7. $4.68
   $\times$ 5

8. $3.29
   $\times$ 7

9. $3.82
   $\times$ 2

Find the product in dollars and cents.

10. $6.75
    $\times$ 4

11. $2.87
    $\times$ 8

12. $1.67
    $\times$ 9

13. $8.43
    $\times$ 6

14. $4.71
    $\times$ 3

15. $7.65
    $\times$ 4

16. $2.48
    $\times$ 9

17. $1.82
    $\times$ 6

18. $4 \times \$5.08 = \blacksquare$
19. $\blacksquare = 7 \times \$3.94$
20. $3 \times \$6.48 = \blacksquare$
21. $\blacksquare = 5 \times \$7.31$

22. $8 \times \$2.65 = \blacksquare$
23. $\blacksquare = 9 \times \$3.84$
24. $\blacksquare = 2 \times \$4.73$
25. $6 \times \$9.24 = \blacksquare$

26. Mario buys 3 dozen pens. Each dozen costs $7.85. About how much money does he need?

27. ✏️ **Write a problem** that uses $4 \times \$5.72$. Solve the problem.

─ *Mixed Review and Test Prep* ─

Divide. (p. 220)

28. $7\overline{)49}$ 29. $6\overline{)48}$ 30. $5\overline{)45}$

31. What multiplication sentence is shown with an array made of 4 rows of 5? (p. 120)

32. **TEST PREP** Trevor bought a book for $12.95 and a video for $15.95. How much did he spend in all? (p. 46)

   A $27.80   C $28.80
   B $27.90   D $28.90

Quick Review

Find each product.

1.	46 × 3	2.	73 × 4	3.	28 × 5
4.	37 × 2	5.	89 × 6		

▶ Learn

ON THE ROAD AGAIN Mr. Wilson drove 1,953 miles from San Francisco to Houston. Three days later, he drove back to San Francisco along the same route. How many miles did Mr. Wilson drive in all?

$2 \times 1,953 = $ ▨

Estimate. Round 1,953 to the nearest thousand.
$2 \times 2,000 = 4,000$

STEP 1 Multiply the ones.	**STEP 2** Multiply the tens.	**STEP 3** Multiply the hundreds.	**STEP 4** Multiply the thousands.
1,953 × 2 ——— 6	1 1,953 × 2 ——— 06	1 1 1,953 × 2 ——— 906	1 1 1,953 × 2 ——— 3,906

So, Mr. Wilson drove 3,906 miles. The estimate was 4,000, so the product is reasonable.

Examples

A 1 1 1,625 × 3 ——— 4,875	**B** 2 41 1,482 × 5 ——— 7,410	**C** 3 3 $20.89 × 4 ——— $83.56

▶ Check

1. **Tell** why there are no tens to regroup in Example A.

CALIFORNIA STANDARDS ⊶NS 2.4 Solve simple problems involving multiplication of multidigit numbers by one-digit numbers. **MR 2.3** Use a variety of methods, such as words, numbers, symbols, charts, graphs, tables, diagrams, and models, to explain mathematical reasoning. *also* ⊶NS 1.3, NS 2.0, ⊶NS 3.3, MR 2.1, **MR 2.4, MR 2.6, MR 3.0, MR 3.1, MR 3.2, MR 3.3**

Find the product. Estimate to check.

2.	1,186 × 8	3.	3,245 × 3	4.	1,514 × 6	5.	4,692 × 2

▶ Practice and Problem Solving

Find the product. Estimate to check.

6.	1,231 × 3	7.	2,843 × 6	8.	2,418 × 4	9.	$30.22 × 3

10.	4,395 × 2	11.	1,587 × 6	12.	$31.25 × 3	13.	3,974 × 2

14.	1,279 × 7	15.	$43.25 × 2	16.	746 × 4

17. $9 \times 642 = $ ■   18. ■ $= 5 \times 1,653$ 19. ■ $= 2 \times 3,742$

20. A pilot flew back and forth from San Francisco to Denver 3 times. The distance between the cities is 1,305 miles. How many miles did the pilot fly?

21. **Write About It** Find the missing digit. Explain how you found it.

$$\begin{array}{r} 2{,}1\blacksquare5 \\ \times \phantom{2{,}1}4 \\ \hline 8{,}700 \end{array}$$

22. **$\frac{a+b}{c}$ Algebra** Use estimation to find the missing factor. Then explain how you found the missing number. $\$9.30 \times$ ■ $= \$37.20$

Mixed Review and Test Prep

23. Write a rule for the table. Then copy and complete the table. (p. 168)

Tricycles	1	2	3	4
Wheels	3	6	■	■

Write the value of each blue digit. (p. 6)

24. 1,0**7**3   25. 4,8**6**1   26. 9,5**2**4

27. **TEST PREP** Dolores spent $30 on 6 boxes of pens. How much did each box of pens cost? (p. 224)

A $5   C $24

B $6   D $180

Extra Practice page H49, Set E

Review/Test

✔ CHECK CONCEPTS

Copy and complete. Use patterns and mental math to help. (pp. 306–307)

1. $6 \times 1 = \blacksquare$
$6 \times 10 = \blacksquare$
$6 \times 100 = \blacksquare$
$6 \times 1,000 = \blacksquare$

2. $5 \times 3 = \blacksquare$
$5 \times 30 = \blacksquare$
$5 \times 300 = \blacksquare$
$5 \times 3,000 = \blacksquare$

3. $8 \times 4 = \blacksquare$
$8 \times 40 = \blacksquare$
$8 \times \blacksquare = 3,200$
$\blacksquare \times 4,000 = 32,000$

✔ CHECK SKILLS

Estimate the product. (pp. 310–311)

4. $\begin{array}{r} 45 \\ \times\ 6 \\ \hline \end{array}$

5. $\begin{array}{r} 323 \\ \times\ \ 3 \\ \hline \end{array}$

6. $\begin{array}{r} 386 \\ \times\ \ 5 \\ \hline \end{array}$

7. $\begin{array}{r} 24 \\ \times\ 8 \\ \hline \end{array}$

8. $\begin{array}{r} 592 \\ \times\ \ 7 \\ \hline \end{array}$

Find the product. (pp. 312–319)

9. $\begin{array}{r} 201 \\ \times\ 4 \\ \hline \end{array}$

10. $\begin{array}{r} 63 \\ \times\ 5 \\ \hline \end{array}$

11. $\begin{array}{r} \$2.07 \\ \times\ \ \ 9 \\ \hline \end{array}$

12. $\begin{array}{r} \$9.31 \\ \times\ \ \ 7 \\ \hline \end{array}$

13. $\begin{array}{r} 1,285 \\ \times\ \ \ 8 \\ \hline \end{array}$

14. $\begin{array}{r} 64 \\ \times\ 2 \\ \hline \end{array}$

15. $\begin{array}{r} 531 \\ \times\ 4 \\ \hline \end{array}$

16. $\begin{array}{r} \$7.49 \\ \times\ \ \ 3 \\ \hline \end{array}$

17. $\begin{array}{r} \$21.96 \\ \times\ \ \ \ 4 \\ \hline \end{array}$

18. $\begin{array}{r} 4,238 \\ \times\ \ \ 5 \\ \hline \end{array}$

19. $4 \times 2,052 = \blacksquare$ **20.** $8 \times 527 = \blacksquare$ **21.** $3 \times \$6.15 = \blacksquare$ **22.** $6 \times \$42.17 = \blacksquare$

✔ CHECK PROBLEM SOLVING

USE DATA For 23–25, use the bar graph. (pp. 308–309)

23. About how much would 4 warthogs weigh?

24. About how much would 3 zebras weigh?

25. Would 2 zebras weigh more than or less than 4 gnu? Explain.

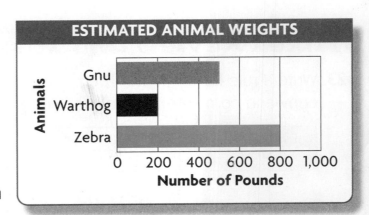

ESTIMATED ANIMAL WEIGHTS

Animals: Gnu, Warthog, Zebra
Number of Pounds: 0, 200, 400, 600, 800, 1,000

Cumulative Review

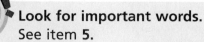

Look for important words.
See item **5.**

An important word is *rounded*. When you are asked to round a number, it is important to find the place to which you need to round.

Also see problem **2**, p. H63.

For 1–10, choose the best answer.

1. 128
 × 5

 A 640   C 540
 B 600   D 510

2. Which number makes this equation true?

■ × 20 = 180

 F 160   H 80
 G 90   J 9

3. $2.52
 × 9

 A $23.68   C $22.58
 B $22.68   D $22.50

4. 3,516
 × 4

 F 1,444   H 14,064
 G 12,064   J NOT HERE

5. What is 1,456 rounded to the nearest hundred?

 A 2,000   C 1,400
 B 1,500   D 1,000

6. Which is the best estimate for this product?

 178
 × 6

 F 1,200   H 800
 G 1,000   J 600

7. Tickets to the amusement park cost $12.25 each. How much will 5 tickets cost?

 A $51.25   C $60.25
 B $60.05   D $61.25

8. Mario had $3.25. He spent $1.97. Then he earned $2.50 for feeding his neighbor's cat. How much money did Mario have then?

 F $1.28   H $5.75
 G $3.78   J $7.72

9. A cruise ship can carry 2,000 passengers on each cruise. There are 4 cruises in a month. How many passengers can the ship carry in a month?

 A 4,000   C 8,000
 B 6,000   D NOT HERE

10. What is the value of the 3 in 93,406?

 F 30   H 3,000
 G 300   J 30,000

Divide by 1-Digit Numbers

The Wrigley Building is a famous landmark in Chicago. The clock that you see on the Wrigley Building has an hour hand that is 6 feet 4 inches long and a minute hand that is 9 feet 2 inches long. Suppose the classes shown on the bar graph are going to see the Wrigley Building. If the same number of students ride on each of 4 buses, how many students will be on each bus?

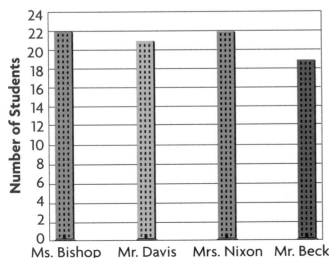

FIELD TRIP

Number of Students (y-axis: 0 to 24)

Classes: Ms. Bishop, Mr. Davis, Mrs. Nixon, Mr. Becker

The Wrigley Building

CHECK WHAT YOU KNOW ✓

Use this page to help you review and remember
important skills needed for Chapter 19.

✓ VOCABULARY

Choose the best term from the box.

1. In the example, the number 8 is the __?__ .

2. In the example, the number 32 is the __?__ .

3. In the example, the number 4 is the __?__ .

Example:

$$8\overline{)32}^{\,4}$$

| dividend |
| divisor |
| quotient |
| sum |

✓ SUBTRACTION FACTS (See p. H6.)

Subtract.

4. $7 - 5 = $ ■
5. $12 - 5 = $ ■
6. $14 - 6 = $ ■
7. ■ $= 13 - 5$

8. $8 - 2 = $ ■
9. ■ $= 15 - 9$
10. $18 - 9 = $ ■
11. $8 - 7 = $ ■

✓ DIVISION FACTS (See p. H21.)

Divide.

12. $81 \div 9 = $ ■
13. $21 \div 3 = $ ■
14. ■ $= 48 \div 8$
15. $18 \div 6 = $ ■

16. $24 \div 6 = $ ■
17. ■ $= 25 \div 5$
18. $16 \div 8 = $ ■
19. $12 \div 3 = $ ■

20. $2\overline{)14}$
21. $5\overline{)40}$
22. $6\overline{)42}$
23. $7\overline{)56}$

24. $9\overline{)36}$
25. $4\overline{)16}$
26. $3\overline{)24}$
27. $7\overline{)49}$

✓ REGROUP HUNDREDS AND TENS (See p. H22.)

Regroup.

28. $50 = 4$ tens ■ ones

29. $60 = $ ■ tens 10 ones

30. $73 = 5$ tens ■ ones

31. 3 tens ■ ones $= 44$

32. 5 hundreds ■ tens $= 610$

33. 2 hundreds ■ tens $= 340$

HANDS ON
Divide with Remainders

Quick Review

1. $3\overline{)9}$ 2. $6\overline{)12}$

3. $4\overline{)24}$ 4. $5\overline{)35}$

5. $7\overline{)28}$

▶ **Explore**

Noah collected all 19 dinosaur toys from Crispy Crunch cereal. He wants to keep an equal number of toys in each of 3 shoe boxes. Can Noah divide them equally among the 3 boxes? Why or why not?

Use counters to find $19 \div 3$.

VOCABULARY
remainder

MATERIALS
counters

> **STEP 1**   Use 19 counters. Draw 3 circles.

Remember

$$30 \div 6 = 5$$
$\uparrow$ $\uparrow$ $\uparrow$
dividend divisor quotient

> **STEP 2**   Divide the 19 counters into 3 equal groups by putting them in the circles.

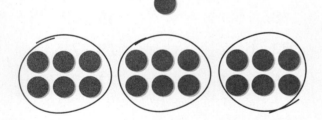

I'm putting 11 counters into 2 equal groups. Will there be any left over?

So, there will be 6 toys in each shoe box, with 1 toy left over.

Try It

Use counters to make equal groups. Draw a picture of the model you made.

a. $11 \div 2$ **b.** $15 \div 4$ **c.** $26 \div 3$

CALIFORNIA STANDARDS NS 2.0 Students calculate and solve problems involving addition, subtraction, multiplication, and division. **MR 2.3** Use a variety of methods, such as words, numbers, symbols, charts, graphs, tables, diagrams, and models, to explain mathematical reasoning. *also* **MR 2.4, MR 3.2**

► Connect

In division, the amount left over when a number cannot be divided evenly is called the **remainder**.

Find 20 ÷ 6.

TECHNOLOGY LINK

More Practice: Use E-Lab, *Dividing with Remainders*.

www.harcourtschool.com/elab2002

STEP 1 Use 20 counters. Draw 6 circles.

STEP 2 Divide the 20 counters into 6 equal groups.

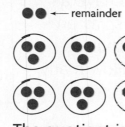

The quotient is 3.
The remainder is 2.

► Practice

Use counters to find the quotient and remainder.

1. $9 \div 2 = \blacksquare$

2. $17 \div 4 = \blacksquare$

3. $10 \div 3 = \blacksquare$

4. $20 \div 3 = \blacksquare$

5. $4\overline{)18}$

6. $5\overline{)19}$

7. $5\overline{)27}$

8. $6\overline{)21}$

Find the quotient and remainder. You may use counters or draw a picture to help.

9. $16 \div 3 = \blacksquare$

10. $21 \div 5 = \blacksquare$

11. $2\overline{)15}$

12. $4\overline{)22}$

13. **Write About It** What is the greatest remainder you could have when you divide by 5? Explain.

14. **? What's the Question?** Darnell has 12 Beastie Buddies in his collection. He puts an equal number in each of 3 drawers. The answer is 4 Beastie Buddies.

Mixed Review and Test Prep

Multiply. (p. 294)

15. $2 \times 13 = \blacksquare$

16. $3 \times 16 = \blacksquare$

17. $5 \times 14 = \blacksquare$

18. $4 \times 19 = \blacksquare$

19. **TEST PREP** Judy bought 3 posters that cost $2.79 each. How much did Judy pay for the 3 posters? (p. 316)

A $8.37   **C** $8.09

B $8.17   **D** $7.39

Model Division of 2-Digit Numbers

Quick Review

1. $16 \div 4 = \blacksquare$
2. $14 \div 7 = \blacksquare$ 3. $15 \div 3 = \blacksquare$
4. $32 \div 4 = \blacksquare$ 5. $18 \div 6 = \blacksquare$

▶ Learn

COOL COLLECTIONS   Suppose you have a collection of 63 hockey cards. You display an equal number of cards on each of 5 shelves. How many cards are on each shelf? How many cards are left over?

Activity

$63 \div 5 = \blacksquare$ $5\overline{)63}$

Materials: base-ten blocks

STEP 1	STEP 2	STEP 3
Divide the tens first. Put an equal number of tens in each group.	One ten is left. Regroup it as 10 ones. Put an equal number of ones in each group.	Use the letter *r* to show the remainder. $63 \div 5 = 12 \text{ r}3$

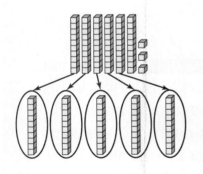

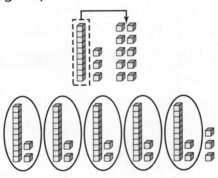

$$63 \div 5 = 12 \text{ r}3$$

remainder

quotient

So, there are 12 cards on each shelf. There are 3 cards left over.

• In Step 2, why was the 1 ten regrouped into 10 ones?

▶ Check

1. **Explain** why $63 \div 5$ has a remainder.

CALIFORNIA STANDARDS NS 2.5 Solve division problems in which a multidigit number is evenly divided by a one-digit number. **MR 2.3** Use a variety of methods, such as words, numbers, symbols, charts, graphs, tables, diagrams, and models, to explain mathematical reasoning. *also* AF 1.2, MR 2.4, MR 3.2

Use the model. Write the quotient and remainder.

2.

21 ÷ 2

3.

44 ÷ 3

▶ Practice and Problem Solving

Use the model. Write the quotient and remainder.

4.

62 ÷ 4

5.

71 ÷ 5

Divide. You may use base-ten blocks to help.

6. 23 ÷ 2 = ■   **7.** 25 ÷ 2 = ■   **8.** 39 ÷ 3 = ■   **9.** 43 ÷ 4 = ■

10. 2)37   **11.** 3)42   **12.** 2)51   **13.** 4)55

14. Matt needs 42 cans of juice for a party. How many 6-packs of juice should he buy?

15. $\frac{a+b}{c}$ **Algebra** Find the missing digit.

$$\begin{array}{r} 15 \ r2 \\ 3)4\blacksquare \end{array}$$

16. **? What's the Error?** This is how Cheryl modeled 32 ÷ 3.

Describe Cheryl's mistake. Draw the correct model.

17. Mitsugi has 4 shelves of hockey cards. Each shelf has 13 cards. How many hockey cards does he have?

Mixed Review and Test Prep

For 18–20, find the product. (p. 294)

18. 12
      × 9

19. 14
      × 3

20. 27
      × 2

21. 9,427 − 5,613 = ■ (p. 62)

22. **TEST PREP** Amy has 4 boxes with 8 toys in each box. She gives 3 toys to her sister. How many toys does Amy have left? (p. 172)

A 35   **C** 29

B 33   **D** 25

Record Division of 2-Digit Numbers

▶ **Learn**

TAKE YOUR PICK   Brad picked 58 apples at an orchard. He put an equal number of apples in 3 bags. How many were in each bag? How many were left over?

$$58 \div 3 = \blacksquare$$

Read: 58 divided by 3   **Write**: $3\overline{)58}$

STEP 1   Divide the 5 tens first. $3\overline{)5}$
Think: $3 \times 1 = 3$

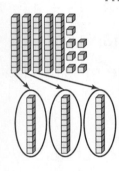

$$
\begin{array}{r}
1 \\
3\overline{)58} \\
-3 \\
\hline
2
\end{array}
$$

1 ten in each group.

Divide. $3\overline{)5}$
Multiply. 3×1
Subtract. $5 - 3$
Compare. $2 < 3$

The difference, 2, must be less than the divisor, 3.

STEP 2   Bring down the 8 ones.

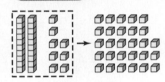

$$
\begin{array}{r}
1 \\
3\overline{)58} \\
-3\downarrow \\
\hline
28
\end{array}
$$

Regroup 2 tens 8 ones as 28 ones.

STEP 3   Divide the 28 ones. $3\overline{)28}$
Think: $3 \times 9 = 27$

9 ones in each group.

$$
\begin{array}{r}
19\ \text{r}1 \\
3\overline{)58} \\
-3 \\
\hline
28 \\
-27 \\
\hline
1
\end{array}
$$

Divide. $3\overline{)28}$
Multiply. 3×9
Subtract. $28 - 27$
Compare. $1 < 3$

STEP 4   Multiply to check your answer.

$$
\begin{array}{r}
19 \leftarrow \text{quotient} \\
\times\ 3 \leftarrow \text{divisor} \\
\hline
57 \\
\\
57 \\
+\ 1 \leftarrow \text{Add the remainder.} \\
\hline
58 \leftarrow \text{This should equal} \\
\text{the dividend.}
\end{array}
$$

So, there were 19 apples in each bag and 1 apple was left over.

CALIFORNIA STANDARDS ○━ℿNS 2.3 Use the inverse relationship of multiplication and division to compute and check results. **NS 2.5** Solve division problems in which a multidigit number is evenly divided by a one-digit number. *also* **NS 2.8,** ○━ℿ**AF 1.1, MR 2.3, MR 2.4, MR 3.2**

More Division

Tyler bought 35 pumpkins for the school's Fall Festival. He puts 4 pumpkins in each box. How many boxes does he use? How many pumpkins are left over?

$$35 \div 4 = \blacksquare$$

STEP 1 Since there are not enough tens to divide, start with ones.

$$\blacksquare$$
$$4\overline{)35}$$ 4 > 3, so place the first digit in the ones place.

STEP 2 Divide the 35 ones.

$$\begin{array}{r} 8\ r3 \\ 4\overline{)35} \\ -32 \\ \hline 3 \end{array}$$

Divide. $4\overline{)35}$
Multiply. 4×8
Subtract. $35 - 32$
Compare. $3 < 4$

STEP 3 Multiply to check your answer.

$$\begin{array}{r} 8 \\ \times 4 \\ \hline 32 \end{array}$$ ← quotient
← divisor

$$\begin{array}{r} 32 \\ + 3 \\ \hline 35 \end{array}$$ ← Add the remainder.
← dividend

So, Tyler uses 8 boxes. There are 3 pumpkins left over.

Examples

A
$$\begin{array}{r} 14 \\ 7\overline{)98} \\ -7 \\ \hline 28 \\ -28 \\ \hline 0 \end{array}$$

Check:
$$\begin{array}{r} 14 \\ \times 7 \\ \hline 98 \end{array}$$

B
$$\begin{array}{r} 22\ r1 \\ 3\overline{)67} \\ -6 \\ \hline 07 \\ -6 \\ \hline 1 \end{array}$$

Check:
$$\begin{array}{r} 22 \\ \times 3 \\ \hline 66 \end{array} \quad \begin{array}{r} 66 \\ + 1 \\ \hline 67 \end{array}$$

MATH IDEA The steps to record divison are divide, multiply, subtract, compare.

- Why couldn't you write a 6 in the ones place in Step 2 above?

▶ Check

1. **Describe** why you do not need to add a remainder when you check Example A.

Divide and check.

2. $29 \div 2 = \blacksquare$ **3.** $26 \div 3 = \blacksquare$ **4.** $52 \div 4 = \blacksquare$ **5.** $38 \div 3 = \blacksquare$

LESSON CONTINUES

▶ Practice and Problem Solving

Divide and check.

6. $33 \div 2 = $ ▨   **7.** $45 \div 4 = $ ▨   **8.** $49 \div 5 = $ ▨   **9.** $41 \div 3 = $ ▨

10. $42 \div 2 = $ ▨   **11.** $50 \div 3 = $ ▨   **12.** $67 \div 6 = $ ▨   **13.** $61 \div 5 = $ ▨

14. $38 \div 4 = $ ▨   **15.** $84 \div 7 = $ ▨   **16.** $23 \div 4 = $ ▨   **17.** $76 \div 3 = $ ▨

18. $4\overline{)77}$ **19.** $6\overline{)52}$ **20.** $5\overline{)73}$ **21.** $3\overline{)66}$

22. $5\overline{)48}$ **23.** $2\overline{)71}$ **24.** $7\overline{)92}$ **25.** $9\overline{)89}$

26. $2\overline{)87}$ **27.** $7\overline{)34}$ **28.** $8\overline{)99}$ **29.** $6\overline{)39}$

Write the *check* step for each division problem.

30. $4\overline{)22}$ (5 r2)   **31.** $6\overline{)43}$ (7 r1)   **32.** $3\overline{)86}$ (28 r2)   **33.** $2\overline{)55}$ (27 r1)

Decide whether you start by dividing the tens or the ones. Write *tens* or *ones*.

34. $5\overline{)22}$ **35.** $6\overline{)81}$ **36.** $2\overline{)15}$ **37.** $3\overline{)32}$

38. $9\overline{)72}$ **39.** $4\overline{)34}$ **40.** $7\overline{)61}$ **41.** $2\overline{)52}$

42. ? What's the Error? Lou began to solve a problem like this:

$$2\overline{)53} \quad \begin{array}{r} 1 \\ -2 \\ \hline 3 \end{array}$$

Describe how he must use the *compare* step to fix his error. Solve the problem.

43. a+b/c Algebra Mrs. Talbert divided 25 markers equally among 6 students. Write a number sentence to show how many markers each student got and how many were left over.

44. REASONING Alyssa divided 79 jelly beans equally into 6 jars. She kept 1 jar and the leftover jelly beans. How many jelly beans did she keep?

45. Marshal bought 16 packs of gum. Each pack holds 3 baseball cards. How many cards did he get?

46. Hitoshi has 97 strawberries. He puts an equal number of strawberries into 3 baskets. How many strawberries are in each basket? How many are left over?

47. It takes Grant 12 minutes to get to his friend's house. It takes him 4 times as long to get to his grandfather's house. How long does it take to get to his grandfather's?

Mixed Review and Test Prep

Find the product. (p. 298)

48. 21
    × 2

49. 34
    × 4

50. 55
    × 3

51. **TEST PREP** Mrs. Martin kept a lunch box she bought in 1971. She was 9 years old when she bought the lunch box. How old was she in 2000? (p. 62)

A 38   **B** 28   **C** 19   **D** 9

Compare. Write <, >, or = for each ●.

52. 3 × 4 ● 6 × 2 (p. 138)

53. 54 ÷ 9 ● 18 ÷ 2 (p. 220)

54. **TEST PREP** Mr. Warner filled some bags with 9 crayons each. He had 4 crayons left. If Mr. Warner started with 67 crayons, how many bags did he fill? (p. 226)

F 9   **G** 8   **H** 7   **J** 6

LINKUP to Social Studies

The steps for making crayons have not changed much since 1903. Wax in different colors is heated to 240°F. Workers pour the hot wax across a table that holds crayon molds. In $7\frac{1}{2}$ minutes, the wax cools into 72 rows of 8 crayons.

USE DATA For 1–2, use the bar graph.

1. How many crayons are in the pack shown in the bar graph?

2. **Write a problem** about the bar graph. Use division. Trade with a partner and solve.

3. How many crayons are made in the 72 rows of 8 crayon molds?

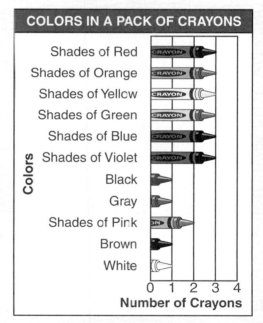

COLORS IN A PACK OF CRAYONS

Colors:
Shades of Red
Shades of Orange
Shades of Yellow
Shades of Green
Shades of Blue
Shades of Violet
Black
Gray
Shades of Pink
Brown
White

Number of Crayons: 0 1 2 3 4

4. The 64-pack of crayons has 4 equal rows. How many crayons are in each row?

Practice Division

▶ Learn

TOO MANY TOYS Ellen has collected 87 plastic toys. She puts an equal number of toys on 6 shelves in her room. How many toys are on each shelf? How many toys are not on shelves?

$$87 \div 6 = \blacksquare$$

Read: 87 divided by 6   **Write:** $6\overline{)87}$

STEP 1 Divide the 8 tens.

$$
\begin{array}{r}
1 \\
6\overline{)87} \\
-6 \\
\hline
2
\end{array}
$$

Divide. $6\overline{)8}$
Multiply. 6×1
Subtract. $8 - 6$
Compare. $2 < 6$

STEP 2 Bring down the 7 ones. Divide the 27 ones.

$$
\begin{array}{r}
14\ r3 \\
6\overline{)87} \\
-6\downarrow \\
\hline
27 \\
-24 \\
\hline
3
\end{array}
$$

Divide. $6\overline{)27}$
Multiply. 6×4
Subtract. $27 - 24$
Compare. $3 < 6$

STEP 3 Multiply to check your answer.

$$
\begin{array}{r}
14 \\
\times\ 6 \\
\hline
84
\end{array}
$$
← quotient
← divisor

$$
\begin{array}{r}
84 \\
+\ 3 \\
\hline
87
\end{array}
$$
← remainder
← dividend

So, there are 14 toys on each shelf and 3 toys not on shelves.

- Could you begin by writing a 1 over the ones place in Step 1? Why or why not?

▶ Check

1. **Tell** why you must multiply 6×4 in Step 2.

Divide.

2. $38 \div 3 = \blacksquare$

3. $29 \div 4 = \blacksquare$

4. $35 \div 2 = \blacksquare$

5. $65 \div 5 = \blacksquare$

CALIFORNIA STANDARDS O—n NS 2.3 Use the inverse relationship of multiplication and division to compute and check results. **NS 2.5** Solve division problems in which a multidigit number is evenly divided by a one-digit number. *also* **NS 2.0, O—n AF 1.1, MR 2.3, MR 2.4, MR 3.2**

Divide.

6. $26 \div 5 = $ ■

7. $52 \div 2 = $ ■

8. $55 \div 4 = $ ■

9. $49 \div 3 = $ ■

10. $6\overline{)69}$

11. $4\overline{)77}$

12. $5\overline{)72}$

13. $3\overline{)64}$

14. $3\overline{)80}$

15. $7\overline{)91}$

16. $4\overline{)65}$

17. $9\overline{)79}$

Divide and check.

18. $70 \div 8 = $ ■

19. $77 \div 2 = $ ■

20. $81 \div 3 = $ ■

21. $88 \div 5 = $ ■

22. $5\overline{)78}$

23. $4\overline{)49}$

24. $8\overline{)36}$

25. $6\overline{)45}$

26. $4\overline{)50}$

27. $3\overline{)65}$

28. $2\overline{)34}$

29. $4\overline{)13}$

30. $34 \div 5$

31. $45 \div 3$

32. $29 \div 7$

33. $55 \div 7$

Write the *check* step for each division problem.

34. $4\overline{)96}^{\,24}$

35. $2\overline{)91}^{\,45\ r1}$

36. $6\overline{)76}^{\,12\ r4}$

37. $3\overline{)20}^{\,6\ r2}$

38. **What if** Ellen put 87 toys on 7 shelves? How many toys would be on each shelf? How many toys would be left over?

39. **MENTAL MATH** How can you use the basic fact $12 \div 4 = 3$ to find $13 \div 4$ in your head?

40. $\frac{a+b}{c}$ **Algebra** Ida had 64 marbles. She gave an equal number to 4 friends. Write a number sentence to show how many each friend got.

41. **REASONING** Suppose you find that $42 \div 3$ has no remainder. Can you tell without dividing whether $45 \div 3$ has a remainder? Explain.

Mixed Review and Test Prep

Find the product. (p. 170)

42. $3 \times 3 \times 5 = $ ■

43. $4 \times 2 \times 6 = $ ■

Add or subtract. (p. 88)

44. $\$5.01 - \$3.35 = $ ■

45. $\$7.89 + \$2.45 = $ ■

46. At the zoo, Cathy counted 74 animals. Sam counted half that number. How many animals did the two friends count? (p. 326)

   A 74   **B** 111   **C** 148   **D** 174

47. **TEST PREP** Which is the missing factor? $6 \times $ ■ $ = 48$ (p. 142)

   F 2   **G** 6   **H** 8   **J** 12

Problem Solving Skill
Interpret the Remainder

Understand ▶ Plan ▶ Solve ▶ Check

PICNIC PLANS   Clare needs to take 50 cans of juice to the picnic. The juice is sold in packages of 6 cans. How many packages must she buy?

Since you need to know how many groups of 6 are in 50, you divide.

Find $50 \div 6$.

$$
\begin{array}{r}
8 \text{ r2} \\
6\overline{)50} \\
-48 \\
\hline
2
\end{array}
$$

The 8 packages of juice will have only 48 cans.

If Clare buys 8 packages of juice, she will still need 2 more cans of juice. So, Clare must buy 9 packages of juice.

Rico is making crispy treats for the picnic. He needs 2 cups of cereal for each pan of treats. If he has 13 cups of cereal, how many pans of treats can he make?

Find $13 \div 2$.

$$
\begin{array}{r}
6 \text{ r1} \\
2\overline{)13} \\
-12 \\
\hline
1
\end{array}
$$

Making 6 pans of treats uses 12 cups of cereal.

If Rico makes 6 pans of treats, he will have only 1 cup of cereal left. This is not enough for another pan of treats. So, he can make 6 pans of treats.

MATH IDEA   When you divide, sometimes you have to decide how to use the remainder to solve the problem.

CALIFORNIA STANDARDS NS 2.0 Students calculate and solve problems involving addition, subtraction, multiplication, and division. **MR 1.1** Analyze problems by identifying relationships, distinguishing relevant from irrelevant information, sequencing and prioritizing information, and observing patterns. *also* NS 2.5, NS 2.8, AF 2.1, MR 1.0, MR 2.3, MR 2.4, MR 2.6, MR 3.1, MR 3.2, MR 3.3

1. **What if** Clare had to take 56 cans of juice? If there are 6 cans in each package, how many packages would she need to buy?

2. Simon has 63 bird stamps in a collection. He can fit 8 stamps on a page. How many pages does he need?

3. Beth is making bows to decorate a float in a parade. It takes 5 feet of ribbon to make a bow. She has 89 feet of ribbon. How many bows can Beth make?

4. Dimitri worked on a math puzzle for 32 minutes. If each part of the puzzle took 5 minutes to solve, how many parts did he solve?

A class of 28 students is going on a field trip. Each van can hold 8 students. How many vans does the class need?

5. Which number sentence can help solve the problem?

   A $28 - 28 = 0$
   B $28 \times 8 = 224$
   C $28 \div 8 = 3 \text{ r}4$
   D $8 \div 8 = 1$

6. How many vans does the class need?

   F 3 vans
   G 4 vans
   H 8 vans
   J 28 vans

Mixed Applications

USE DATA For 7–9, use the price list.

7. Tomas bought 1 lb of peanuts and 1 lb of cashews. He paid with $5. How much change did he get?

8. Sheila has $5.36. Does she have enough to buy 1 pound of raisins, 1 pound of peanuts, and 1 pound of cashews? Explain.

9. Cole bought 2 pounds of raisins. The checker charged him $1.19. Is this price reasonable? Explain.

PRICE LIST	
Raisins	$0.85 per lb
Peanuts	$1.99 per lb
Cashews	$2.50 per lb

Review/Test

✔ CHECK VOCABULARY AND CONCEPTS

Choose the best term from the box.

| divisor |
| remainder |

1. In division, the amount left over is called the __?__. (p. 325)

Find the quotient and remainder. You may use counters or draw a picture to help. (pp. 324-325)

2. $11 \div 2 = \blacksquare$ 3. $26 \div 4 = \blacksquare$ 4. $38 \div 5 = \blacksquare$ 5. $14 \div 5 = \blacksquare$

✔ CHECK SKILLS

Use the model. Write the quotient and remainder. (pp. 326-327)

6.
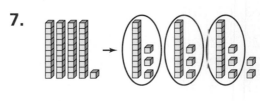

$27 \div 2 = \blacksquare$

7.

$41 \div 3 = \blacksquare$

Divide. (pp. 328-333)

8. $22 \div 3 = \blacksquare$ 9. $35 \div 2 = \blacksquare$ 10. $50 \div 4 = \blacksquare$ 11. $59 \div 5 = \blacksquare$

12. $2\overline{)53}$ 13. $5\overline{)72}$ 14. $6\overline{)82}$ 15. $3\overline{)78}$

16. $8\overline{)69}$ 17. $3\overline{)94}$ 18. $9\overline{)95}$ 19. $4\overline{)71}$

Write the *check* step for each division problem. (pp. 328-333)

20. $3\overline{)25}$ $\overset{8\ r1}{}$ 21. $6\overline{)43}$ $\overset{7\ r1}{}$ 22. $4\overline{)51}$ $\overset{12\ r3}{}$ 23. $2\overline{)44}$ $\overset{22}{}$

✔ CHECK PROBLEM SOLVING

Solve. (pp. 334-335)

24. A class of 33 students is going to the science museum. If each van holds 9 students, how many vans does the class need?

25. Caroline needs 4-inch pieces of yarn for an art project. She has 58 inches of yarn. How many 4-inch pieces can she make?

Cumulative Review

TIP!

Choose the answer.
See item **5.**

All the answer choices show expressions with multiplication. Think how the order property for multiplication relates to equivalent expressions.

Also see problem **6,** p. H65.

For 1–10, choose the best answer.

1. $8\overline{)92}$

   A 10 r2   C 11 r4
   B 11 r2   D 12

2. $58 \div 5 = \blacksquare$

   F 9   H 11
   G 10 r3   J 11 r3

3. There are 47 students going by car to visit the Space Museum. Each car can carry 4 students. How many cars are needed?

   A 7   C 11
   B 10   D 12

4. $0.47
   $+\$0.35$

   F $0.82   H $0.12
   G $0.72   J NOT HERE

5. Which expression is equivalent to 38×9?

   A 8×39 C 9×39
   B 9×38 D 18×9

6. $6\overline{)93}$

   F 10 r3   H 15 r3
   G 15   J 16

7. $30 \div 4 = \blacksquare$

   A 7 r2   C 9
   B 8 r2   D 10

8. David began working on his homework at 4:35 and finished at 6:00. How long did he work?

   F 2 hours 35 minutes
   G 1 hour 25 minutes
   H 1 hour 20 minutes
   J 25 minutes

9. Mr. King's class painted 26 pictures to decorate 3 bulletin boards. Each bulletin board has space for 9 pictures. How many spaces will there be left to fill?

   A 0   C 3
   B 1   D 9

10. Mrs. Carson baked 26 rice cakes. She divided them equally among the 6 members of her family. How many rice cakes were left over?

   F 0   H 2
   G 1   J NOT HERE

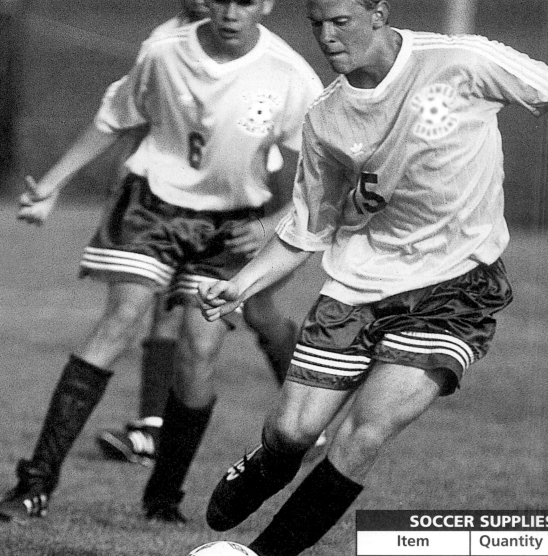

SOCCER SUPPLIES		
Item	Quantity	Cost
Corner flags	8	$64
Goal nets	4	$120
Soccer balls	9	$144
Air pumps	3	$60
Shinguards	6 pairs	$72

Soccer is the world's most popular sport. Millions of people in more than 140 countries play soccer. Soccer as it is played today developed in England in the 1800s. The table lists the costs of some of the equipment used in soccer. Suppose a soccer coach orders 6 pairs of shinguards. How much does each shinguard cost? Explain how to find the answer.

CHECK WHAT YOU KNOW ✓

Use this page to help you review and remember
important skills needed for Chapter 20.

✓ MULTIPLICATION AND DIVISION FACTS (See p. H22.)

Find the product.

1. $6 \times 6 = \blacksquare$ **2.** $\blacksquare = 3 \times 9$ **3.** $6 \times 8 = \blacksquare$ **4.** $7 \times 3 = \blacksquare$

5. $6 \times 7 = \blacksquare$ **6.** $5 \times 5 = \blacksquare$ **7.** $\blacksquare = 2 \times 9$ **8.** $8 \times 8 = \blacksquare$

9. $\blacksquare = 9 \times 7$ **10.** $6 \times 4 = \blacksquare$ **11.** $5 \times 6 = \blacksquare$ **12.** $\blacksquare = 9 \times 9$

13. $8 \times 6 = \blacksquare$ **14.** $7 \times 7 = \blacksquare$ **15.** $4 \times 9 = \blacksquare$ **16.** $8 \times 7 = \blacksquare$

Find the quotient.

17. $12 \div 6 = \blacksquare$ **18.** $63 \div 7 = \blacksquare$ **19.** $\blacksquare = 24 \div 3$ **20.** $30 \div 6 = \blacksquare$

21. $\blacksquare = 35 \div 5$ **22.** $48 \div 6 = \blacksquare$ **23.** $16 \div 4 = \blacksquare$ **24.** $\blacksquare = 81 \div 9$

25. $40 \div 8 = \blacksquare$ **26.** $\blacksquare = 20 \div 5$ **27.** $42 \div 7 = \blacksquare$ **28.** $64 \div 8 = \blacksquare$

29. $8\overline{)56}$ **30.** $9\overline{)45}$ **31.** $4\overline{)32}$ **32.** $7\overline{)49}$

33. $6\overline{)54}$ **34.** $5\overline{)40}$ **35.** $6\overline{)42}$ **36.** $8\overline{)48}$

✓ REGROUP HUNDREDS AND TENS (See p. H22.)

Write the missing number to regroup.

37. 7 hundreds 18 tens = $\blacksquare$ hundreds 8 tens

38. 5 hundreds 23 tens = 7 hundreds $\blacksquare$ tens

39. 1 hundred $\blacksquare$ tens = 2 hundreds 3 tens

40. 4 hundreds 34 tens = $\blacksquare$ hundreds 4 tens

Mental Math: Patterns in Division

Quick Review

1. 2×3
2. 20×3
3. 200×3
4. 300×2
5. What number times 3 equals 900?

▶ Learn

PATTERN SEARCH Ms. Ward showed these problems to her students. She asked them to look for a pattern.

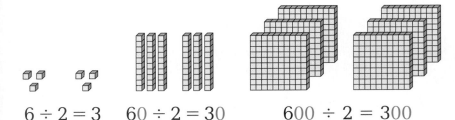

$6 \div 2 = 3$ $60 \div 2 = 30$ $600 \div 2 = 300$

Mallory said the number of zeros in the dividend is the same as in the quotient. Do you agree?

Remember

$14 \div 7 = 2$

$\uparrow$ $\uparrow$ $\uparrow$

dividend   divisor   quotient

Examples

A $25 \div 5 = 5$
$250 \div 5 = 50$
$2,500 \div 5 = 500$

B $12 \div 6 = 2$
$120 \div 6 = 20$
$1,200 \div 6 = 200$

C $40 \div 8 = 5$
$400 \div 8 = 50$
$4,000 \div 8 = 500$

• In Example C, why is there one more zero in the dividend than in the quotient?

MATH IDEA You can use mental math and patterns to help you divide multiples of 10 and 100.

▶ Check

1. **Explain** what happens to the number of zeros in the quotient as the number of zeros in the dividend increases.

 CALIFORNIA STANDARDS NS 2.5 Solve division problems in which a multidigit number is evenly divided by a one-digit number. **MR 1.1** Analyze problems by identifying relationships, distinguishing relevant from irrelevant information, sequencing and prioritizing information, and observing patterns. *also* **AF 1.2, NS 2.0, MR 2.0, MR 2.2, MR 2.3**

Copy and complete. Use patterns and mental math.

2. $8 \div 8 = \blacksquare$

$80 \div 8 = \blacksquare$

$800 \div 8 = \blacksquare$

3. $12 \div 3 = \blacksquare$

$120 \div 3 = \blacksquare$

$1,200 \div 3 = \blacksquare$

4. $20 \div 4 = \blacksquare$

$200 \div 4 = \blacksquare$

$2,000 \div 4 = \blacksquare$

▶ Practice and Problem Solving

Copy and complete. Use patterns and mental math.

5. $35 \div 5 = \blacksquare$

$350 \div 5 = \blacksquare$

$3,500 \div 5 = \blacksquare$

6. $28 \div \blacksquare = 4$

$280 \div 7 = \blacksquare$

$\blacksquare \div 7 = 400$

7. $30 \div 6 = \blacksquare$

$\blacksquare \div 6 = 50$

$3,000 \div \blacksquare = 500$

Use mental math and a basic fact to find the quotient.

8. $700 \div 7$

9. $360 \div 6$

10. $1,600 \div 4$

11. $800 \div 2$

12. $4,800 \div 8$

13. $400 \div 5$

14. $1,200 \div 4$

15. $30 \div 3$

16. $420 \div 7$

USE DATA For 17–18, use the graph.

17. Grace bought a package of balloons. If each balloon costs 5 cents, how much does the package cost?

18. Jorge said he likes the red and blue balloons best. How many more red and blue balloons are in the package than orange and yellow?

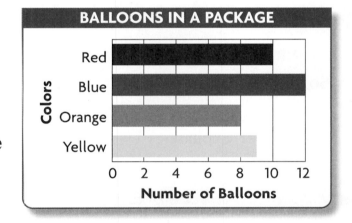

19. $\frac{a+b}{c}$ **Algebra** A grapefruit is 3 times as heavy as an apple. Together they weigh 24 ounces. What does each one weigh?

Mixed Review and Test Prep

20. $16 \div 2 = \blacksquare$

21. $64 \div 8 = \blacksquare$

22. $63 \div 9 = \blacksquare$

23. $45 \div 5 = \blacksquare$ (p. 220)

24. **TEST PREP** Leah started eating lunch at 11:55 A.M. It took her 30 minutes. When did Leah finish eating? (p. 98)

A 12:25 P.M. **B** 12:30 P.M. **C** 12:35 P.M. **D** 12:55 P.M.

Extra Practice page H51, Set A

Quick Review

1. 700 ÷ 7   2. 90 ÷ 3
3. 200 ÷ 4   4. 1,800 ÷ 2
5. 120 ÷ 3

▶ **Learn**

PUNCH LINE Mrs. Allison is making 135 cups of punch. If she divides the punch evenly among 3 punch bowls, about how many cups will be in each bowl?

When you do not need an exact answer, you can *estimate* to find the quotient.

Estimate. $135 \div 3$ or $3\overline{)135}$

STEP 1	**STEP 2**	**STEP 3**
Look at the first two digits.	Think of a basic fact that is close to $13 \div 3$.	Then use the pattern to estimate.
$135 \div 3 = \blacksquare$	$12 \div 3 = 4$	$12 \div 3 = 4$
		$120 \div 3 = 40$

So, about 40 cups of punch will be in each bowl.

• Will the exact answer be more than or less than 40 cups? Explain.

▶ **Check**

1. **Tell** what estimate you would get if you used $15 \div 3 = 5$ as the basic fact to find the number of cups of punch.

Estimate each quotient. Write the basic fact you used to find the estimate.

2. $181 \div 2 = \blacksquare$ 3. $501 \div 5 = \blacksquare$ 4. $374 \div 6 = \blacksquare$

5. $8\overline{)490}$ 6. $3\overline{)268}$ 7. $7\overline{)223}$

CALIFORNIA STANDARDS NS 2.5 Solve division problems in which a multidigit number is evenly divided by a one-digit number. **MR 2.1** Use estimation to verify the reasonableness of calculated results. also **NS 2.0, MR 1.1, MR 2.0, MR 2.2**

Estimate each quotient. Write the basic fact you used to find the estimate.

8. $143 \div 5 = \blacksquare$

9. $174 \div 9 = \blacksquare$

10. $161 \div 2 = \blacksquare$

11. $253 \div 6 = \blacksquare$

Estimate the quotient.

12. $365 \div 5 = \blacksquare$

13. $116 \div 4 = \blacksquare$

14. $493 \div 8 = \blacksquare$

15. $618 \div 6 = \blacksquare$

16. $784 \div 2 = \blacksquare$

17. $407 \div 4 = \blacksquare$

18. $3\overline{)876}$

19. $7\overline{)644}$

20. $9\overline{)716}$

21. $5\overline{)442}$

22. $8\overline{)331}$

23. $8\overline{)552}$

For 24–26, find an estimate to solve each problem.

24. The students raised $226 to buy 3 new trees for school. About how much can they spend for each tree?

25. Kathy put 102 photos into her album. She put 3 photos on each page. About how many pages have photos?

26. There are 4 large tables at the sports banquet. There are 163 team members. About how many team members will be at each table?

27. REASONING Masao saved $1.50 more this week than he saved last week. He saved $2.25 last week. How much money in all has he saved in the two weeks?

28. **?** **What's the Error?** A clerk has 196 shirts to display on 6 racks. She wants to display about the same number of shirts on each rack. She said she can put about 20 shirts on each rack. What's her error?

Mixed Review and Test Prep

Complete. (p. 188)

29. $\blacksquare \times 3 = 30 \div 5$

30. $24 \div \blacksquare = 4 \times 2$

31. $2 \times 2 = \blacksquare \div 4$

32. $14 \div 2 = 1 \times \blacksquare$

33. **TEST PREP** Doug's number pattern starts with 3, 9, 15, 21, 27. Which number will NOT be in Doug's pattern? (p. 136)

A 51   **C** 45

B 49   **D** 39

Place the First Digit in the Quotient

▶ **Learn**

ROLL CALL   There are 6 classes of third graders. Each class has the same number of students. There are 192 third graders in all. How many students are in each class?

Since each class has an equal number of students, you can divide to find the answer.

$$192 \div 6 = \blacksquare \quad \text{or} \quad 6\overline{)192}$$
$$\downarrow$$

Estimate. $180 \div 6 = 30$

STEP 1	STEP 2	STEP 3
Decide where to place the first digit in the quotient.	Divide the 19 tens by 6.	Bring down the 2 ones. Divide the 12 ones.

STEP 1

Decide where to place the first digit in the quotient.

$6\overline{)192}$ $1 < 6$, so look at the tens.

$6\overline{)192}$ $19 > 6$, so use 19 tens.

STEP 2

Divide the 19 tens by 6.

$$\begin{array}{r} 3 \\ 6\overline{)192} \\ -18 \\ \hline 1 \end{array}$$

Multiply.
Subtract.
Compare. $1 < 6$
The difference must be less than the divisor.

STEP 3

Bring down the 2 ones. Divide the 12 ones.

$$\begin{array}{r} 32 \\ 6\overline{)192} \\ -18\downarrow \\ \hline 12 \\ -12 \\ \hline 0 \end{array}$$

Multiply.
Subtract.
Compare. $0 < 6$

So, each class has 32 students.

Since 32 is close to the estimate of 30, the quotient is reasonable.

• How can you use multiplication to check your answer?

MATH IDEA When you divide, you must decide where to place the first digit in the quotient.

CALIFORNIA STANDARDS NS 2.5 Solve division problems in which a multidigit number is evenly divided by a one-digit number. **MR 1.1** Analyze problems by identifying relationships, distinguishing relevant from irrelevant information, sequencing and prioritizing information, and observing patterns. *also* **NS 2.0, MR 2.0, MR 2.1, MR 2.2**

1. **Explain** how an estimate can help you decide how many digits will be in the quotient.

Copy. Place an X where the first digit in the quotient should be.

2. $5\overline{)65}$　　3. $3\overline{)324}$　　4. $4\overline{)56}$　　5. $6\overline{)138}$

▶ **Practice and Problem Solving**

Copy. Place an X where the first digit in the quotient should be.

6. $4\overline{)92}$　　　7. $8\overline{)104}$　　　8. $6\overline{)650}$　　　9. $7\overline{)98}$

Find the quotient.

10. $5\overline{)155}$　　11. $2\overline{)72}$　　12. $3\overline{)84}$　　13. $4\overline{)92}$

14. $8\overline{)680}$　　15. $4\overline{)136}$　　16. $7\overline{)294}$　　17. $5\overline{)755}$

18. $3\overline{)72}$　　19. $4\overline{)472}$　　20. $9\overline{)819}$　　21. $6\overline{)294}$

22. $7\overline{)154}$　　23. $2\overline{)78}$　　24. $5\overline{)90}$　　25. $9\overline{)558}$

26. **Write a problem** in which the dividend has 3 digits and the first digit of the quotient is in the tens place.

27. **? What's the Question?** Ruth's dogs eat 5 pounds of dog food in a week. The answer is 12 weeks.

28. Bonnie has 235 pennies. John has the same amount of money in nickels. How many nickels does John have?

Mixed Review and Test Prep

What time does each clock show? (p. 94)

29.

30.

31.

32.

33. **TEST PREP** Max went to soccer practice 3 days a week for 9 weeks. How many practices did he go to? (p. 122)

　A 3　　　　B 24　　　　C 27　　　　D 30

Extra Practice page H51, Set C

Practice Division of 3-Digit Numbers

Quick Review

1. $56 \div 7$
2. $45 \div 5$
3. $24 \div 6$
4. $72 \div 9$
5. How many twos are in 15? What is left over?

▶ **Learn**

FAIR TRADE Roger and his sister have 358 trading cards. They want to divide them evenly. How many cards does each one get?

$$358 \div 2 = \blacksquare \text{ or } 2\overline{)358}$$

Estimate.　　$400 \div 2 = 200$

STEP 1	**STEP 2**	**STEP 3**	**STEP 4**
Decide where to place the first digit in the quotient.	Divide 3 hundreds by 2.	Bring down the 5 tens. Divide 15 tens.	Bring down the 8 ones. Divide 18 ones.
$\blacksquare$ $2\overline{)358}$ $\quad 3 > 2$, so divide the hundreds.	$\begin{array}{r} 1 \\ 2\overline{)358} \\ -2 \\ \hline 1 \end{array}$ Multiply. Subtract. Compare. $1 < 2$	$\begin{array}{r} 17 \\ 2\overline{)358} \\ -2\downarrow \\ \hline 15 \\ -14 \\ \hline 1 \end{array}$ Multiply. Subtract. Compare. $1 < 2$	$\begin{array}{r} 179 \\ 2\overline{)358} \\ -2 \\ \hline 15 \\ -14\downarrow \\ \hline 18 \\ -18 \\ \hline 0 \end{array}$ Multiply. Subtract. Compare. $0 < 2$

So, each one gets 179 trading cards.

• Is the quotient reasonable? Explain.

Examples

A
$$\begin{array}{r} 85 \\ 5\overline{)425} \\ -40 \\ \hline 25 \\ -25 \\ \hline 0 \end{array}$$

B
$$\begin{array}{r} 116 \\ 6\overline{)696} \\ -6 \\ \hline 09 \\ -6 \\ \hline 36 \\ -36 \\ \hline 0 \end{array}$$

C
$$\begin{array}{r} 93 \\ 3\overline{)279} \\ -27 \\ \hline 09 \\ -9 \\ \hline 0 \end{array}$$

D
$$\begin{array}{r} 138 \\ 7\overline{)966} \\ -7 \\ \hline 26 \\ -21 \\ \hline 56 \\ -56 \\ \hline 0 \end{array}$$

CALIFORNIA STANDARDS NS 2.0 Students calculate and solve problems involving addition, subtraction, multiplication, and division. **NS 2.5** Solve division problems in which a multidigit number is evenly divided by a one-digit number. *also* **MR 2.0, MR 2.1**

▶ Check

TECHNOLOGY LINK

More Practice:
Use Mighty Math
Calculating Crew, *Nick Knack*, Level Q.

1. Discuss how you can decide where to place the first digit in a quotient.

Find the quotient.

2. $5\overline{)365}$ **3.** $3\overline{)231}$ **4.** $4\overline{)336}$ **5.** $2\overline{)568}$

▶ Practice and Problem Solving

Find the quotient.

6. $7\overline{)448}$ **7.** $6\overline{)276}$ **8.** $5\overline{)335}$ **9.** $2\overline{)384}$

10. $5\overline{)145}$ **11.** $7\overline{)847}$ **12.** $6\overline{)486}$ **13.** $4\overline{)644}$

14. $6\overline{)324}$ **15.** $9\overline{)819}$ **16.** $7\overline{)294}$ **17.** $4\overline{)904}$

18. Katie delivers 497 newspapers in 7 days. She delivers the same number each day. How many papers does she deliver each day?

19. **Write About It** Explain why it is important to know the basic division facts when you are dividing 3-digit dividends.

20. **? What's the Error?** Raul wrote this problem to show his sister. What did he do wrong? Correct his error.

$$
\begin{array}{r}
122 \text{ r}4 \\
4\overline{)492} \\
-4 \\
\hline
09 \\
-8 \\
\hline
12 \\
-8 \\
\hline
4
\end{array}
$$

21. REASONING Julia bought some trading cards for $3.00. She paid for them with the same number of dimes as nickels. How many of each coin did she use?

Mixed Applications

22. 305 (p. 42)
$+225$

23. 482 (p. 42)
$+437$

24. $8{,}374$ (p. 62)
$-2{,}196$

25. $5{,}607$ (p. 62)
$-3{,}424$

26. **TEST PREP** Coach Sloan bought 8 baseballs for $32. What was the cost for one baseball? (p. 224)

A $8   **C** $4

B $6   **D** $2

Extra Practice page H51, Set D

HANDS ON
Divide Amounts of Money

MATERIALS
play money

▶ **Explore**

Rafi spent $3.74 for 2 toy cars. Each car was the same price. How much did each car cost?

Use division to find the cost of one car.

$$\$3.74 ÷ 2 = ■ \quad \text{or} \quad 2\overline{)\$3.74}$$

↓

Estimate. $4.00 ÷ 2 = $2.00

STEP 1 Model the amount Rafi spent.

Dollars   **Dimes**

Pennies

STEP 2 Divide the dollars into 2 groups. Regroup 1 dollar as 10 dimes.

STEP 3 Divide the dimes. Regroup 1 dime as 10 pennies.

STEP 4 Divide the pennies.

$1.87 $1.87

So, Rafi spent $1.87 on each car.

Try It

Use play money to model each problem. Record the numbers as you complete each step.

a. $5.26 ÷ 2 = ■   **b.** $3.64 ÷ 4 = ■

You can divide money amounts like whole numbers.
Place the dollar sign and decimal point in the quotient.

Divide. $4.92 ÷ 3 = ■ or 3)$\overline{4.92}$

STEP 1	**STEP 2**	**STEP 3**	**STEP 4**
Divide the 4 dollars.	Bring down the 9 dimes. Divide 19 dimes.	Bring down the 2 pennies. Divide the 12 pennies.	Write the quotient with a dollar sign and a decimal point.
$\begin{array}{r} 1 \\ 3\overline{)\$4.92} \\ -3 \\ \hline 1 \end{array}$ Multiply. Subtract. Compare. $1 < 3$	$\begin{array}{r} 16 \\ 3\overline{)\$4.92} \\ -3\downarrow \\ \hline 19 \\ -18 \\ \hline 1 \end{array}$ Multiply. Subtract. Compare. $1 < 3$	$\begin{array}{r} 164 \\ 3\overline{)\$4.92} \\ -3 \\ \hline 19 \\ -18 \\ \hline 12 \\ -12 \\ \hline 0 \end{array}$ Multiply. Subtract. Compare. $0 < 3$	$\begin{array}{r} \$1.64 \\ 3\overline{)\$4.92} \\ -3 \\ \hline 19 \\ -18 \\ \hline 12 \\ -12 \\ \hline 0 \end{array}$

- Why is it important to keep each digit in the quotient in the correct place-value position?

Find the quotient.

1. 2)$\overline{\$1.76}$ **2.** 3)$\overline{\$8.04}$ **3.** 6)$\overline{\$8.28}$ **4.** 2)$\overline{\$9.26}$

5. 9)$\overline{\$5.13}$ **6.** 8)$\overline{\$9.36}$ **7.** 3)$\overline{\$1.95}$ **8.** 4)$\overline{\$8.48}$

9. REASONING A 5-ounce basket of cherries costs $1.95. Ali's mother said not to buy the cherries unless they are less than $0.50 per ounce. Should Ali buy the cherries? Explain.

10. **? What's the Error?** Bob paid $3.00 for a bag of popcorn that he shared with 4 friends. Bob said each friend owes him $0.06. What's his error?

Mixed Review and Test Prep

Find the missing addend. (p. 68)

11. 25 + ■ = 53 **12.** ■ + 32 = 47

13. ■ + 19 = 38 **14.** 84 + ■ = 101

15. **TEST PREP** What is the missing factor in 3 × ■ × 4 = 24? (p. 142)

A 5   **B** 4   **C** 3   **D** 2

Problem Solving Strategy
Solve a Simpler Problem

Understand ▶ Plan ▶ Solve ▶ Check

PROBLEM Sunrise Elementary School is having its annual field day. There are 640 students in the school. The students are placed in 8 groups to participate in the events. How many students are in each group?

 Understand

• What do you know?

• What are you asked to find?

Plan

• What strategy can you use to solve the problem?

  You can use *solve a simpler problem.*

Solve

• How can you use the strategy *solve a simpler problem*?

  Think: $640 \div 8 = \blacksquare$.
  You can use the basic fact $64 \div 8$ to find $640 \div 8$.
  $64 \div 8 = 8$, so $640 \div 8 = 80$.

So, the students are in 8 groups of 80.

Check

• How can you decide if your answer is correct?

 CALIFORNIA STANDARDS MR 1.2 Determine when and how to break a problem into simpler parts.
NS 2.5 Solve division problems in which a multidigit number is evenly divided by a one-digit number.
also **NS 2.0, MR 1.0, MR 2.0, MR 2.1**

▶ Problem Solving Practice

PROBLEM SOLVING STRATEGIES

Draw a Diagram or Picture
Make a Model or Act It Out
Make an Organized List
Find a Pattern
Make a Table or Graph
Predict and Test
Work Backward
Solve a Simpler Problem
Write an Equation
Use Logical Reasoning

For 1–4, *solve a simpler problem.*

1. **What if** there were 720 students in the school and they needed to be placed into 8 groups? How many students would be in each group?

2. Betty has $4.00 in nickels. How many nickels does she have?

José has $12.00. He wants to buy model plane kits that cost $3.89 each. How many kits can he buy?

3. What simpler problem can José use to help him decide about how many kits he can buy?

   A $12 - 3 = 9$ **C** $12 \times 3 = 36$
   B $3 + 3 = 6$ **D** $12 \div 4 = 3$

4. How many kits can José buy?

   F 2   **G** 3   **H** 4   **J** 5

Mixed Strategy Practice

USE DATA For 5–7, use the table.

5. The Thomas family rented 1 canoe and 2 fishing poles for 3 days each. How much did they spend in all?

6. The Davis family rented a canoe and a raft for 4 days. How much did they spend?

7. **REASONING** The Murray family rented a canoe and 3 fishing poles. They paid $71.00 in all. For how many days did they rent the equipment?

HAPPY HOLLOW CAMP
Daily Rentals

Canoe	$25.00
Raft	$6.25
Fishing pole	$3.50

8. The Clarks arrived home at 5:00 P.M. They had spent 2 hours fishing, 45 minutes driving to the camp, and 45 minutes driving home. At what time did they leave home?

Review/Test

✓ CHECK CONCEPTS

Copy and complete. Use patterns and mental math to help. (pp. 340–341)

1. $6 \div 6 = \blacksquare$
   $60 \div 6 = \blacksquare$
   $600 \div 6 = \blacksquare$

2. $9 \div 3 = \blacksquare$
   $\blacksquare \div 3 = 30$
   $900 \div \blacksquare = 300$

3. $\blacksquare \div 4 = 3$
   $120 \div 4 = \blacksquare$
   $1{,}200 \div \blacksquare = 300$

✓ CHECK SKILLS

Estimate the quotient. (pp. 342–343)

4. $325 \div 4 = \blacksquare$

5. $573 \div 7 = \blacksquare$

6. $268 \div 3 = \blacksquare$

7. $8\overline{)497}$

8. $5\overline{)509}$

9. $9\overline{)362}$

Find the quotient. (pp. 344–349)

10. $2\overline{)24}$

11. $6\overline{)360}$

12. $8\overline{)312}$

13. $2\overline{)180}$

14. $3\overline{)213}$

15. $5\overline{)255}$

16. $4\overline{)252}$

17. $9\overline{)747}$

18. $3\overline{)744}$

19. $6\overline{)678}$

20. $5\overline{)\$2.40}$

21. $8\overline{)\$9.52}$

22. $2\overline{)\$9.78}$

23. $3\overline{)\$8.82}$

✓ CHECK PROBLEM SOLVING

Solve. (pp. 350–351)

24. Ms. Jones stopped after work to buy apples for school lunches. She had $6.00 and wanted to buy 8 apples that cost $0.69 each. Did she have enough money? Explain.

25. The hobby shop ordered 120 model cars. The models come packaged with 4 cars in a box. How many boxes of cars will the shop receive?

Cumulative Review

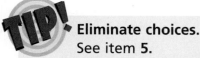

Eliminate choices.
See item **5.**

After you estimate your answer you can eliminate unreasonable choices. Only one answer choice is greater than $1.00. Use your estimate to decide if this is the most reasonable amount.

Also see problem **5,** p. H64.

For 1–10, choose the best answer.

1. 6)‾600‾

    A 10   **C** 100
    B 60   **D** 120

2. 7)‾945‾

    F 135   **H** 149
    G 145   **J** NOT HERE

3. $28 \div 3 = \blacksquare$

    A 8 r3   **C** 9 r1
    B 9   **D** 9 r2

4. 8)‾184‾

    F 20   **H** 25
    G 24   **J** NOT HERE

5. Brandon needs to buy 18 erasers for some friends. Each eraser costs 9 cents. How much money does Brandon need?

    A $0.02   **C** $0.27
    B $0.09   **D** $1.62

6. Jerry bought 8 markers. He paid $6.40 in all. How much did he pay for each marker?

    F $0.80   **H** $8.80
    G $0.88   **J** $80.00

7. Akira has 270 pennies. Suzanne has an equivalent amount in nickels. How many nickels does Suzanne have?

    A 27   **C** 135
    B 54   **D** 270

8. 5)‾$9.35‾

    F $1.05   **H** $1.85
    G $1.80   **J** $1.87

9. Which is the best estimate for this quotient?

               8)‾736‾

    A 9   **C** 900
    B 90   **D** 9,000

10. An airplane has 140 seats. Each row of the plane has 4 seats. How many rows of seats does the airplane have?

    F 35   **H** 60
    G 40   **J** 560

MATH DETECTIVE

Mystery Signs

REASONING Use the clues to help you find the mystery signs. Some cases may have more than one solution.

Hints: Use each number or operation only once.

Complete the operation in ( ) first.

Examples

A Write +, −, ×, or ÷ for each ● so the number sentence is true.

(7 ● 3) ● 8 = 29

Answer:

(7 × 3) + 8 = 29

B Write 3, 42, or 2 for each ■ to make the number sentence true.

(■ ÷ ■) + ■ = 16

Answer:

(42 ÷ 3) + 2 = 16

Cases 1–6

Write +, −, ×, or ÷ for each ● so the number sentence is true.

1. (32 ● 4) ● 1 = 7
2. (40 ● 5) ● 11 = 211
3. (9 ● 39) ● 8 = 6
4. (90 ● 6) ● 1 = 15
5. 49 ● (4 ● 3) = 7
6. 5 ● (50 ● 10) = 25

Cases 7–9

Write a number for each ■ to make the number sentence true.

7. (■ × ■) + ■ = 26   Use 2, 4, and 6.
8. (■ − ■) ÷ ■ = 25   Use 2, 8, and 58.
9. ■ + (■ × ■) = 65   Use 2, 3, and 21.

STRETCH YOUR THINKING Make the number sentence true. Write + or × for each ●. Write 6, 7, or 5 for each ■.

(■ ● ■) ● ■ = 41

CASE CLOSED

UNIT 6

Challenge

Prime Numbers

You can show factors of a number by using square tiles.

Use 16 square tiles. Make arrays to show all the factors of 16.

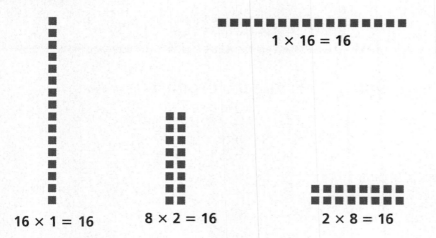

1 × 16 = 16

16 × 1 = 16   **8 × 2 = 16**   **2 × 8 = 16**   **4 × 4 = 16**

So, the number 16 has five factors: 1, 2, 4, 8, and 16.

The number 16 is a **composite number** because it has more than two factors.

Use 7 square tiles. Make arrays to show all the factors of 7.

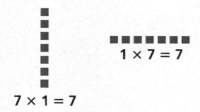

1 × 7 = 7

7 × 1 = 7

The number 7 has only two factors: 1 and 7. So, 7 is a prime number. **Prime numbers** have only 2 factors, 1 and the number itself.

Try It

Use square tiles. Make arrays to show all the factors for each. Write *prime* or *composite*.

1. 6   **2.** 3   **3.** 15   **4.** 20   **5.** 13

Study Guide and Review

VOCABULARY

Choose the best term from the box.

1. In division, the amount left over is called the __?__.

   (p. 324)

> array
> divisor
> remainder

STUDY AND SOLVE

Chapter 17

Multiply 2-digit numbers by 1-digit numbers.

Find the product. $9 \times 36 = \blacksquare$

$$\begin{array}{r} {\scriptstyle 5} \\ 36 \\ \times\ 9 \\ \hline 324 \end{array}$$

- Multiply the ones. Regroup into tens and ones.
- Multiply the tens. Add the regrouped tens.
- Regroup the tens as hundreds and tens.

So, $9 \times 36 = 324$.

Find each product. (pp. 294–299)

2. $\begin{array}{r} 15 \\ \times\ 2 \\ \hline \end{array}$ 3. $\begin{array}{r} 26 \\ \times\ 5 \\ \hline \end{array}$ 4. $\begin{array}{r} 38 \\ \times\ 4 \\ \hline \end{array}$

5. $\begin{array}{r} 42 \\ \times\ 8 \\ \hline \end{array}$ 6. $\begin{array}{r} 67 \\ \times\ 6 \\ \hline \end{array}$ 7. $\begin{array}{r} 88 \\ \times\ 9 \\ \hline \end{array}$

8. There are 24 hours in a day. How many hours are there in 9 days?

Chapter 18

Multiply 3-digit numbers by 1-digit numbers.

Find the product. $7 \times 482 = \blacksquare$

$$\begin{array}{r} {\scriptstyle 51} \\ 482 \\ \times\ 7 \\ \hline 3{,}374 \end{array}$$

- Multiply the ones. Regroup if needed.
- Multiply the tens. Regroup if needed.
- Multiply the hundreds. Regroup if needed.

So, $7 \times 482 = 3{,}374$.

Find each product. (pp. 312–319)

9. $\begin{array}{r} 450 \\ \times\ 2 \\ \hline \end{array}$ 10. $\begin{array}{r} \$2.16 \\ \times\ \ \ 5 \\ \hline \end{array}$

11. $\begin{array}{r} 412 \\ \times\ 3 \\ \hline \end{array}$ 12. $\begin{array}{r} \$6.55 \\ \times\ \ \ 4 \\ \hline \end{array}$

13. $\begin{array}{r} 428 \\ \times\ 7 \\ \hline \end{array}$ 14. $\begin{array}{r} 391 \\ \times\ 1 \\ \hline \end{array}$

Chapter 19

Divide 2-digit dividends.

Divide and check. $97 \div 8 = \blacksquare$

$$
\begin{array}{r}
12 \text{ r1} \\
8)\overline{97} \\
-8\downarrow \\
\hline
17 \\
-16 \\
\hline
1
\end{array}
$$

- Divide the 9 tens.
- Bring down the 7 ones.
- Divide the 17 ones.
- Write the remainder beside the quotient.

So, $97 \div 8 = 12$ r1.
Check: $8 \times 12 = 96$; $96 + 1 = 97$

Divide and check. (pp. 328–333)

15. $54 \div 6 = \blacksquare$ **16.** $17 \div 5 = \blacksquare$

17. $25 \div 4 = \blacksquare$ **18.** $42 \div 3 = \blacksquare$

19. $2)\overline{37}$ **20.** $6)\overline{90}$

21. $4)\overline{45}$ **22.** $5)\overline{74}$

23. $3)\overline{99}$ **24.** $8)\overline{67}$

Chapter 20

Divide 3-digit dividends.

$147 \div 3 = \blacksquare$

$$
\begin{array}{r}
49 \\
3)\overline{147} \\
-12\downarrow \\
\hline
27 \\
-27 \\
\hline
0
\end{array}
$$

$318 \div 2 = \blacksquare$

$$
\begin{array}{r}
159 \\
2)\overline{318} \\
-2\downarrow \\
\hline
11 \\
-10\downarrow \\
\hline
18 \\
-18 \\
\hline
0
\end{array}
$$

So, $147 \div 3 = 49$. So, $318 \div 2 = 159$.

Divide. (pp. 344–349)

25. $2)\overline{166}$ **26.** $3)\overline{291}$

27. $5)\overline{\$2.45}$ **28.** $7)\overline{434}$

29. $6)\overline{306}$ **30.** $8)\overline{736}$

31. $4)\overline{\$4.68}$ **32.** $8)\overline{544}$

33. $9)\overline{801}$ **34.** $3)\overline{\$7.23}$

PRACTICE AND PROBLEM SOLVING

Solve. (pp. 300–301, 334–335)

35. Paul bought 3 slices of pizza for $1.49 each. He had a $10 bill. What operations would you use to find out how much change Paul received? What was Paul's change?

36. Mrs. Ramsey bakes pies at a bakery. She has 52 cups of apples. If 6 cups of apples are needed for one pie, how many pies can she make? Explain.

California Connections

HIGHWAYS

Interstate highways often provide the shortest and quickest way to drive from one place to another.

▲ California has more than 2,300 miles of interstate highways.

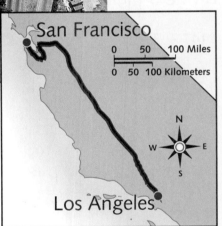

▲ The red line shows the quickest route from San Francisco to Los Angeles.

The driving distance from San Francisco to Los Angeles on interstate highways is 380 miles and takes about 6 hours.

1. How many miles will you travel if you drive on interstate highways from San Francisco to Los Angeles and then back to San Francisco?

2. If you drive 65 miles per hour, how far will you travel in 3 hours?

3. Suppose the driver and three passengers share the driving time equally for the trip from San Francisco to Los Angeles. How many miles will each one drive?

4. **What if** you pay $1.65 for a gallon of gasoline? How much will it cost to put 9 gallons of gasoline in the car?

THE SCENIC BYWAYS

People who are not in a hurry and who like to see the sights prefer to drive on the more scenic byways from one place to another.

The driving distance from San Francisco to Los Angeles on scenic byways along the coast is 460 miles.

▲ **The blue line shows the scenic route from San Francisco to Los Angeles.**

1. Suppose you drive the same number of miles each day and take 4 days to travel from San Francisco to Los Angeles on the scenic byways. How many miles will you drive each day?

2. Santa Cruz is 77 miles from San Francisco. If you drive 40 miles each hour, can you drive from San Francisco to Santa Cruz in 2 hours? Explain.

3. The mission at San Luis Obispo is 254 miles from San Francisco. How many miles is it from this mission to Los Angeles?

4. **What if** you made 3 round trips along the scenic byways from San Francisco to Los Angeles? How many miles would you have driven in all?

Often the scenic byway has many bridges, winding roads, and scenic views. ▼

Solid and Plane Figures

Collecting model trains is a popular hobby. The largest model railway in the world is in New Jersey. It has 8 miles of track and 135 model trains. Look carefully at the photo for shapes that you can identify. Name the different shapes that you see.

CHECK WHAT YOU KNOW ✓

Use this page to help you review and remember
important skills needed for Chapter 21.

✓ IDENTIFY SOLID FIGURES (See p. H23.)

Choose the best term from the box.

SOLID FIGURES
cone
cube
cylinder
square pyramid
rectangular prism
sphere

1.

2.

3.

4.

5.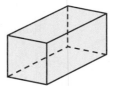

6.

7.

8.

9.

✓ IDENTIFY PLANE FIGURES (See p. H23.)

Choose the best term from the box.

PLANE FIGURES
circle
rectangle
square
triangle

10.

11.

12.

13.

14.

15.

16.

17.

18.

Solid Figures

 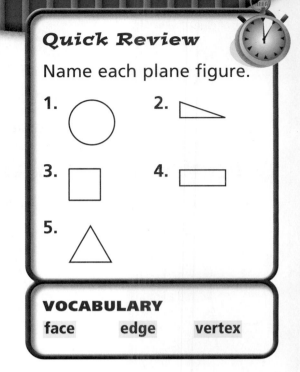
► Learn

FIGURE IT OUT Use names of solid figures to describe objects around you.

cube

rectangular prism

sphere

cylinder

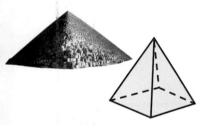
square pyramid

cone

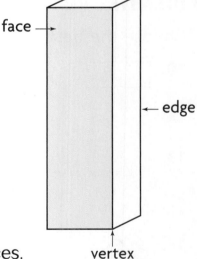
face

edge

vertex

A **face** is a flat surface of a solid figure.

An **edge** is the line segment formed where two faces meet.

A **vertex** is a corner where three or more edges meet. Two or more corners are called vertices.

A rectangular prism has 6 faces, 12 edges, and 8 vertices.

• How many edges does a cube have? a sphere?

• **REASONING** Which solid figures will roll? Explain how you know.

CALIFORNIA STANDARDS MG 2.0 Students describe and compare the attributes of plane and solid geometric figures and use their understanding to show relationships and solve problems. **MG 2.5** Identify, describe, and classify common three-dimensional geometric objects. *also* **MR 1.0, MR 1.1, MR 2.0, MR 2.3, MR 2.4, MR 3.0, MR 3.3**

Tracing Faces

Use names of plane figures to describe the faces of solid figures.

Activity

Materials: solid figures (square pyramid, rectangular prism, cube), paper, crayons

Trace the faces of several solid figures. Then name the faces that make up each solid figure.

STEP 1

On a large sheet of paper, make a chart like the one below. Trace the faces of each solid figure.

Name of Figure	Faces	Names and Number of Faces
Square pyramid	□ △ △ △ △	

STEP 2

Record the names and number of faces for each solid figure.

Name of Figure	Faces	Names and Number of Faces
Square pyramid	□ △ △ △ △	1 square 4 triangles

- **REASONING** Use the words *all*, *some,* or *none* to describe the faces of the solid figures you traced.

MATH IDEA Some solid figures have faces, edges, and vertices. Faces of solid figures are plane figures such as squares, rectangles, and triangles.

TECHNOLOGY LINK

More Practice:
Use E-Lab, *Tracing and Naming Faces.*

www.harcourtschool.com/
elab2002

▶ Check

1. **Explain** how you know a sphere has no faces.

Name the solid figure that each object looks like.

2.

3.

4.

5.

LESSON CONTINUES ▶

Name the solid figure that each object looks like.

6.

7.

8.

9.

Which solid figure has the faces shown? Write *a*, *b*, or *c*.

10.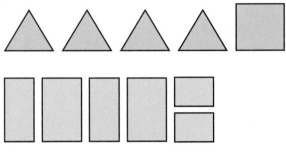

11.

> a. **rectangular prism**
> b. **square pyramid**
> c. **cube**

12.

Copy and complete the table.

	FIGURE	FACES	EDGES	VERTICES
13.	Rectangular prism	■	■	■
14.	Sphere	■	■	■
15.	Cube	■	■	■
16.	Square pyramid	■	■	■

17. REASONING An analogy is a comparison of similar features of objects. For example, *day* is to *light* as *night* is to *darkness*. Complete each analogy.

   a. A cereal *box* is to a *rectangular prism* as a *ball* is to a __?__ .

   b. A *square* is to a *cube* as a *rectangle* is to a __?__ .

18. **Write About It** List objects you might find at a grocery store that look like each of the following solid figures. Think of at least two objects for each figure.

   a. sphere

   b. rectangular prism

   c. cylinder

19. Josh painted a box shaped like a rectangular prism. Each face was a different color. How many colors did Josh use?

20. How are a cube and a rectangular prism alike? How are they different?

21. Write a problem about a solid figure. Give clues about the figure. Exchange with a classmate and decide what the figure is.

Mixed Review and Test Prep

22. $7 \times 4 = $ ▮
(p. 134)

23. $8 \times 0 = $ ▮
(p. 132)

24. $3 \times $ ▮ $= 15$
(p. 142)

25. $4 \times $ ▮ $= 8$
(p. 142)

Find each sum or difference. (p. 88)

26. $\$10.48$
  $+\$\ 6.97$

27. $\$8.36$
  $+\$4.52$

28. $\$9.41$
  $-\$3.73$

29. $\$10.00$
  $-\$\ 1.25$

30. **TEST PREP** How much is one $5 bill, 8 quarters, 2 dimes, 2 nickels, and 1 penny? (p. 84)

A $8.31   **C** $8.06

B $8.26   **D** $7.31

31. **TEST PREP** Wendy says, "The movie starts at ten minutes after six." At what time does the movie start?
(p. 94)

F 5:50   **H** 6:10

G 6:01   **J** 10:06

Thinker's Corner

VISUAL THINKING You can use what you know about the faces of solid figures to model solid figures.

Materials: paper, pencil, scissors, tape

a. Trace the shape at the right onto a piece of paper. Be sure to trace the dotted lines.

b. Cut out the shape along the solid lines.

c. Fold each of the 4 points up on the dotted lines.

d. Tape the edges of the figure together.

  1. Name the figure you made.

  2. Name the faces of the figure.

Combine Solid Figures

▶ Learn

PUT IT ALL TOGETHER Some objects are made up of two or more solid figures put together. Look at the house on the right. What solid figures make up the shape of the house?

Look at each part of the house separately. Think about the solid figures you know.

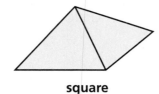

cube   square pyramid   rectangular prism

So, the house is made up of a cube, a square pyramid, and a rectangular prism.

MATH IDEA Solid figures can be combined to make different solid objects.

Examples

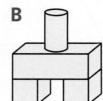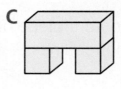

A   B   C

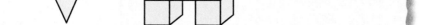

• What solid figures are used to make Object A?

• What solid figures are used to make Object B?

▶ Check

1. Explain how you can make Object B look like Object C.

366

CALIFORNIA STANDARDS MG 2.5 Identify, describe, and classify common three-dimensional geometric objects. **MG 2.6** Identify common solid objects that are the components needed to make a more complex solid object. *also MG 2.0, MR 2.0, MR 2.4, MR 3.0, MR 3.3*

Name the solid figures used to make each object.

2.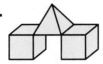

3.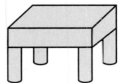

4.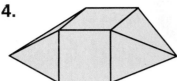

▶ **Practice and Problem Solving**

Name the solid figures used to make each object.

5.

6.

7.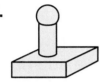

Each pair of objects should be the same. Name the solid figure that is missing.

8.

9.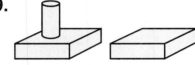

10.

11.

12.

13.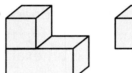

14. Name at least two different faces that make up the solid object at the right.

15. What solid figures could you make if you cut a cube in half?

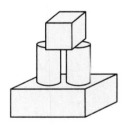

Mixed Review and Test Prep

16. Round 765 to the nearest hundred. (p. 28)

17. Round 1,080 to the nearest thousand. (p. 30)

18. 343 (p. 42)
+239

19. 751 (p. 58)
−390

20. **TEST PREP** Four friends shared 32 crackers equally. How many crackers did each friend get? (p. 202)

A 4   **C** 8
B 7   **D** 28

Extra Practice page H52, Set B

Line Segments and Angles

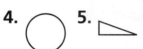

Quick Review

Write the number of sides each figure has.

1. ▭ 2. ☐ 3. △

4. ○ 5. ◁

▶ **Learn**

POINT TO POINT Victor drew this plane figure by connecting points on grid paper. The sides of his figure are made up of line segments.

VOCABULARY

line	ray
point	angle
line segment	right angle

A **line** is straight. It continues in both directions. It does not end.

A **point** is an exact position or location.

point

A **line segment** is straight. It is the part of a line between two points, called endpoints.

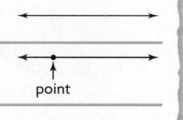

- What plane figure did Victor draw?

- How many sides does the figure have? How many line segments?

- **REASONING** How are lines and line segments alike? How are they different?

CALIFORNIA STANDARDS MG 2.0 Students describe and compare the attributes of plane and solid geometric figures and use their understanding to show relationships and solve problems. **MG 2.4** Identify right angles in geometric figures or in appropriate objects and determine whether other angles are greater or less than a right angle. *also* **⊶MG 2.2, MR 1.1, MR 2.0, MR 2.3, MR 2.4, MR 3.0, MR 3.3**

Identifying Angles

TECHNOLOGY LINK

To learn more about lines, watch the Harcourt Math Newsroom Video, *Georgia Dome*.

A **ray** is a part of a line. It has one endpoint. It is straight and continues in one direction.

An **angle** is formed by two rays with the same endpoint.

A **right angle** is a special angle. It forms a square corner.

Some angles are *less than* a right angle.

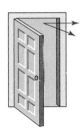  

Some angles are *greater than* a right angle.

 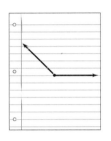 

Look at the photo of Victor.

- What plane figure did Victor draw?

- How many angles does the figure have? How many right angles?

MATH IDEA You can identify line segments and angles in plane figures. You can tell if an angle is a right angle or if it is greater than or less than a right angle.

▶ Check

1. **Draw** a picture of an angle that is greater than a right angle. Explain how you know it is greater than a right angle.

Name each figure.

2.

3.

4.

5.

LESSON CONTINUES ▶

Name each figure.

6.

7. ←——————→

8. •↓

9. ←———•↓

Use a corner of a piece of paper to tell whether each angle is a *right angle, greater than* a right angle, or *less than* a right angle.

10.

11.

12.

13.

14.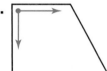

15.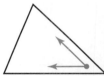

Copy and complete the table.

	Figure	Number of Line Segments	Number of Angles	Number of Right Angles
16.	▢	■	■	■
17.	◯	■	■	■
18.	△	■	■	■
19.	▭	■	■	■
20.	◺	■	■	■
21.	▱	■	■	■

22. Compare the number of line segments and the number of angles for each figure in the table. What pattern do you notice?

23. **Write About It** Draw a right angle and a triangle that has a right angle. Describe the parts of each figure.

24. REASONING Blanca says that a figure with 2 right angles can be a triangle. Do you agree or disagree? Draw a picture to help.

25. List at least 3 objects in the room that contain right angles. How can you be sure the angles are right angles?

Mixed Review and Test Prep

Find each quotient. (p. 220)

26. $16 \div 2 = \blacksquare$ **27.** $36 \div 4 = \blacksquare$

28. $25 \div 5 = \blacksquare$ **29.** $21 \div 3 = \blacksquare$

Find each product. (p. 31)

30. $5 \times 416 = \blacksquare$

31. $3 \times 178 = \blacksquare$

32. $8 \times 409 = \blacksquare$

33. $7 \times 380 = \blacksquare$

34. **TEST PREP** What is the value of the blue digit in 3,495? (p. 6)

A 4,000   C 40

B 400   D 4

35. **TEST PREP** Will bought a magazine for $2.79 and 3 pieces of candy for 9¢ each. How much money did he spend? (p. 172)

F $2.88   H $3.06

G $2.96   J $5.49

THINKER's CORNER

CLOCKS AND ANGLES Use what you learned in this lesson to describe the angles made by the hands of a clock.

Write whether each angle is a *right angle, greater than* a right angle, or *less than* a right angle.

1.

7:45

2.
11:15

3.

9:00

4.

3:30

5. REASONING The hour hand is between 3 and 4. The minute hand is pointing to the 5. What time is it? Describe the angle made by the hands.

6. Joy left for school at 7:50 A.M. It took her 15 minutes to walk to school. What time is it? Describe the angle made by the hands.

Types of Lines

▶ Learn

GET IN LINE Here are some ways to describe the relationships between lines.

Lines that cross are **intersecting lines** . Intersecting lines form angles.

Some intersecting lines cross to form right angles.

Lines that never cross are **parallel lines** . Since parallel lines never cross, they do not form angles.

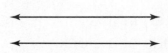

You may see intersecting and parallel lines in the world around you.

REASONING Are the angles in the climbing dome right angles? How can you check?

MATH IDEA You can classify lines as intersecting or parallel. Some intersecting lines cross to form right angles.

▶ Check

1. **Explain** how to tell whether the swing chains are intersecting or parallel.

CALIFORNIA STANDARDS MG 2.0 Students describe and compare the attributes of plane and solid geometric figures and use their understanding to show relationships and solve problems. **MG 2.4** Identify right angles in geometric figures or in appropriate objects and determine whether other angles are creater or less than a right angle. *also* **MR 1.1, MR 2.0, MR 2.3, MR 2.4, MR 3.0, MR 3.3**

Describe the lines. Write *parallel* or *intersecting*.

2.

3. ↑ ↑

4. ⤬

Practice and Problem Solving

Describe the lines. Write *parallel* or *intersecting*.

5. ⤬

6. ⤬

7. ⫽

Describe the intersecting lines. Write *form right angles* or *do not form right angles*.

8.

9.

10.

USE DATA For 11–13, use the map at the right.

11. Which street does not cross Oak Street?

12. Describe the angle the bicycle trail makes at the intersection of Oak Street and Pine Street.

13. **? What's the Question?** The answer is Maple Street.

14. **REASONING** Use what you know about line relationships to describe the sides of a rectangle.

15. **Write About It** Draw and label sets of intersecting and parallel lines. Use a ruler to help. Include lines that cross to form right angles. Describe the angles in each of your drawings.

Mixed Review and Test Prep

Find the quotient. (p. 220)

16. $8\overline{)40}$

17. $7\overline{)49}$

18. $9\overline{)72}$

19. $10\overline{)30}$

20. **TEST PREP** Talia is paid $4 to baby-sit for 1 hour. How much will she earn in 5 hours? (p. 134)

A $9   C $25

B $20   D $30

Extra Practice page H52, Set D

HANDS ON
Circles

▶ **Explore**

Draw two different circles. Follow
Steps 1–4 using large and small paper clips.

Quick Review

Name each figure.

1. •———•   2. •——▶

3. ◀——▶   4. ↑
                   ◀———•

5. •

VOCABULARY

center   diameter
radius

MATERIALS large and
small paper clips, two
pencils, ruler

STEP 1

Draw a point.   •

STEP 2

Draw a circle by placing pencils
in the ends of the paper clip.
The pencil on the point should
not move.

STEP 3

Use a ruler to draw a line segment
from the point to anywhere on
the circle.

STEP 4

Use a ruler to draw a line across the
circle and through the point.

*These are the circles I drew.
How are my circles alike?*

Talk About It

• How are the circles you drew alike?
  How are they different?

CALIFORNIA STANDARDS MG 2.0 Students describe and compare the attributes cf plane and solid geometric
figures and use their understanding to show relationships and solve problems. **MR 2.3** Use a variety of methods, such
as words, numbers, symbols, charts, graphs, tables, diagrams, and models, to explain mathematical reasoning.
also **MR 2.0, MR 2.4, MR 3.0, MR 3.3**

The **center** of a circle is a point in the middle of a circle. It is the same distance from anywhere on the circle.

A **radius** of a circle is a line segment. Its endpoints are the center of the circle and any point on the circle.

A **diameter** of a circle is a line segment. It passes through the center. Its endpoints are points on the circle.

▶ **Practice**

For 1–4, use the circle at the right.

1. What part of the circle is red?

2. What part of the circle is blue?

3. What part of the circle is yellow?

4. Describe the relationship between the radius and the diameter.

5. **? What's the Error?** Ty says that a circle with a diameter of 5 inches is larger than a circle with a radius of 3 inches. Describe his error. Draw a picture to help.

6. **REASONING** Josh makes 3 cuts in a pie. Each cut is a different diameter of the pie. How many pieces does Josh have?

Mixed Review and Test Prep

7. $3 \times 7 = $ ■ (p. 122) **8.** $9 \times 8 = $ ■ (p. 164)

9. $\begin{array}{r} 4,623 \\ +3,957 \\ \hline \end{array}$ (p. 46)

10. $\begin{array}{r} 8,017 \\ -6,459 \\ \hline \end{array}$ (p. 62)

11. **TEST PREP** A sandwich costs $3.55. Don pays with a $5 bill. How much change should Don get? (p. 88)

A $1.45   **C** $2.45

B $1.55   **D** $3.45

Problem Solving Strategy
Break Problems into Simpler Parts

Understand → Plan → Solve → Check

PROBLEM   The patterns on the faces of the Soma Cube puzzle contain squares. How many squares are on all faces of the Soma Cube?

• What are you asked to find?

• What information will you use?

• Is there information you will not use? If so, what?

• What strategy can you use to solve the problem?

You can break the problem into simpler parts.

The Soma Cube puzzle was discovered in 1936 by the Danish poet and scientist Piet Hein.

• How can you use the strategy to solve the problem?

Count the number of squares on one face of the Soma Cube.

Add to find the number of squares on all 6 faces of the Soma Cube.

$14 + 14 + 14 + 14 + 14 + 14 = 84$

So, there are 84 squares on all faces of the Soma Cube.

• How can you decide whether your answer is correct?

CALIFORNIA STANDARDS MR 1.2 Determine when and how to break a problem into simpler parts. **MR 2.2** Apply strategies and results from simpler problems to more complex problems. *also* **MG 2.0, MR 1.0, MR 1.1, MR 2.0, MR 2.3, MR 2.4, MR 3.0, MR 3.1**

Problem Solving Practice

PROBLEM SOLVING STRATEGIES

Draw a Diagram or Picture
Make a Model or Act It Out
Make an Organized List
Find a Pattern
Make a Table or Graph
Predict and Test
Work Backward
▶ **Solve a Simpler Problem**
Write an Equation
Use Logical Reasoning

Break problems into simpler parts to solve.

1. **What if** you count only the smallest squares on the Soma Cube? How many small squares are on all faces of the Soma Cube?

2. The faces of the Soma Cube also contain rectangles like the one below. How many of these rectangles are on all faces of the Soma Cube?

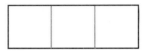

This square pyramid is at the Louvre Museum in Paris, France. The patterns on the faces contain triangles.

3. How many triangles are on one face of the pyramid? Hint: The triangles are not all the same size.

   A 1 triangle   **C** 5 triangles
   B 4 triangles   **D** 10 triangles

4. Which number sentence shows how to find the number of triangles on 4 faces of the pyramid?

   F $10 \times 5 = 50$ **H** $4 \times 4 = 16$
   G $10 + 10 + 10 + 10 = 40$ **J** $1 + 1 + 1 + 1 = 4$

The pyramids were designed by the Chinese American architect Ieoh Ming Pei. The glass lets light into underground halls and rooms of the museum.

Mixed Strategy Practice

USE DATA For 5–7, use the bar graph.

5. Ms. Colmery's class filled 5 rows and 2 extra chairs in the museum auditorium. How many chairs were in each row?

6. **REASONING** Describe at least two ways chairs can be arranged in equal rows for Mr. Leong's class.

7. How many people visited the museum in all? Write a number sentence and solve.

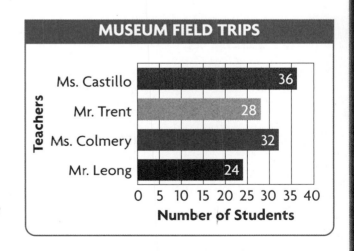

MUSEUM FIELD TRIPS

Ms. Castillo — 36
Mr. Trent — 28
Ms. Colmery — 32
Mr. Leong — 24

Teachers

0 5 10 15 20 25 30 35 40
Number of Students

Review/Test

✓ CHECK VOCABULARY AND CONCEPTS

For 1–2, choose the best term from the box.

face
edge
radius

1. A flat surface of a solid figure is a __?__. (p. 362)

2. A line segment whose endpoints are the center of a circle and any point on the circle is a __?__. (p. 375)

3. A solid figure has 6 square faces. What is it? (pp. 362–365)

4. A solid figure has 6 rectangular faces. What is it? (pp. 362–365)

✓ CHECK SKILLS

5. Name the solid figures used to make this object. (pp. 366–367)

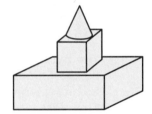

6. Write if each angle is a *right angle, greater than* a right angle, or *less than* a right angle. (pp. 368–371)

a.

b.

Name each figure. (pp. 368–371)

7.   8.   9.   10.

Describe the lines. Write *intersecting* or *parallel*. (pp. 372–373)

11.   12.   13.

✓ CHECK PROBLEM SOLVING

For 14–15, use the figure at the right. (pp. 376–377)

14. Each face of this cube has a checkerboard pattern of blue and white squares. How many blue squares are on all faces of the cube?

15. How many blue and white squares are on all faces of the cube? Explain how you know.

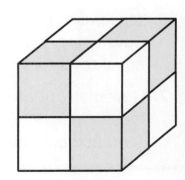

Cumulative Review

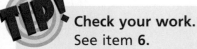 **Check your work.**
See item **6.**
Draw a picture to check your work. Be sure you draw exactly what is described.
Also see problem **7,** p. H65.

For 1–10, choose the best answer.

1. Which solid figure has 6 faces, 12 edges, and 8 vertices?

    A cylinder

    B sphere

    C cone

    D rectangular prism

2. Which solid figure has no faces, no edges, and no vertices?

    F cylinder   **H** cone

    G sphere   **J** cube

3. Which solid figures could describe a box of cereal and a can of peaches?

    A cone and cube

    B cone and sphere

    C rectangular prism and cylinder

    D sphere and cylinder

4. How many angles does a rectangle have?

    F 2   **H** 4

    G 3   **J** 5

5. Which is a name for a plane figure that has exactly 3 line segments?

    A triangle   **C** cube

    B square   **D** cylinder

6. Darren drew a circle. Then he drew a line segment from the center of the circle to a point on the circle. Which word best describes the line segment he drew?

    F center   **H** diameter

    G radius   **J** side

7. Which number sentence shows one way to find the number of faces on 2 cubes?

    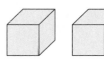

    A $8 \times 2 = 16$

    B $6 + 8 + 6 + 8 = 28$

    C $6 + 6 = 12$

    D $2 + 4 + 8 = 14$

8. George and Joey saved money to buy a gift for their mother. Together they have $17.42. George saved $9.58. How much did Joey save?

    F $27.00   **H** $7.84

    G $8.16   **J** NOT HERE

9. How is 5,016 written in expanded form?

    A $5,000 + 10 + 6$

    B $5,000 + 100 + 6$

    C $5,000 + 160$

    D $500 + 10 + 6$

10. $48 \div 6 = $ ▇

    F 6   **H** 9

    G 7   **J** NOT HERE

Polygons

Traffic signs help drivers know things about the roads they are driving on. Different shapes are used for different types of signs. The stop sign is the only sign in the shape of an octagon. The signs for school zones are in the shape of a pentagon. Yield signs are triangles. Make a list of the different signs you see on your way to school. Tell what geometric shapes are used for the signs.

CHECK WHAT YOU KNOW ✓

Use this page to help you review and remember
important skills needed for Chapter 22.

✓ SIDES AND CORNERS

Tell the number of sides and corners in each figure. (See p. H24.)

1.

2.

3.

4.

5.

6.

7.

8.

9.

✓ GEOMETRIC PATTERNS (See p. H24.)

Name the missing figure in each pattern. Describe
the pattern.

10. △ △ □ △ _?_ □ △ △ □

11. ▭ ○ ▭ ○ ▭ _?_ ▭ ○

12. △ □ □ △ □ □ △ ▭ _?_ △ ▭ □

13. ▨ ⬤ ▨ ⬤ ▨ _?_ ▨ ⬤

14. ▱ △ ▱ ▱ △ △ ▱ ▱ _?_ △ △ △

Polygons

▶ **Learn**

SHAPE UP! A closed figure begins and ends at the same point. An open figure has ends that do not meet. A **polygon** is a closed plane figure with straight sides. The sides of a polygon are line segments.

VOCABULARY

polygon	quadrilateral
pentagon	hexagon
octagon	

polygons

A  B

not polygons

C   D

• Explain why figures C and D are *not* polygons.

MATH IDEA You can name and sort polygons by the number of *sides* or *angles* they have.

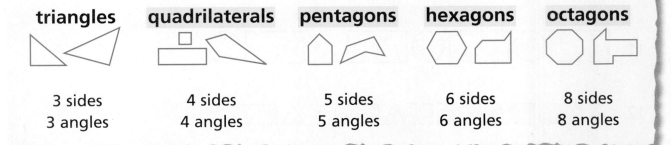

triangles	quadrilaterals	pentagons	hexagons	octagons
3 sides	4 sides	5 sides	6 sides	8 sides
3 angles	4 angles	5 angles	6 angles	8 angles

• What do you notice about the number of sides and the number of angles in polygons?

▶ **Check**

1. **Explain** why a circle is *not* a polygon.

CALIFORNIA STANDARDS O—n **MG 2.1** Identify, describe, and classify polygons (including pentagons, hexagons, and octagons). **MG 2.0** Students describe and compare the attributes of plane and solid geometric figures and use their understanding to show relationships and solve problems. *also* **MG 2.4, MR 2.3, MR 2.4, MR 3.0, MR 3.3**

Tell if each figure is a polygon. Write *yes* or *no*.

2.
3.
4.
5.
6.

▶ Practice and Problem Solving

Tell if each figure is a polygon. Write *yes* or *no*.

7.
8.
9.
10.
11.

Write the number of sides and angles each polygon has. Then name the polygon.

12.
13.
14.
15.
16.

For 17–19, write the letters of the figures that answer the questions.

17. Which are polygons?

18. Which is a quadrilateral?

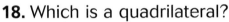
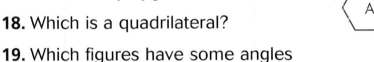

19. Which figures have some angles that are less than a right angle?

20. **REASONING** What is another name for a rectangle? Explain how you know.

21. **Write About It** Draw polygons with 3, 4, 5, 6, and 8 sides. Then label each polygon.

Mixed Review and Test Prep

22. $\begin{array}{r} 466 \\ +527 \end{array}$ (p. 42)

23. $\begin{array}{r} 684 \\ +359 \end{array}$ (p. 42)

24. $\begin{array}{r} 700 \\ -293 \end{array}$ (p. 58)

25. $\begin{array}{r} 900 \\ -564 \end{array}$ (p. 58)

26. **TEST PREP** Dawson got to the soccer field at 3:35 P.M. The game started 40 minutes later. What time did the game start? (p. 98)

   A 3:55 P.M.   **C** 4:15 A.M.
   B 4:00 P.M.   **D** 4:15 P.M.

Extra Practice page H53, Set A

20 feet. Make a bar graph to show the same information.

6. Andrea wants to know how far a garden snail can move in 1 hour. Write a number sentence and solve.

7. **What's the Question?** The answer is 5 times as fast.

8. How long would it take the sloth to move 40 feet?

| Garden Snail | 🕐🕐🕐🕐🕐🕐🕐🕐🕐🕐 |
| Sloth | 🕐🕐 |

Key: Each 🕐 = 2 minutes.

Problem Solving Strategy
Find a Pattern

Understand → Plan → Solve → Check

PROBLEM Karl is making a design with pattern blocks. What will be the next 4 pattern blocks in his design?

CHAPTER 22

Review/Test

✓ CHECK VOCABULARY

Choose the best term from the box.

1. A repeating pattern formed by figures that cover a surface without overlapping or leaving any space between them is a __?__. (p. 388)

2. Any polygon that has 4 sides and 4 angles is a __?__. (p. 382)

3. A closed plane figure with straight sides is a __?__. (p. 382)

> polygon
> quadrilateral
> pentagon
> tessellation

✓ CHECK SKILLS

Write the number of sides and angles each polygon has. Then name the polygon. (pp. 382–383)

4.   5.   6.   7.   8.

Tell if each figure will tessellate. Write *yes* or *no*. (pp. 388–389)

9.   10.   11.   12.

13. How many lines of symmetry does the letter A have? (pp. 384–387)

✓ CHECK PROBLEM SOLVING

Find a pattern to solve. (pp. 390–391)

14. Stan made this pattern. What will be the next 3 shapes in his pattern? Describe his pattern.

15. Alicia wrote this number pattern. Describe the pattern. Then write the next 3 numbers in her pattern.

50, 42, 34, 26, ■, ■, ■

Cumulative Review

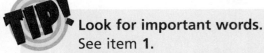

Look for important words.
See item **1.**

Not is an important word. Instead of finding a polygon, you need to look for the one that is **not** a polygon. That means it is **not** a plane figure with straight sides.

Also see problem **2**, p. H63.

For 1–9, choose the best answer.

1. Which figure is **not** a polygon?

A   B   C   D

2. Which describes a quadrilateral?

   F 3 sides, 3 angles
   G 4 sides, 4 angles
   H 5 sides, 5 angles
   J 6 sides, 6 angles

3. Which figure is congruent to this figure?

A   B   C   D

4. Which numbers would be next in this pattern?

25, 21, 17, 13

   F 12, 11, 10   **H** 9, 7, 5
   G 11, 9, 7   **J** 9, 5, 1

5. How many lines of symmetry does this figure have?

   A 1   **C** 3
   B 2   **D** 4

6. Which letter does **not** have a line of symmetry?

   F P   **H** O
   G M   **J** V

7. Stanley is going to the basketball game. He leaves his house at 6:45 P.M. The game starts at 7:30 P.M. How much time will Stanley have to get to the game?

   A 35 minutes   **C** 45 minutes
   B 40 minutes   **D** 1 hour

8. The Redbrook School had a book fair. The sales were $43.76 the first day, $21.98 the second day, and $56.32 the last day. What was the total amount of sales?

   F $122.96   **H** $121.06
   G $122.06   **J** NOT HERE

9. What is five thousand, fifteen written in standard form?

   A 5,015   **C** 15,500
   B 5,050   **D** NOT HERE

Triangles and Quadrilaterals

A quilt is made by sewing together pieces of cloth to form a design or pattern. In pioneer days, blankets were very expensive and not easy to find. Many women used their leftover pieces of cloth to make quilts. Look at the quilt pictured here. How many shapes do you see? Copy and complete the table.

SHAPES ON THE QUILT

Shape	Number
Square	■
Triangle	■

CHECK WHAT YOU KNOW

Use this page to help you review and remember
important skills needed for Chapter 23.

✓ CONGRUENT FIGURES (See p. H25.)

Tell whether the two figures are congruent. Write
yes or *no*.

1.

2.

3.

4.

5.

6.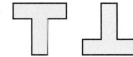

7.

8.

9.

✓ TYPES OF ANGLES (See p. H25.)

Write if each angle is a *right angle, greater than* a
right angle, or *less than* a right angle.

10.

11.

12.

13.

14.

15.

1 Triangles

▶ Learn

TRIPLE PLAY   Polygons with 3 sides and 3 angles are triangles.

triangles   **not triangles**

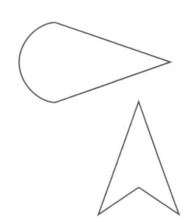

You can describe any angle in a triangle as a right angle, greater than a right angle, or less than a right angle.

Quick Review

Write the number of sides and angles each polygon has.

1.   2.

3.   4.

5.

Remember
A right angle forms a square corner.

Examples

A

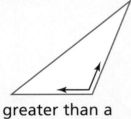

greater than a
right angle

B

less than a
right angle

C

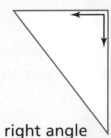

right angle

• How can you check if an angle is greater than or less than a right angle?

▶ Check

1. **Describe** the other two angles in Example C above.

CALIFORNIA STANDARDS ○┐ **MG 2.2** Identify attributes of triangles. **MG 2.4** Identify right angles in geometric figures or in appropriate objects and determine whether other angles are greater or less than a right angle. *also* **MG 2.0, MR 2.3, MR 2.4**

Write if each angle is a *right angle, greater than* a right angle, or *less than* a right angle.

2.
3.
4.
5.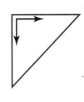
6.

▶ Practice and Problem Solving

Write if each angle is a *right angle, greater than* a right angle, or *less than* a right angle.

7.
8.
9.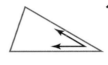
10.
11.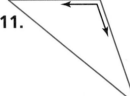

For 12, use the photo at the right.

12. This shape was folded from a piece of paper. How many triangles do you see on the shape?

13. **Write About It** Name 2 ways triangles A and B are alike. Name 2 ways they are different.

14. Joan folded a square piece of paper in half to make a triangle. Describe what you know about the triangle.

Mixed Review and Test Prep

15. Andy went on a trip from September 2 to September 23. How many weeks was his trip? (p. 102)

September						
Sun	Mon	Tue	Wed	Thu	Fri	Sat
1	2	3	4	5	6	7
8	9	10	11	12	13	14
15	16	17	18	19	20	21
22	23	24	25	26	27	28
29	30					

16. $\begin{array}{r} 18 \\ \times\ 2 \\ \hline \end{array}$
(p. 294)

17. $\begin{array}{r} 80 \\ \times\ 4 \\ \hline \end{array}$
(p. 294)

18. $\begin{array}{r} 239 \\ \times\ 3 \\ \hline \end{array}$
(p. 312)

19. **TEST PREP** James had 8 roses. He put an equal number in 8 vases. Which number sentence shows how to find the number of roses in each vase? (p. 194)

A $8 - 8 = 0$ **C** $8 \div 1 = 8$

B $8 \div 8 = 1$ **D** $8 + 8 = 16$

Extra Practice page H54, Set A

▶ **Learn**

TIME FOR TRIANGLES Beverly and Armando sorted these triangles in different ways.

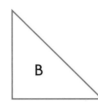
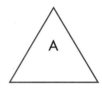

This is how Beverly sorted the triangles.

Quick Review

Write if each angle is a *right angle, greater than a right angle,* or *less than a right angle.*

M.R.A

1. L.R.A   2.

3. R.A   4.

5.

L.L.A

R.A

VOCABULARY

| equilateral | scalene |
| isosceles | right triangle |

This is how Beverly sorted the triangles.

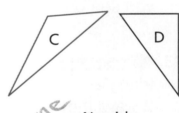

All sides   Two sides   scalene   No sides
are equal.   are equal.   are equal.

This is how Armando sorted the triangles.

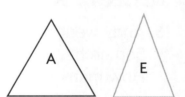

One angle is   One angle is greater   All angles are less
a right angle.   than a right angle.   than a right angle.

• How did Beverly sort the triangles? How did Armando sort the triangles?

• How would Beverly sort triangle F? How would Armando sort it?

CALIFORNIA STANDARDS ⊙━┓ **MG 2.2** Identify attributes of triangles. **MG 2.4** Identify right angles in geometric figures or in appropriate objects and determine whether other angles are greater or less than a right angle. *also* **MG 2.0, MR 1.1, MR 2.3, MR 2.4**

Name Triangles

You can name triangles by the number of equal sides they have.

equilateral

2 cm G 2 cm

2 cm

3 equal sides

isosceles

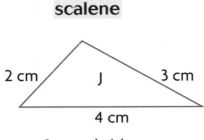

3 cm H 3 cm

2 cm

2 equal sides

scalene

2 cm J 3 cm

4 cm

0 equal sides

You can sort triangles by their angles.

right triangle

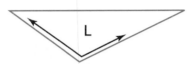

K

1 right angle

L

1 angle greater than a right angle

M

3 angles less than a right angle

- How are triangles J and M alike? How are they different?

MATH IDEA You can sort triangles by their sides or their angles.

▶ Check

1. **Describe** triangle N by its sides. Then describe it by its angles.

right Angle

N

For 2–5, use the triangles at the right. Write *O, P,* and *Q.*

2. Which triangles have 3 sides and 3 angles?

3. Which triangle is isosceles?

4. Which triangle is a right triangle?

5. Which triangles have 3 angles that are less than a right angle?

L, R

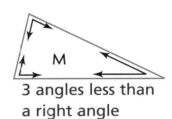

3 cm O 5 cm

4 cm

isosceles

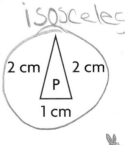

2 cm P 2 cm

1 cm

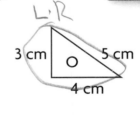

2 cm Q 2 cm

2 cm

TECHNOLOGY LINK

More Practice: Use Mighty Math Number Heroes, *Geoboard,* Level Q.

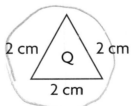

LESSON CONTINUES ▶

► Practice and Problem Solving

For 6–8, use the triangles at the right.
Write A, B, or C.

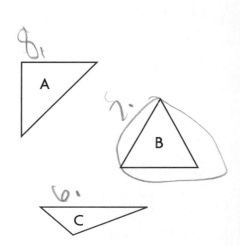

6. Which triangle is scalene?

7. Which triangles have at least 2 equal sides?

8. Which triangle has 1 angle that is greater than a
   right angle?

**For 9–12, write one letter from each box to
describe each triangle.**

a. equilateral	**d.** It has 1 right angle.
b. isosceles	**e.** It has 1 angle greater than a right angle.
c. scalene	**f.** All angles are less than a right angle.

9.
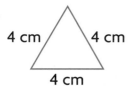

10.
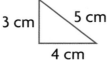

11.

12.
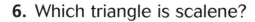

For 13–18, name each triangle. Write _equilateral, isosceles,_ or _scalene._

13.
4 cm   4 cm
4 cm

14.
4 cm   6 cm
8 cm

15.
2 cm   2 cm
3 cm

16.
3 cm
3 cm   3 cm

17.
4 cm   4 cm
3 cm

18.
3 cm   2 cm
4 cm

USE DATA For 19–20, use the diagram at the right.

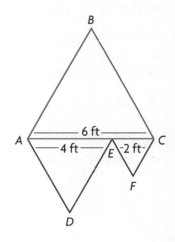

19. Mrs. Liu has a garden with paths that are equilateral
    triangles. The shortest path from A to C is the
    green path. Which path is longer: the blue path or
    the red path? Explain.

20. **? What's the Question?** The answer is 4 feet longer.

21. A right triangle has sides that are 3 inches, 4 inches, and 5 inches. Is this triangle equilateral, isosceles, or scalene? Explain.

22. Draw a triangle with 2 equal sides and one right angle. You may use grid paper to help. Name the triangle.

Mixed Review and Test Prep

23.
$$6 \\ \times 8$$
(p. 148)

24.
$$9 \\ \times 7$$
(p. 164)

25.
$$3 \\ \times 9$$
(p. 122)

26. $5\overline{)35}$
(p. 200)

27. $8\overline{)48}$
(p. 214)

28. $7\overline{)49}$
(p. 214)

For 29–30, use the bar graph. (p. 252)

29. **TEST PREP** Which number sentence shows how to find how many more people went to the Summer Concert than to the Storytelling Festival?

A $120 - 90 = 30$ **C** $120 - 40 = 80$
B $90 - 40 = 50$ **D** $90 + 40 = 130$

30. **TEST PREP** About 40 more people are expected to attend the Summer Concert next year. Choose the best number of seats to set up for the concert.

F 50   **H** 120
G 80   **J** 130

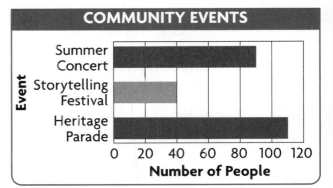

COMMUNITY EVENTS

--LINKUP---
to Art

Many artists use triangles in their works. The Native American blanket at the right uses different kinds of triangles. Find and name as many triangles on the blanket as you can.

Materials: grid paper, pencil, crayons

1. Draw and color a blanket design on grid paper. Use different kinds of triangles in your design.

2. Trade designs with a classmate and describe the triangles you see.

Quadrilaterals

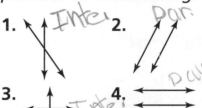
▶ Learn

LIMIT OF FOUR Polygons with 4 sides and
4 angles are quadrilaterals.

quadrilaterals   **not quadrilaterals**

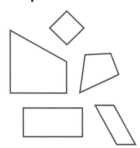 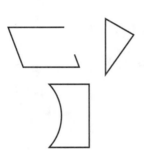 

An angle in a quadrilateral can be a right angle
or greater than or less than a right angle. The
sides of a quadrilateral can be parallel.

Examples

A

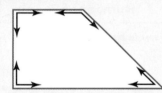

2 right angles;
1 angle is less than a right angle;
1 angle is greater than a right angle;
1 pair of parallel sides.

B

4 right angles;
2 pairs of parallel sides.

• How can you check if two sides of a
  quadrilateral are parallel?

▶ Check

1. Draw two different quadrilaterals that have
2 parallel sides. You may use grid paper to help.
Explain how the quadrilaterals are different.

CALIFORNIA STANDARDS O⎯π**MG 2.3** Identify attributes of quadrilaterals. **MG 2.4** Identify right angles in
geometric figures or in appropriate objects and determine whether other angles are greater than or less than a right
angle. *also* **MG 2.0, MR 2.3, MR 2.4**

Describe the angles and sides of each quadrilateral.

2.

~~ISOS~~

3.

ISOS

4.

ISOS

▶ Practice and Problem Solving

Describe the angles and sides of each quadrilateral.

5.

quadrilateral

6.

Scalene

7.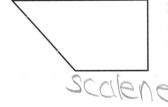

scalene

For 8–11, use the quadrilaterals at the right.
Write *true* or *false* for each statement.

8. All of the quadrilaterals have 4 sides and 4 angles.

9. None of the quadrilaterals have right angles.

10. Some of the quadrilaterals have 6 sides.

11. Some of the quadrilaterals have parallel sides.

12. ? **What's the Error?** Mandy says that the blue figure is a quadrilateral. Describe Mandy's error. There's only side that are even

13. ✏️ Write a problem using the figure below. Trade problems with a classmate and solve.

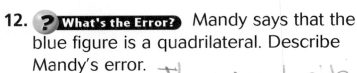

14. Use grid paper. Draw a quadrilateral with only one angle greater than a right angle. (HINT: include at least one right angle.)

─ *Mixed Review and Test Prep* ─

Find the quotient and remainder.

(p. 324)

15. 7)45

6 R3
45
42
03

16. 6)52

8 R6
8 × 6
54

17. 57 ÷ 8 = 7 R1

18. 39 ÷ 4 = 9 R7

19. **TEST PREP** Find the missing factor.

■ × 7 = 56 (p. 142)

A 7   C 9

Ⓑ 8   D 49

Extra Practice page H54, Set C

Sort Quadrilaterals

Quick Review

Describe the intersecting lines. Write *form right angles* or *do not form right angles.*

1.   2.

3.   4.

5.

▶ Learn

FOUR BY FOUR   Any polygon with 4 sides and 4 angles is a quadrilateral. There are special names for some kinds of quadrilaterals.

Activity

Materials: quadrilateral worksheet, scissors
- Cut out the quadrilaterals from your worksheet.

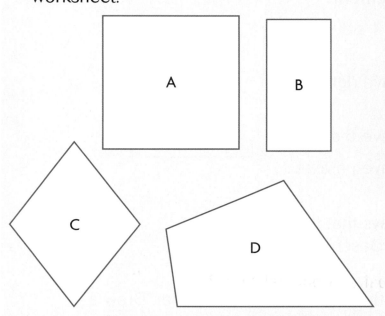

- Sort the quadrilaterals into the groups listed below. Some quadrilaterals may be in more than one group.
  a. have right angles
  b. do not have right angles
  c. have parallel sides
  d. do not have parallel sides

- Which quadrilaterals are in more than one group? Explain.

VOCABULARY
parallelogram
rhombus

 CALIFORNIA STANDARDS ○━□**MG 2.3** Identify attributes of quadrilaterals. **MG 2.4** Identify right angles in geometric figures or in appropriate objects and determine whether other angles are greater or less than a right angle. *also* **MG 2.0, MR 1.1, MR 2.3, MR 2.4**

Name Quadrilaterals

Here are some names of quadrilaterals with parallel sides.

parallelograms	**rhombuses or rhombi**	**rectangles**	**squares**

  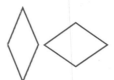 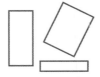 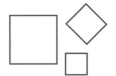

2 pairs of parallel sides 2 pairs of equal sides	2 pairs of parallel sides 4 equal sides	2 pairs of parallel sides 2 pairs of equal sides 4 right angles	2 pairs of parallel sides 4 equal sides 4 right angles

- How are a rectangle and a square alike? How are they different?

MATH IDEA You can name and sort quadrilaterals by looking at their sides and angles.

▶ Check

1. **Describe** the sides and angles of this quadrilateral. What is another name for it?

For 2–4, use the quadrilaterals at the right.
Write *E, F, G, H,* and *J.*

2. Which quadrilaterals have 2 pairs of parallel sides?

3. Which quadrilaterals have 2 or more right angles?

4. How are quadrilateral E and quadrilateral G alike? How are they different?

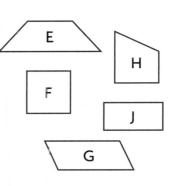

Write as many names as you can for each quadrilateral.

5.

6.

7.

8.

9.

10.

LESSON CONTINUES ▶

**For 11–13, use the quadrilaterals at the right.
Write A, B, C, D, and E.**

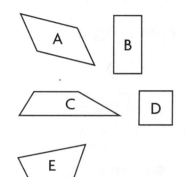

11. Which quadrilaterals have 2 pairs of equal sides?

12. Which quadrilaterals have no right angles?

13. How are quadrilateral A and quadrilateral D alike?
How are they different?

**For 14–19, write as many names as you can for each
quadrilateral.**

14. 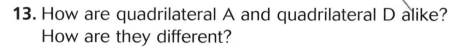 **15.**   **16.**

17.   **18.**   **19.**

**For 20–23, write *all* the letters that describe
each quadrilateral.**

20.   **21.**

22. 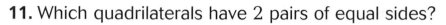 **23.**

a. It has 4 equal sides.

b. It has 2 pairs of parallel sides.

c. It has 4 right angles.

d. It has 2 pairs of equal sides.

24. REASONING How is the blue figure
like the figures to its right?

25. I have 4 equal sides and 4 right
angles. What am I?

26. I have 5 sides and 5 angles.
What am I?

27. REASONING Can a quadrilateral
have no equal sides and no
parallel sides? Draw a picture to
help you explain.

28. **Write About It** Draw and label
4 different quadrilaterals on grid
paper. Explain how each is
different.

29. Akemi sees a tile with 4 right angles. She says it must be a square. Do you agree or disagree? Explain.

30. Dante drew a quadrilateral with 4 right angles and 2 pairs of parallel sides. What could he have drawn?

Mixed Review and Test Prep

Which number is greater? (p. 20)

31. 9,362 or 9,529

32. 5,108 or 5,018

Which number is less? (p. 20)

33. 2,814 or 2,148   **34.** 8,730 or 998

35. Round 6,398 to the nearest thousand. (p. 30)

36. Maurice will meet his mother in 30 minutes. What time will it be? (p. 94)

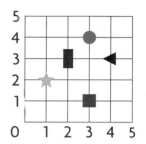

For 37–38, use the grid. (p. 260)

37. **TEST PREP** Which is the ordered pair for the triangle?

A (1,2) **B** (2,3) **C** (3,4) **D** (4,3)

38. **TEST PREP** Which shape is found at (3,4) on the grid?

F square   **H** rectangle
G circle   **J** triangle

--LINKUP---
to Reading

STRATEGY • USE GRAPHIC AIDS Graphic aids such as charts, diagrams, and maps help display information visually. Sometimes the information needed to solve a problem is given in a graphic aid. The diagram at the right shows that all squares are quadrilaterals.

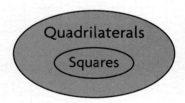

1. What does this diagram show?

2. What does this diagram show?

3. Erin says a square is always a rectangle but a rectangle is not always a square. Do you agree or disagree? Explain.

4. Make a diagram to show the relationship among polygons, quadrilaterals, and parallelograms.

Extra Practice page H54, Set D

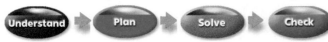

LESSON **5**

Problem Solving Skill

Identify Relationships

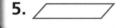

Understand → Plan → Solve → Check

WHAT'S IN A NAME? Sam drew the polygon below. What are all the ways to name the polygon? What is the best name for the polygon?

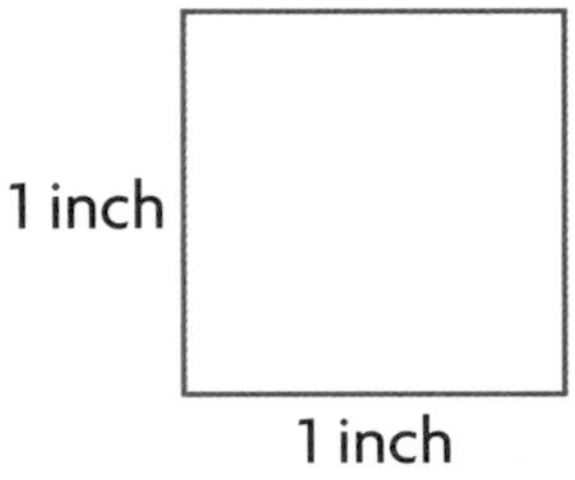

Quick Review

Write the best name for each quadrilateral.

1. 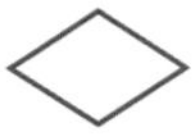

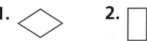

2. 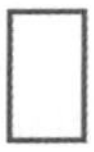

3. 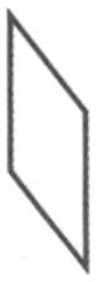

4. 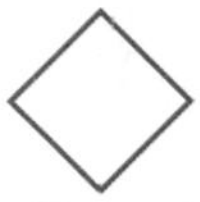

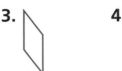

5. 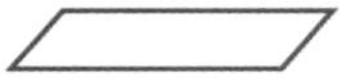

MATH IDEA You can identify relationships among figures by comparing their characteristics.

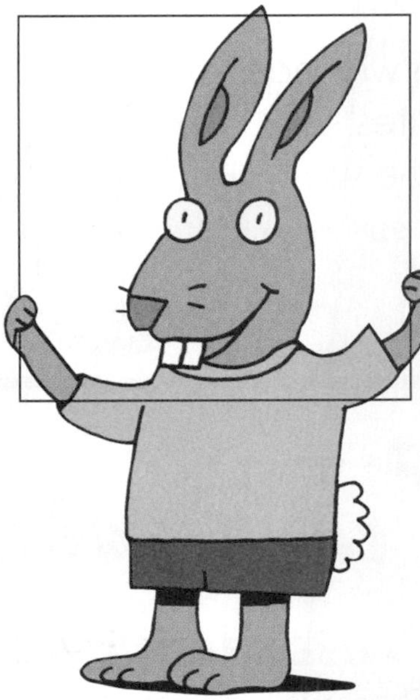

CHARACTERISTICS	NAME
4 sides, 4 angles	quadrilateral
2 pairs of parallel sides	parallelogram
4 right angles	rectangle
4 equal sides	square

So, the polygon is a quadrilateral, a parallelogram, a rectangle, and a square. The best name for the polygon is square.

Talk About It

- Explain how you know that *square* is the best name for Sam's figure.
- **REASONING** Every rectangle is a parallelogram. Is every parallelogram a rectangle? Explain.

CALIFORNIA STANDARDS MR 1.1 Analyze problems by identifying relationships, distinguishing relevant from irrelevant information, sequencing and prioritizing information, and observing patterns. **MG 2.3** Identify attributes of quadrilaterals. *also* **MG 2.0, MG 2.4, MR 1.0, MR 2.0, MR 2.3, MR 2.4, MR 3.0, MR 3.3**

1. What are all the ways to name the polygon at the right? What is the best name for the polygon?

2. **Write About It** Draw a polygon with at least 2 equal sides. Label your figure with as many names as you can. Circle the best name for your figure and explain why you chose it.

For 3–4, use the figures at the right.

3. Choose the best names for the figures.

   A quadrilateral and square
   B quadrilateral and rhombus
   C parallelogram and rectangle
   D square and parallelogram

4. Which statement is *not true*?

   F They are quadrilaterals.
   G They are polygons.
   H They are parallelograms.
   J They are triangles.

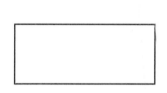

Mixed Applications

USE DATA For 5–6, use the line plot at the right.

5. The X's on the line plot stand for students. How many students saw more than 4 movies last year?

6. What is the range for this set of data? What is the mode?

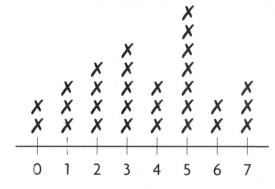

Movies Seen Last Year

7. There were 438 people on a train. At the station, 113 people got off and 256 people got on. How many people are on the train now?

8. Mimi bought a sandwich for $2.75 and a carton of milk for $1.25. She paid with a $5 bill. How much change should she get?

Review/Test

✓ CHECK VOCABULARY

Choose the best term from the box.

1. A triangle with 3 equal sides is __?__ . (p. 399)

equilateral
scalene
isosceles

✓ CHECK SKILLS

Write if each angle is a *right angle, greater than* a
right angle, or *less than* a right angle. (pp. 396–397)

2.   3.   4.   5.   6.

Name each triangle. Write *equilateral, isosceles,* or
scalene. (pp. 398–401)

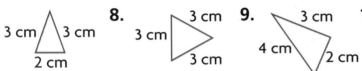

7.   3 cm   3 cm / 2 cm
8.   3 cm   3 cm / 3 cm
9.   3 cm   4 cm / 2 cm
10.   3 cm 2 cm / 3 cm
11.   3 cm   4 cm / 5 cm

Describe the angles and sides of each quadrilateral. (pp. 402–403)

12.   13.   14.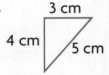

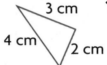

Write the name for each quadrilateral. (pp. 404–407)

15.   16.   17.   18.   19.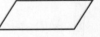

✓ CHECK PROBLEM SOLVING

Solve. (pp. 408–409)

20. What are all the ways to name the polygon at the
right? What is the best name for the polygon?

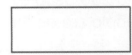

Cumulative Review

Eliminate choices.
See item **3**.

Eliminate any figures that do not have 4 sides. Then eliminate any figure that does not always have all sides equal.

Also see problem **5**, p. H64.

For 1–10, choose the best answer.

1. In which figure is each angle less than a right angle?

A

C

B

D

2. Which figure always has 2 pairs of parallel sides?

  F triangle   **H** quadrilateral
  G circle   **J** square

3. Which figure always has 4 equal sides?

  A rectangle   **C** triangle
  B rhombus   **D** parallelogram

4. Margaret ate breakfast at 7:15 A.M. She left the house at 7:40 A.M. and arrived at school 20 minutes later. What time did Margaret get to school?

  F 7:35 A.M.   **H** 8:00 A.M.
  G 7:55 A.M.   **J** 8:05 A.M.

5. What is the value of the 5 in 35,268?

  A 500   **C** 5,000
  B 1,000   **D** 50,000

6. Which angle is greater than a right angle?

F   H

G   J

7. Pablo was at summer camp every day for 6 weeks. How many days did Pablo spend at camp?

  A 60   **C** 30
  B 42   **D** NOT HERE

8. Jamie spent $17 on a book bag. He still had $9 when he got home. How much money did Jamie have before he bought the book bag?

  F $36   **H** $8
  G $26   **J** NOT HERE

9. Which best describes the figure?

  A parallelogram   **C** rhombus
  B rectangle   **D** square

10. Which figure could never be named a quadrilateral?

  F triangle   **H** rectangle
  G parallelogram   **J** square

Study Guide and Review

VOCABULARY

Choose the best term from the box.

1. A closed plane figure with straight sides is a __?__ . (p. 382)

2. A corner of a solid figure where three or more edges meet is called a __?__ . (p. 362)

> face
> polygon
> vertex

STUDY AND SOLVE

Chapter 21

Describe solid figures.

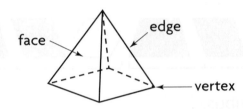

This square pyramid has 5 faces, 8 edges, and 5 vertices. One face is a square, and the other four faces are triangles.

For 3–5, use the figure. (pp. 362–365)

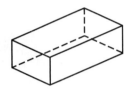

3. How many faces, edges, and vertices does the figure have?

4. What shape are the faces of this figure?

5. What is this figure called?

Tell if an angle is a right angle, greater than a right angle, or less than a right angle.

This angle is a **right angle**. It forms a square corner.

Describe the angles below.

 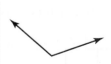

This angle is *less than* a right angle. | This angle is *greater than* a right angle.

Write whether the angle is a *right angle, greater than* a right angle, or *less than* a right angle. (pp. 368–371)

6.   7.

8.   9.

Chapter 22

Identify polygons.

What is this polygon?

It is an **octagon**.
It has 8 sides and 8 angles.

Write the number of sides and angles each polygon has. Then name the polygon. (pp. 382–383)

10.   11.

Chapter 23

Classify triangles and quadrilaterals.

Which figure is an equilateral triangle?

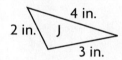

Figure K is an equilateral triangle. It has 3 equal sides.

Which figure is a parallelogram?

Figure Q is a parallelogram. It has 2 pairs of equal sides.

For 12–15, use the figures. (pp. 396–407)

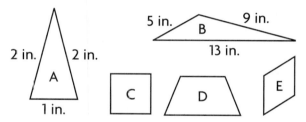

12. Which is a scalene triangle?

13. Which is an isosceles triangle?

14. Which quadrilaterals have 2 pairs of parallel sides?

15. How many right angles does figure C have? figure D?

PROBLEM SOLVING PRACTICE

Solve. (pp. 376–377, 408–409)

16. Jeff put a different letter on each small square of this cube. How many letters did he use?

17. Compare a rectangle and a parallelogram. Describe their sides and angles to tell how they are alike and different.

California Connections

LANDMARKS

Buildings often suggest geometric figures. Look for examples of solid figures in these California landmarks.

▲ **The bell tower at the University of California at Berkeley**

▲ **The Transamerica Pyramid in San Francisco**

1. Name a solid figure suggested by each building.

2. Choose one of the figures you named. Tell how many faces, vertices, and edges the solid figure has.

3. Draw each of the plane figures that are faces of the solid figure you chose. Name each face.

4. Choose one of the faces from above. Write three sentences to describe the figure. Use as many of the words in the box below as you can.

triangle	parallelogram	line segment	right angle	square	point
polygon	quadrilateral	rectangle	angle	intersecting	

MISSIONS

Mission Santa Barbara and Mission San Carlos Borroméo del Rio Carmelo are 2 of the 21 missions built along the California coast between 1769 and 1823.

◀ **Mission Santa Barbara, called Queen of the Missions, is famous for its bell towers.**

◀ **Mission San Carlos Borroméo del Rio Carmelo, called Carmel Mission, has a museum and beautiful gardens.**

1. Look at the diagrams of the fronts of the missions. Which diagram shows symmetry? Trace the diagram and draw the line of symmetry.

2. **REASONING** Look for congruent figures on each diagram. Trace a pair of congruent figures from one of the diagrams. Tell how you know the figures are congruent.

3. Look for symmetry around you. Choose a building in your community that has symmetry. Make a simple sketch of the front of the building and draw the line or lines of symmetry.

Customary Units

On July 21, 1969, Neil Armstrong became the first person to set foot on the moon's surface. While there, Armstrong and fellow astronaut Edwin "Buzz" Aldrin collected 44 pounds of moon rocks and other materials to bring back to Earth. On Earth, objects weigh 6 times as much as they weigh on the moon. This means that 44 pounds of moon material weighs about 264 pounds on Earth. On the moon, Aldrin weighed 60 pounds with his space suit on. How much did he weigh on Earth? Look at the table. What would the backpack of equipment weigh on Earth?

Edwin "Buzz" Aldrin

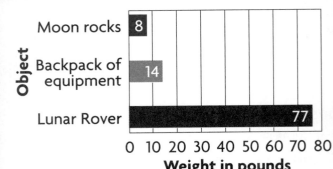

WEIGHTS OF OBJECTS ON THE MOON

Object

Moon rocks	8
Backpack of equipment	14
Lunar Rover	77

0 10 20 30 40 50 60 70 80
Weight in pounds

CHECK WHAT YOU KNOW

Use this page to help you review and remember important skills needed for Chapter 24.

✔ VOCABULARY

Choose the correct term from the box.

inch
sum

1. An __?__ is a customary unit used to measure length.

✔ USE A CUSTOMARY RULER (See p. H26.)

For 2–3, use the picture at the right.

2. Is nail A less than or greater than 2 inches long?

3. Which nail is about 3 inches long?

A   B

✔ MEASURE TO THE NEAREST INCH (See p. H26.)

Write the length to the nearest inch.

4.

5.

inches   1   2   3   4   5

✔ FIND A RULE (See p. H27.)

Write a rule for the table. Then copy and complete the table.

6.

gloves	1	2	3	4	5	6
fingers	5	10	15	■	■	■

7.

rings	1	2	3	4	5	6
cost	$3	$6	$9	■	■	■

8.

butterflies	1	5	3	4	7	9
wings	2	10	6	■	■	■

9.

packs	3	2	7	5	1	8
crackers	21	14	49	■	■	■

Length

Quick Review

1. $7 \times \blacksquare = 42$

2. $\blacksquare \times 5 = 45$ **3.** $\blacksquare \times 8 = 32$

4. $6 \times \blacksquare = 48$ **5.** $9 \times \blacksquare = 63$

▶ Learn

HOW LONG? An estimate is an answer that is close to the actual answer. You can estimate length by using an item close to 1 inch, like a small paper clip or your knuckle.

Activity

MATERIALS: paper clips, ruler

STEP 1

Copy the table.

LENGTHS OF RIBBONS		
Color	Estimate	Measure
green	about 1 inch	1 inch
blue		
purple		
orange		
red		

STEP 2

Use paper clips to estimate the length of the blue ribbon. Record your estimate in the table.

STEP 3

Use a ruler. Measure the length of the blue ribbon to the nearest inch. Record your measurement in the table.

STEP 4

Repeat Steps 2 and 3 for the purple, orange, and red ribbons.

Remember
To use a ruler:

- line up one end of the object with the end of the ruler.
- find the inch mark closest to the object's other end.

The ribbon is 2 inches long to the nearest inch.

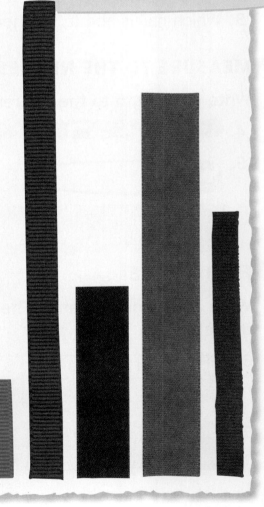

- Why does it make sense to use a small paper clip to estimate inches?

CALIFORNIA STANDARDS MG 1.0 Students choose and use appropriate units and measurement tools to quantify the properties of objects. **MG 1.1** Choose the appropriate tools and units (metric and U.S.) and estimate and measure the length, liquid volume, and weight/mass of given objects. *also* **MR 2.1, MR 2.3, MR 2.4, MR 2.5**

Measuring

You can also measure to the nearest half inch.

Example

What is the length of this crayon to the nearest half inch?

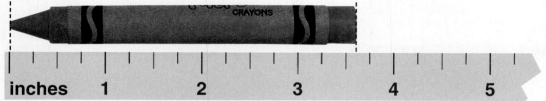

To measure to the nearest half inch:

STEP 1

Line up one end of the crayon with the left side of the ruler.

STEP 2

Find the $\frac{1}{2}$-inch mark that is closest to the other end of the crayon.

So, the length of the crayon to the nearest half inch is $3\frac{1}{2}$ inches.

TECHNOLOGY LINK

To learn more about customary measurement, watch the **Harcourt Math Newsroom Video** *Blue Whale Tracking.*

▶ **Check**

1. **Describe** where you find $\frac{1}{2}$-inch marks on a ruler.

Estimate the length in inches. Then use a ruler to measure to the nearest inch.

2.

3.

4.

5.

Measure the length to the nearest half inch.

6.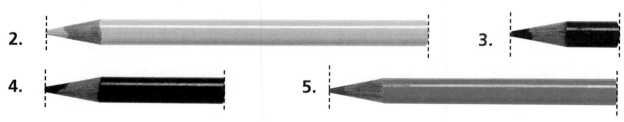

7.

8.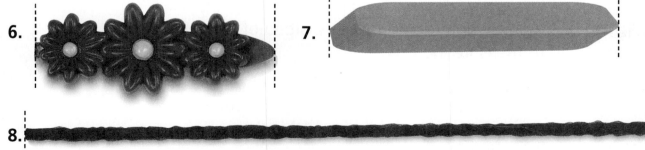

LESSON CONTINUES

► Practice and Problem Solving

Estimate the length in inches. Then use a ruler to measure to the nearest inch.

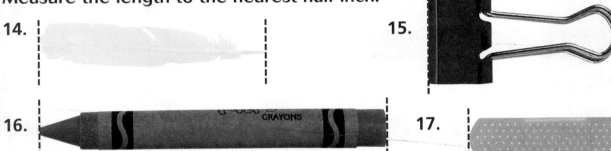

9.

10.

11.

Measure the length to the nearest inch.

12.

13.

Measure the length to the nearest half inch.

14.

15.

16.

17.

18.

Use a ruler. Draw a line for each length.

19. 1 inch

20. $2\frac{1}{2}$ inches

21. 4 inches

22. $5\frac{1}{2}$ inches

23. **REASONING** Why do you use a ruler instead of your knuckle to measure inches?

24. **? What's the Question?** Lori had 2 rolls of ribbon. Each roll had 12 inches of ribbon. After she used some ribbon for a costume, she had 18 inches left. The answer is 6 inches.

25. Joyce used 72 beads to make 9 necklaces with an equal number of beads on each. How many beads were on 2 necklaces?

26. A brush measures $6\frac{1}{2}$ inches. Between which two inch marks does the end of the brush lie?

27. Without using a ruler, draw a line about 8 inches long. Measure the line you drew to the nearest inch. How close was your line to 8 inches?

28. Suppose you need at least 5 inches of yarn for an art project. Is this yellow piece of yarn long enough? Explain.

Mixed Review and Test Prep

29. How many sides does the solid figure have? (p. 362)

30. **TEST PREP** Noah had 8 rows of 6 toy cars. He gave 5 cars to his brother. How many cars does Noah have left? (p. 172)

A 58 **B** 53 **C** 43 **D** 19

31. $3 \times 1 \times 4 = $ ■ (p.170)

32. $5 \times 2 \times 4 = $ ■ (p.170)

33. **TEST PREP** The movie begins at 11:15 A.M. It is 2 hours and 10 minutes long. What time will it end? (p. 98)

F 11:25 A.M. **H** 1:15 P.M.
G 12:25 P.M. **J** 1:25 P.M.

Thinker's Corner

Measure the pieces of yarn and break the code! To find the correct letter for each blank, match the measurement and the color of the yarn.

S
W
T
E
T

H
C
R
A
E

What did the mother bird call the baby bird?

| ___ | ___ | ___ | ___ | ___ | ___ | ___ | ___ | ___ | ___ |
| ? | ? | ? | ? | ? | ? | ? | ? | ? | ? |

$1\frac{1}{2}$ inches 2 inches 1 inch 1 inch $1\frac{1}{2}$ inches 2 inches 1 inch $\frac{1}{2}$ inch $\frac{1}{2}$ inch $1\frac{1}{2}$ inches

Extra Practice page H55, Set A

Inch, Foot, Yard, and Mile

Quick Review

1. $41 + 52$
2. $38 + 39$
3. $12 + 34$
4. $27 + 63$
5. $12 + 12 + 12$

VOCABULARY
foot (ft)
yard (yd)
mile (mi)

► Learn

CHOOSE AND USE You know that an inch (in.) is used to measure length and distance. Other customary units used to measure length and distance are the **foot (ft)**, **yard (yd)**, and **mile (mi)**.

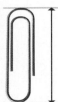

A paper clip is about 1 inch long.

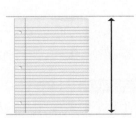

A piece of notebook paper is about 1 foot long.

A baseball bat is about 1 yard long.

You can walk 1 mile in about 20 minutes.

TABLE OF MEASURES
1 foot = 12 inches
1 yard = 3 feet = 36 inches
1 mile = 5,280 feet

- Explain which unit you would use to measure the length of your hand.

► Check

1. Write the names of two objects in your classroom you would measure using feet.

Choose the unit you would use to measure each. Write *inch, foot, yard,* or *mile*.

2. the length of a pencil

3. the length of a butterfly

4. the length of a parking lot

5. the distance a train goes in 30 minutes

A yardstick is 3 times as long as a one foot ruler.

CALIFORNIA STANDARDS MG 1.0 Students choose and use appropriate units and measurement tools to quantify the properties of objects. MG 1.1 Choose the appropriate tools and units (metric and U.S.) and estimate and measure the length, liquid volume, and weight/mass of given objects. *also* MR 2.3, MR 2.4

▶ Practice and Problem Solving

**Choose the unit you would use to measure each.
Write *inch, foot, yard,* or *mile*.**

6. the height of a refrigerator

7. the length of your sneaker

8. the distance between your classroom and the cafeteria

9. the distance you could walk in two hours

10. the length of the school gym

11. the length of a spoon

**Choose the best unit of measure.
Write *inches, feet, yards,* or *miles*.**

12. Sal rides the bus 3 _?_ to school.

13. A football is about 1 _?_ long.

14. The distance between the floor and the doorknob is about 3 _?_ .

15. Sarah's math book is 11 _?_ long.

16. Angie thinks this grasshopper is about 4 inches long. Do you agree with her estimate? Measure and record the length to check.

17. **? What's the Question?** Mitchell had 30 stickers. He gave each of 6 friends an equal number of stickers. The answer is 5 stickers.

18. **Write About It** Estimate the distance from your desk to the classroom door in yards. Then measure the distance. Record your estimate and the actual measurement.

Mixed Review and Test Prep

For 19–22, use the graph. (p. 252)

19. How many votes are for red?

20. Which color has the most votes?

21. How many more votes are there for blue than for green?

22. What kind of graph is this?

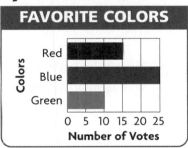

FAVORITE COLORS

23. **TEST PREP** 4,623 − 2,307 (p. 62)

A 2,314 **C** 2,916
B 2,316 **D** 6,930

HANDS ON
Capacity

▶ **Explore**

Capacity is the amount a container can hold.
Cup (c), pint (pt), quart (qt), and gallon (gal)
are customary units for measuring capacity.

VOCABULARY

capacity	quart (qt)
cup (c)	gallon (gal)
pint (pt)	

MATERIALS

cup, pint, quart, and gallon
containers; water, rice, or
beans

cup (c)   pint (pt)   quart (qt)   gallon (gal)

**Copy the table to help find how many cups
are in a pint, a quart, and a gallon.**

NUMBER OF CUPS		
	Estimate	Measure
Cups in a pint?		
Cups in a quart?		
Cups in a gallon?		

STEP 1 Estimate the number of
cups it will take to fill the pint
container. Record your estimate.

STEP 2 Fill a cup and pour it into
the pint container. Repeat until the
pint container is full.

STEP 3 Record the actual
number of cups it took to fill the
pint container.

We are using pints to fill a quart.
How many pints will it take?

STEP 4 Repeat Steps 1–3 for the
quart and the gallon containers.

Try It

a. How many pints does
it take to fill a quart?

b. How many pints does
it take to fill a gallon?

▶ Connect

How are cups, pints, quarts, and gallons related?

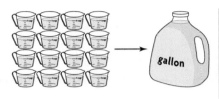

2 cups in 1 pint

4 cups in 1 quart

2 pints in 1 quart

16 cups in 1 gallon

8 pints in 1 gallon

4 quarts in 1 gallon

▶ Practice

Choose the better estimate.

1.

1 cup or
1gallon?

2.

3 cups or
3 quarts?

3.

2 cups or
2 quarts?

4.

10 pints or
10 gallons?

Compare. Write < , > , or = for each ●.

5. 1 cup ● 3 pints   **6.** 1 gallon ● 3 quarts   **7.** 4 pints ● 1 quart

8. Estimation Leticia estimates that she needs about 2 quarts of juice to serve 7 cups. Do you agree? Explain.

9. Write these units of capacity in order from least to greatest: *pint, cup, gallon, quart.*

Mixed Review and Test Prep

Multiply. (p. 148)

10. $6 \times 5 = $ ■   **11.** $6 \times 9 = $ ■

Divide. (p. 214)

12. $56 \div 7 = $ ■   **13.** $49 \div 7 = $ ■

14. TEST PREP Donato gave away 23 marbles. Then he had 345 marbles. How many did he have to begin with? (p. 42)

A 368   **B** 345   **C** 322   **D** 300

HANDS ON
Weight

Quick Review

Compare. Use $<$, $>$, or $=$ for each ●.

1. $16 \div 8$ ● 2×2

2. 10 ● $24 \div 3$

3. $81 \div 9$ ● 9

4. $18 \div 3$ ● $10 \div 2$

5. 13 ● 3×4

▶ **Explore**

An **ounce (oz)** and a **pound (lb)** are customary units for measuring weight.

9 pennies weigh about 1 ounce.

144 pennies weigh about 1 pound.

You can estimate and then weigh objects to decide if they weigh about 1 ounce or about 1 pound.

VOCABULARY
ounce (oz)   pound (lb)

MATERIALS
simple balance
pennies

STEP 1 Place 9 pennies on one side of a balance to show 1 ounce. Find two objects that you think weigh about 1 ounce each. Weigh them to check.

STEP 2 Record what your objects are and whether they weigh more than, less than, or the same as 1 ounce.

45 pennies weigh about 5 oz. What things in your classroom weigh about 5 oz?

STEP 3 Place 144 pennies on one side of the balance to show 1 pound. Find two objects that you think weigh about 1 pound each. Weigh them to check.

STEP 4 Record what your objects are and whether they weigh more than, less than, or the same as 1 pound each.

• Give an example of a small object that is heavier than a large object.

Try It

Name an object that weighs about each amount.

a. 5 ounces   **b.** 3 pounds

CALIFORNIA STANDARDS MG 1.0 Students choose and use appropriate units and measurement tools to quantify the properties of objects. **MG 1.1** Choose the appropriate tools and units (metric and U.S.) and estimate and measure the length, liquid volume, and weight/mass of given objects. *also* **NS 2.8, MG 1.4, MR 2.1, MR 2.3, MR 2.4, MR 2.5**

▶ Connect

How are pounds and ounces related?

These things each weigh about 1 ounce.

A loaf of bread weighs about 1 pound.

16 ounces = 1 pound

▶ Practice

Choose the unit you would use to weigh each.
Write *ounce* or *pound*.

1.

2.

3.

4.

Choose the better estimate.

5.

6.

7.

8.

5. 1 ounce or 1 pound?

6. 3 ounces or 3 pounds?

7. 2 ounces or 2 pounds?

8. 2 ounces or 2 pounds?

9. **REASONING** Bill has 24 cookies divided equally into 6 bags. How many are in 3 bags?

10. Mr. Reynolds cooked three 6-ounce hamburgers. Did he use more than one pound of hamburger? Explain.

11. 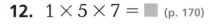 **Write a problem** about items from your home. Use pounds and ounces in your problem.

Mixed Review and Test Prep

12. $1 \times 5 \times 7 = $ ▇ (p. 170)

13. $3 \times 2 \times 8 = $ ▇ (p. 170)

14. $2 \times 3 \times 3 = $ ▇ (p. 170)

15. $12 + 16 + 8 = $ ▇ (p. 36)

16. **TEST PREP** Which shape is shown? (p. 382)

   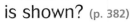

   A square   **C** circle
   B pentagon   **D** rectangle

Ways to Change Units

▶ **Learn**

CHANGE IT   The students in Mrs. Lopez's class need 32 cups of juice for a picnic. How many quarts of juice should they buy?

To change cups into quarts, they must know how these units are related.

- A quart is larger than a cup.

- 4 cups = 1 quart

cup (c)   quart (qt)

Jake and Theresa used different ways to change cups into quarts.

Remember
Table of Measures
Length
12 inches = 1 foot
3 feet = 1 yard
Capacity
2 cups = 1 pint
4 cups = 1 quart
2 pints = 1 quart
8 pints = 1 gallon
4 quarts = 1 gallon

Jake
I'll draw 32 cups. I'll circle groups of 4 to show quarts.

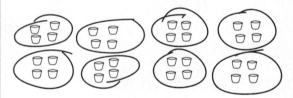

There are 8 groups of 4 cups. So, 32 cups equals 8 quarts.

Theresa
I'll make a table.

Quarts	1	2	3	4	5	6	7	8
Cups	4	8	12	16	20	24	28	32

The table shows that 8 quarts equals 32 cups.

So, they should buy 8 quarts of juice.

MATH IDEA To change one unit into another, first decide how the units are related.

- Explain why Jake drew circles around groups of 4 cups.

CALIFORNIA STANDARDS ○─┐ **AF 2.1** Solve simple problems involving a functional relationship between two quantities. **MG 1.4** Carry out simple unit conversions within a system of measurement. *also* **AF 1.0, AF 1.4, AF 2.0, AF 2.2, MR 2.3, MR 2.4, MR 3.3**

▶ Check

1. **Write** how many pints are in 1 quart. Use the Table of Measures on page 430 to help.

Copy and complete. Use the Table of Measures to help.

2. Change gallons to pints.
   larger unit __?__

   1 gallon = ▦ pints

3. Change feet to inches.
   larger unit __?__

   1 foot = ▦ inches

▶ Practice and Problem Solving

Copy and complete. Use the Table of Measures to help.

4. Change yards to feet.
   larger unit __?__

   1 yard = ▦ feet

5. Change quarts to gallons.
   larger unit __?__

   ▦ quarts = 1 gallon

Change the units. Use the Table of Measures to help.

6. ▦ cups = 1 pint

   12 cups = ▦ pints

7. ▦ inches = 1 foot

feet	1
inches	12

   ▦ inches = 3 feet

8. Callie has 23 inches of yarn. Is this more than or less than 2 feet? Explain.

9. **? What's the Error?** Dylan is finding how many gallons equal 24 pints. He drew this model. Draw the correct model.

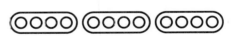

Mixed Review and Test Prep

Multiply. (p. 164)

10. $9 \times 4 =$ ▦   11. $9 \times 7 =$ ▦

Divide. (p. 202)

12. $12 \div 4 =$ ▦   13. $28 \div 4 =$ ▦

14. **TEST PREP** Yolanda had $1.19. She lost a quarter. How much does she have now? (p. 88)

   A $0.84   C $1.09
   B $0.94   D $1.44

Algebra: Rules for Changing Units

▶ Learn

WHAT'S THE RULE?   You can use rules to help you change units. To make a rule for changing units, you must first decide how the units are related.

Examples

A Use a rule to find the number of quarts in 5 gallons.

Remember: 4 quarts = 1 gallon

To change a larger unit into a smaller unit, multiply.

gallons	1	2	3	4	5
quarts	4	8	12	16	20

Rule:
Multiply the number of gallons by 4.

$\blacksquare$ quarts = 5 gallons

$\blacksquare = 5 \times 4$

$20 = 5 \times 4$

So, there are 20 quarts in 5 gallons.

B Use a rule to find the number of gallons in 12 quarts.

Remember: 4 quarts = 1 gallon

To change a smaller unit into a larger unit, divide.

quarts	4	8	12
gallons	1	2	3

Rule:
Divide the number of quarts by 4.

$\blacksquare$ gallons = 12 quarts

$\blacksquare = 12 \div 4$

$3 = 12 \div 4$

So, there are 3 gallons in 12 quarts.

▶ Check

1. **Discuss** why you multiply to change a larger unit into a smaller unit.

CALIFORNIA STANDARDS ⊶ **AF 2.1** Solve simple problems involving a functional relationship between two quantities. **AF 2.2** Extend and recognize a linear pattern by its rules. *also* **AF 1.4, AF 2.0, MG 1.4, MR 2.3, MR 2.4, MR 3.3**

Use the rules to change the units. (2 pints = 1 quart)

2. How many pints are in 5 quarts?

Rule: Multiply the number of quarts by 2.

$\blacksquare = 5 \times 2$

$\blacksquare$ pints = 5 quarts

3. How many quarts are in 8 pints?

Rule: Divide the number of pints by 2.

$\blacksquare = 8 \div 2$

$\blacksquare$ quarts = 8 pints

▶ Practice and Problem Solving

Use the rules to change the units. (4 cups = 1 quart)

4. How many quarts are in 24 cups?

Rule: Divide the number of cups by 4.

$\blacksquare = 24 \div 4$

$\blacksquare$ quarts = 24 cups

5. How many cups are in 7 quarts?

Rule: Multiply the number of quarts by 4.

$\blacksquare = 7 \times 4$

$\blacksquare$ cups = 7 quarts

Write the rule and change the units. You may make a table to help.

6. How many feet are in 3 yards?

$\blacksquare$ feet = 3 yards

7. How many yards are in 12 feet?

$\blacksquare$ yards = 12 feet

8. How many pints are in 16 cups?

$\blacksquare$ pints = 16 cups

9. How many cups are in 6 pints?

$\blacksquare$ cups = 6 pints

10. **REASONING** Gail is 5 feet tall. How many inches tall is she? Copy and complete the table at the right.

feet	1	2	3	4	5
inches	12				

Mixed Review and Test Prep

Divide. (p. 204)

11. $3 \div 1 = \blacksquare$ **12.** $4 \div 4 = \blacksquare$

13. $0 \div 1 = \blacksquare$ **14.** $1 \div 1 = \blacksquare$

15. **TEST PREP** What is the sum of 416 and 295? (p. 42)

A 121   **C** 701

B 611   **D** 711

Extra Practice page H55, Set D

Problem Solving Skill
Use a Graph

Understand → Plan → Solve → Comprueba

Quick Review
1. $3 \times 4 = \blacksquare$
2. $6 \times 4 = \blacksquare$ 3. $4 \times 2 = \blacksquare$
4. $3 \times 2 = \blacksquare$ 5. $5 \times 4 = \blacksquare$

GRAPHING GALLONS This graph shows how quarts and gallons are related. Stacy and Ethan used the graph to decide how many quarts are in 6 gallons.

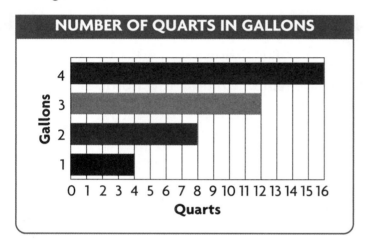

NUMBER OF QUARTS IN GALLONS

Gallons (vertical axis): 4, 3, 2, 1
Quarts (horizontal axis): 0 1 2 3 4 5 6 7 8 9 10 11 12 13 14 15 16

Stacy	Ethan
By looking at the purple bar, I can see that 4 quarts = 1 gallon. I'll write a rule to change gallons to quarts.	I'll look for a pattern in the lengths of the bars.
Rule: Multiply the number of gallons by 4.	1 gallon = 4 quarts
$\blacksquare$ quarts = 6 gallons x 4	2 gallons = 8 quarts
$\blacksquare$ quarts = 6 x 4	3 gallons = 12 quarts
24 = 6 x 4	4 gallons = 16 quarts
So, there are 24 quarts in 6 gallons.	Each bar increases by 4.
	5 gallons is 16 + 4 = 20.
	6 gallons is 20 + 4 = 24.
	So, there are 24 quarts in 6 gallons.

Talk About It

• How could Stacy tell that there are 4 quarts in 1 gallon?

CALIFORNIA STANDARDS O–π **AF 2.1** Solve simple problems involving a functional relationship between two quantities. **MG 1.4** Carry out simple unit conversions within a system of measurement. *also* **NS 1.0, NS 2.8, AF 2.0, AF 2.2, MR 1.1, MR 2.3, MR 2.4, MR 3.3**

▶ Problem Solving Practice

USE DATA For 1–5, use the graph.

1. Explain how you can use the graph to find how many cups are in 1 pint.

2. How many cups are in 4 pints?

3. How many pints are in 6 cups?

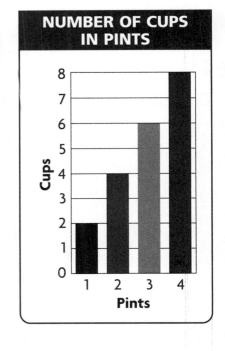

NUMBER OF CUPS IN PINTS

Choose the letter of the correct answer.

4. Which rule could you use to help find how many cups are in 5 pints?

   A Multiply the number of pints by 4.
   B Multiply the number of pints by 2.
   C Multiply the number of gallons by 4.
   D Multiply the number of cups by 3.

5. How many cups are in 5 pints?

   F 5 cups   **H** 15 cups
   G 10 cups   **J** 20 cups

Mixed Applications

Solve.

6. Alan had 6 boxes of cookies. Each box had 8 cookies. He sold 2 boxes. How many cookies did he have left?

7. A yard of fabric costs $3. Faye bought 4 yards of fabric and 11 inches of ribbon. How much did she pay for the fabric?

8. Mr. Castillo delivered 3 cases of pasta to each of 3 stores. Each case had 8 boxes of pasta. How many boxes of pasta did he deliver?

9. Toni went fishing every day for 2 weeks. During the third week she went fishing on Monday and Friday. How many days did Toni go fishing in the three weeks?

Review/Test

✅ CHECK CONCEPTS

Choose the better estimate. (pp. 426–429)

1.

    1 ounce or
    1 pound?

2.

    2 pints or
    2 gallons?

3.

    30 ounces or
    30 pounds?

✅ CHECK SKILLS

Measure the length to the nearest inch. (pp. 420–423)

4.

5.

Choose the unit you would use to measure each.
Write *inch, foot, yard,* or *mile.* (pp. 424–425)

6. the length of a marker

7. the height of a wall

Write the rule and change the units.
You may make a table to help. (pp. 430–433)

8. How many inches are in 2 feet?

    ▨ inches = 2 feet

9. How many pints are in 8 cups?

    ▨ pints = 8 cups

✅ CHECK PROBLEM SOLVING

Use Data For 10–12, use the graph.
(pp. 434–435)

10. How many pints are in 1 quart?

11. How many pints are in 4 quarts?

12. How many quarts are in 4 pints?

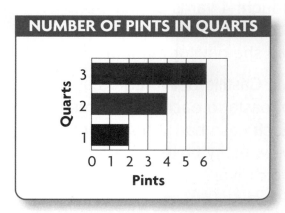

NUMBER OF PINTS IN QUARTS

Cumulative Review

Get the information you need.
See item **3.**

You are expected to know which units are used to measure weight. Think about a soccer ball you have held then find an amount that is reasonable.

Also see problem **3,** p. H63.

For 1–10, choose the best answer.

1. Patsy wants to measure the length of her pet mouse. Which unit should she use?

   A mile   **C** foot
   B yard   **D** inch

2. Use your ruler to measure the length of the string to the nearest inch.

   F 1 inch   **H** 3 inches
   G 2 inches   **J** 4 inches

3. Which is the most reasonable estimate for the weight of a soccer ball?

   A 1 pound   **C** 1 cup
   B 1 ounce   **D** 1 inch

4. How many inches are in 4 feet?

feet	1	2	3	4
inches	12	24	36	■

   F 4 inches   **H** 48 inches
   G 12 inches   **J** 52 inches

5. 4,872 − 3,469 = ■

   A 1,417   **C** 1,403
   B 1,413   **D** NOT HERE

6. Which statement is true?

   F This angle measures less than a right angle.
   G This angle is a right angle.
   H This angle measures greater than a right angle.
   J This angle is a straight angle.

7. There are 8 pints in 1 gallon. How many pints are in 3 gallons?

   A 24 pints   **C** 8 pints
   B 16 pints   **D** NOT HERE

8. How many feet are in 3 yards?

   F 3 ft   **H** 9 ft
   G 6 ft   **J** 12 ft

9. How many quarts are in 8 gallons?

   Rule: Multiply the number of gallons by 4.

   A 32 qt   **C** 4 qt
   B 12 qt   **D** 2 qt

10. Which statement is **not** true?

    F $2,125 < 2,373$
    G $578 > 475$
    H $5,300 + 42 = 5,342$
    J $9,123 > 9,124$

Metric Units

During the summer months, one of the hottest places in the United States is Death Valley. The table shows the average high and low temperatures during July in four cities and in Death Valley. Which city listed is the hottest? How much cooler is the high temperature for that city than Death Valley?

JULY TEMPERATURES IN °C		
Place	Temperature	
	High	Low
Death Valley, CA	46°C	31°C
San Francisco, CA	22°C	12°C
Columbus, OH	29°C	17°C
Philadelphia, PA	30°C	19°C
Key West, FL	33°C	28°C

Death Valley

CHECK WHAT YOU KNOW

Use this page to help you review and remember important skills needed for Chapter 25.

✔ VOCABULARY

Choose the correct term from the box.

1. A metric unit used to measure length is a __?__.

> centimeter
> inch

✔ USE A METRIC RULER (See p. H27.)

For 2–4, use the yarn shown at the right.

2. Which piece of yarn is 3 centimeters long?

3. Which piece of yarn is about 1 centimeter long?

4. Which piece of yarn is about 1 centimeter longer than yarn B?

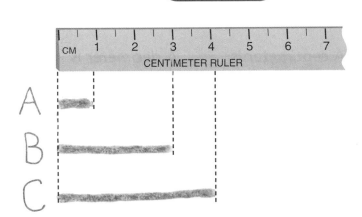

✔ MEASURE TO THE NEAREST CENTIMETER (See p. H28.)

Write the length to the nearest centimeter.

5.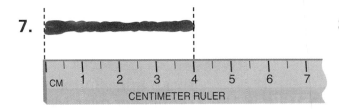

6.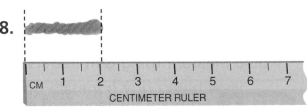

7.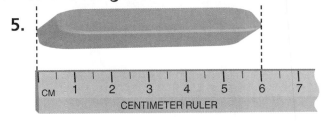

8.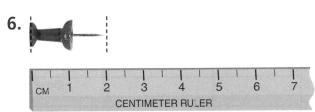

HANDS ON
Mass: Grams and Kilograms

▶ **Explore**

The **gram (g)** and the **kilogram (kg)** are metric units for measuring mass.

A paper clip has a mass of about 1 gram.

A large book has a mass of about 1 kilogram.

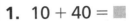
Find the mass of objects in your classroom.

STEP 1

Place 10 paper clips on one side of the simple balance to show 10 g.

STEP 2

Find an object that you think might equal 10 g. Use the balance to check.

STEP 3

Repeat Steps 1 and 2 for 25 g and 1 kg. Use the book to show 1 kg.

A nickel has a mass of about 5 grams. What things in your classroom have a mass of about 5 grams?

Try It

Name an object that has a mass of each amount.

a. 5 grams   **b.** 2 kilograms

CALIFORNIA STANDARDS MG 1.0 Students choose and use appropriate units and measurement tools to quantify the properties of objects. **MG 1.1** Choose the appropriate tools and units (metric and U.S.) and estimate and measure the length, liquid volume, and weight/mass of given objects. also **MG 1.4, MR 2.3, MR 2.4**

▶ Connect

This cat has a mass of 5 kg. How many grams is that?

1,000 grams = 1 kilogram

Kilograms	1	2	3	4	5
Grams	1,000	2,000	3,000	4,000	5,000

So, the cat has a mass of 5,000 g.

▶ Practice

Choose the better estimate.

1.

200 g or 200 kg?

2.

18 g or 18 kg?

3.

6 g or 6 kg?

Choose the tool and unit to measure each.

4. length of a pencil   **5.** mass of a grape

6. capacity of a bucket

Tools	Units	
ruler	cm	g
liter container	kg	mL
simple balance	L	m

7. **Write About It** Do objects of about the same size always have about the same mass? Give an example.

8. Kim's kitten has a mass of 2 kg. How many grams is that?

9. Choose an object you think has a greater mass than your shoe. Then compare the object and your shoe on the balance. Record which has the greater mass.

Mixed Review and Test Prep

For 10–13, subtract. (p. 58)

10. 441 − 78   **11.** 703 − 264

12. 157 − 139   **13.** 500 − 167

14. **TEST PREP** Lino had 15 photos. He put an equal number of photos on each of 3 album pages. How many photos are on each page? (p. 202)

A 45   **B** 10   **C** 5   **D** 3

HANDS ON
Measure Temperature

▶ **Explore**

Degrees Fahrenheit (°F) are customary units of temperature, and **degrees Celsius (°C)** are metric units of temperature.

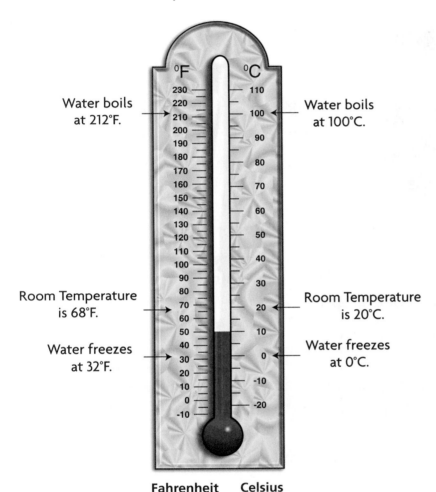

Water boils at 212°F. → Water boils at 100°C.

Room Temperature is 68°F. → Room Temperature is 20°C.

Water freezes at 32°F. → Water freezes at 0°C.

Fahrenheit   Celsius

Quick Review

Skip-count. Write the missing number.

1. 10, 15, 20, 25, ▣

2. 80, 75, ▣, 65, 60

3. 110, 115, 120, ▣, 130

4. 95, ▣, 105, 110, 115

5. 90, ▣, 86, 84, 82

VOCABULARY
degrees Fahrenheit (°F)
degrees Celsius (°C)

MATERIALS Celsius and Fahrenheit thermometers

STEP 1

Estimate the temperature outside the classroom in degrees Celsius and in degrees Fahrenheit. Record your estimates.

STEP 2

Measure the temperature outside the classroom by using thermometers. Record the differences between the actual measurements and your estimates.

CALIFORNIA STANDARDS MG 1.0 Students choose and use appropriate units and measurement tools to quantify the properties of objects. **MR 2.3** Use a variety of methods, such as words, numbers, symbols, charts, graphs, tables, diagrams, and models, to explain mathematical reasoning. *also* **MR 2.4**

Try It

What is the temperature now?

a. The temperature was 71°F.
It dropped 20°.

b. The temperature was 25°C.
It dropped 3°.

▶ Connect

To read a thermometer, find the number at the top of
the red bar. The thermometer on page 450 shows that
the temperature is

50°F   **Read:** fifty degrees Fahrenheit
10°C   **Read:** ten degrees Celsius

▶ Practice

Write each temperature in °F.

1.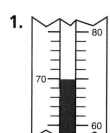

2.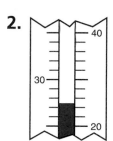

Write each temperature in °C.

3.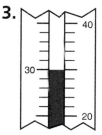

4.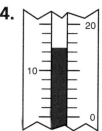

Choose the better estimate.

5.

6.

25°F or 95°F?   56°F or 98°F?

Choose the better estimate.

7.

8.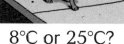

8°C or 25°C?   0°C or 22°C?

9. The temperature was 45°F at 9:00 A.M. It was 62°F at
noon. How many degrees did the temperature rise?

Mixed Review and Test Prep

For 10–13, multiply. (p. 164)

10. $8 \times 9 = $ ■   **11.** $9 \times 9 = $ ■

12. $4 \times 9 = $ ■   **13.** $9 \times 6 = $ ■

14. **TEST PREP** Add 1,276 and 894. (p. 46)

    A 1,060   **C** 2,160
    B 2,070   **D** 2,170

Review/Test

✔ CHECK VOCABULARY AND CONCEPTS

Choose the best term from the box.

> degrees Celsius
> degrees
>   Fahrenheit
> gram
> liter
> meter

1. Capacity can be measured by using metric units such as the __?__ . (p. 446)

2. A metric unit for measuring mass is the __?__ . (p. 448)

3. Customary units of temperature are __?__ . (p. 450)

4. Metric units of temperature are __?__ . (p. 450)

Choose the better estimate. (pp. 446–451)

5.

2 mL or 2 L?

6.

2 g or 2 kg?

7.

30°F or 60°F?

✔ CHECK SKILLS

Estimate the length in centimeters. Then use a ruler to measure to the nearest centimeter. (pp. 440–443)

8. ├────────┤

9.

Choose the unit you would use to measure each. Write *cm, m,* or *km.* (pp. 440–443)

10. the length of a pencil

11. the distance to the moon

12. the length of a caterpillar

13. the distance you run in 10 seconds

✔ CHECK PROBLEM SOLVING

Use *make a table* to solve. (pp. 444–445)

14. Ashley has 2 meters of yarn. How many centimeters of yarn does she have?

15. Ray had 1 meter of string. Then he bought 400 centimeters more. How many meters does he have in all?

Cumulative Review

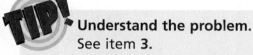

TIP! Understand the problem.
See item **3**.

You need to understand that *most likely* means it will probably happen more often than any of the other choices.

Also see problem **1**, p. H62.

For 1–9, choose the best answer.

1. Which temperature does the thermometer show?

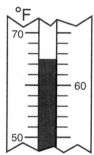

°F
70
60
50

A 60°F   **C** 70°F
B 65°F   **D** 75°F

2. Which is the best estimate for the amount of water in a fish tank?

F 1 mL   **H** 1 L
G 10 mL   **J** 10 L

3. On what color is the pointer *most likely* to land?

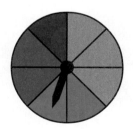

A orange   **C** blue
B green   **D** red

4. Four tapes cost a total of $32. How much would one tape cost?

F $8   **H** $32
G $9   **J** $36

5. Which is the best estimate of the outdoor temperature?

A 0°C   **C** 32°C
B 25°C   **D** 100°C

6. A solid figure has 6 faces. All the faces are squares. Which solid figure is described?

F square pyramid
G sphere
H cone
J cube

7. $51 \div 3 = $ ▪

A 16   **C** 48
B 17   **D** NOT HERE

8. Which is a unit for measuring the *capacity* of this object?

F meter   **H** liter
G gram   **J** centimeter

9. Which is a unit for measuring the *mass* of a tennis raquet?

A meter   **C** gram
B liter   **D** inch

Perimeter, Area, and Volume

People have grown flowers for thousands of years. In the 1600s, tulips in Holland were so valuable that the bulbs were used as money! Suppose you wanted to plant a flower garden and put a fence around it. Look at the garden diagram below. How can you find how many feet of fencing you need to go around the garden?

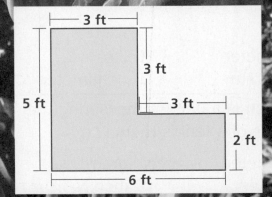

CHECK WHAT YOU KNOW

Use this page to help you review and remember important skills needed for Chapter 26.

✔ COLUMN ADDITION (See p. H10.)

Find each sum.

1. $\begin{array}{r} 3 \\ 7 \\ +5 \\ \hline \end{array}$

2. $\begin{array}{r} 2 \\ 4 \\ +8 \\ \hline \end{array}$

3. $\begin{array}{r} 1 \\ 3 \\ +9 \\ \hline \end{array}$

4. $\begin{array}{r} 36 \\ 12 \\ +9 \\ \hline \end{array}$

5. $\begin{array}{r} 12 \\ 7 \\ +23 \\ \hline \end{array}$

6. $\begin{array}{r} 44 \\ 9 \\ +6 \\ \hline \end{array}$

7. $\begin{array}{r} 245 \\ 65 \\ +72 \\ \hline \end{array}$

8. $\begin{array}{r} 99 \\ 201 \\ +31 \\ \hline \end{array}$

9. $\begin{array}{r} 102 \\ 40 \\ +14 \\ \hline \end{array}$

10. $\begin{array}{r} 57 \\ 400 \\ +3 \\ \hline \end{array}$

11. $\begin{array}{r} 142 \\ 231 \\ +105 \\ \hline \end{array}$

12. $\begin{array}{r} 65 \\ 19 \\ +15 \\ \hline \end{array}$

✔ MULTIPLICATION FACTS (See p. H11.)

Find each product.

13. $7 \times 4 = \blacksquare$

14. $3 \times 6 = \blacksquare$

15. $2 \times 5 = \blacksquare$

16. $4 \times 9 = \blacksquare$

17. $7 \times 2 = \blacksquare$

18. $6 \times 6 = \blacksquare$

19. $9 \times 8 = \blacksquare$

20. $10 \times 3 = \blacksquare$

21. $4 \times 5 = \blacksquare$

22. $\begin{array}{r} 3 \\ \times 7 \\ \hline \end{array}$

23. $\begin{array}{r} 7 \\ \times 8 \\ \hline \end{array}$

24. $\begin{array}{r} 9 \\ \times 6 \\ \hline \end{array}$

25. $\begin{array}{r} 5 \\ \times 5 \\ \hline \end{array}$

26. $\begin{array}{r} 2 \\ \times 8 \\ \hline \end{array}$

27. $\begin{array}{r} 7 \\ \times 6 \\ \hline \end{array}$

28. $\begin{array}{r} 3 \\ \times 3 \\ \hline \end{array}$

29. $\begin{array}{r} 9 \\ \times 5 \\ \hline \end{array}$

30. $\begin{array}{r} 10 \\ \times 2 \\ \hline \end{array}$

31. $\begin{array}{r} 7 \\ \times 9 \\ \hline \end{array}$

32. $\begin{array}{r} 1 \\ \times 6 \\ \hline \end{array}$

33. $\begin{array}{r} 4 \\ \times 4 \\ \hline \end{array}$

34. $\begin{array}{r} 5 \\ \times 6 \\ \hline \end{array}$

35. $\begin{array}{r} 2 \\ \times 9 \\ \hline \end{array}$

36. $\begin{array}{r} 3 \\ \times 5 \\ \hline \end{array}$

37. $\begin{array}{r} 8 \\ \times 6 \\ \hline \end{array}$

38. $\begin{array}{r} 5 \\ \times 8 \\ \hline \end{array}$

39. $\begin{array}{r} 2 \\ \times 6 \\ \hline \end{array}$

HANDS ON
Perimeter

▶ Explore

The distance around a figure is called its **perimeter**.

You can build a rectangle by making an array with square tiles.

1 unit

1 unit

Each side of a square tile has a length of 1 unit.

VOCABULARY
perimeter

MATERIALS
square tiles

Then you can find the perimeter of the rectangle by counting the units.

STEP 1 Make a rectangle by placing 3 rows with 5 square tiles in each row.

STEP 2 Count the number of units around the outside of the rectangle.

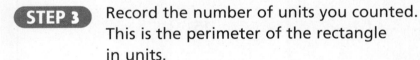

STEP 3 Record the number of units you counted. This is the perimeter of the rectangle in units.

How can I use square tiles to find the perimeter of my math book?

MATH IDEA You can find the perimeter of a figure by counting the number of units around the outside of the figure.

Try It

Use square tiles to find the perimeter of each.

a. your math book

b. your desktop

CALIFORNIA STANDARDS O━━MG **1.3** Find the perimeter of a polygon with integer sides. **MG 1.0** Students choose and use appropriate units and measurement tools to quantify the properties of objects. *also* NS 1.1, MR 2.3, MR 2.4

You can count units to find the perimeters of these plane figures.

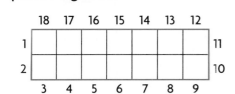

Perimeter = 18 units

Perimeter = 14 units

▶ **Practice**

Find the perimeter of each figure.

1.

2.

3.

4.

5.

6.

7. **?** **What's the Error?** Elise wrote that the perimeter of this rectangle is 6 units. Describe her error. Write the correct perimeter.

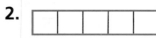

8. Marita made a rectangle with 5 rows of 6 tiles. What is the perimeter of her rectangle?

Mixed Review and Test Prep

9. $27 \div 9 = \blacksquare$ (p. 218)

10. $72 \div 8 = \blacksquare$ (p. 214)

11. $404 - 139 = \blacksquare$ (p. 58)

12. Find the missing factor. (p. 142)

   $7 \times \blacksquare = 42$

13. **TEST PREP** Lin has 2 bags of sandwiches. Each bag holds 2 sandwiches. Each sandwich is cut into 4 pieces. How many pieces of sandwich does Lin have in all? (p. 170)

   A 4   **B** 8   **C** 10   **D** 16

Estimate and Find Perimeter

Quick Review

1. $2 + 3 + 3 = \blacksquare$

2. $5 + 6 + 7 = \blacksquare$

3. $7 + 3 + 6 = \blacksquare$

4. $4 + 2 + 9 = \blacksquare$

5. $10 + 4 + 4 + 2 = \blacksquare$

▶ **Learn**

AROUND AND AROUND You can estimate the perimeter of your math book.

Activity

MATERIALS: toothpicks, paper clips

STEP 1 Copy the table. Estimate the perimeter of your math book in paper clips and in toothpicks. Record your estimates.

STEP 2 Use paper clips. Record how many paper clips it takes to go around all the edges of your math book.

STEP 3 Use toothpicks. Record how many toothpicks it takes to go around all the edges of your math book.

PERIMETER OF MY MATH BOOK	Estimate	Measurement
Number of paper clips		
Number of toothpicks		

• How does your estimate compare with your actual measurement?

• Did it take more paper clips or more toothpicks to measure the perimeter of your math book? Explain.

CALIFORNIA STANDARDS ○┓MG 1.3 Find the perimeter of a polygon with integer sides. **MG 1.1** Choose the appropriate tools and units (metric and U.S.) and estimate and measure the length, liquid volume, and weight/mass of given objects. *also* **MG 1.0, ○┓MG 2.3, MR 2.1, MR 2.3, MR 2.4**

Use a Ruler

You can use a ruler to find the perimeter.

Example A Find the perimeter of the bookmark in centimeters.

Use a ruler to find the lengths of the sides.
Add the lengths of the sides.

10 cm + 2 cm + 10 cm + 2 cm = 24 cm

The perimeter is 24 cm.

You can add the lengths of the sides to find the perimeter.

Example B Find the perimeter of the figure.

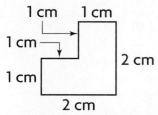

Add the lengths of the sides:
1 cm + 2 cm − 2 cm + 1 cm + 1 cm + 1 cm = 8 cm

The perimeter is 8 cm.

- Why do you need to measure only 2 sides of the bookmark to find its perimeter?

TECHNOLOGY LINK

More Practice: Use Mighty Math Carnival Count-down, *Pattern Block Roundup*, Level N.

Check

1. **Discuss** whether you could measure only 2 sides to find the perimeter in Example B.

Find the perimeter.

2.

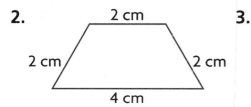

3.

Find the perimeter.

4.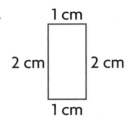
1 cm
2 cm   2 cm
1 cm

5.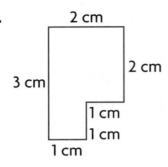
2 cm
3 cm   2 cm
1 cm
1 cm
1 cm

6.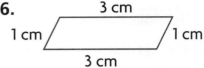
3 cm
1 cm   1 cm
3 cm

7.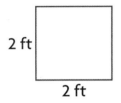
2 ft
2 ft

8.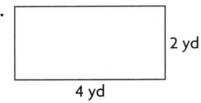
2 yd
4 yd

9.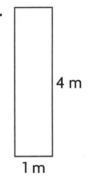
4 m
1 m

Use your centimeter ruler to find the perimeter.

10.

11.

12. REASONING The length of one side of a square is 6 cm. What is the perimeter of the square?

13. A rectangle is 5 inches long and 2 inches wide. What is its perimeter?

14. Use graph paper. Draw a rectangle with a perimeter of 12 units.

15. Jana's beach towel is 5 feet long and 3 feet wide. What is its perimeter?

16. $\frac{a+b}{c}$ **Algebra** This triangle has a perimeter of 8 cm. How long is Side C?

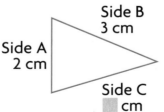
Side B
3 cm
Side A
2 cm
Side C
█ cm

17. **Write About It** Find an object with a perimeter you could measure by using inches. Explain how to estimate and measure its perimeter. Then measure the perimeter with a ruler.

Mixed Review and Test Prep

USE DATA For 18–22, use the graph. (p. 252)

18. How many students voted for cheese?

19. How many more students voted for cheese than for vegetables?

20. How many students voted for crackers and trail mix?

21. **TEST PREP** How many students did NOT vote for popcorn?

    A 21   C 27
    B 24   D 31

22. **TEST PREP** How many students voted in all?

    F 12   H 40
    G 32   J 50

FAVORITE SNACKS

Snacks:
- Popcorn
- Crackers
- Trail Mix
- Cheese
- Vegetables
- Fruit

Number of Votes: 0 2 4 6 8 10

--LiNKUP---
to Social Studies

In 1806, Thomas Jefferson built a house in the Blue Ridge Mountains of Virginia. Jefferson built a special room that is a perfect cube at the center of the house. This room is 20 feet long, 20 feet wide, and 20 feet tall. Jefferson's granddaughter used to draw in this room because a large window called a skylight was in the ceiling.

1. There are 32 panes of glass in the center room skylight. The panes are in 2 rows of 16. How else might Jefferson have arranged the panes in equal rows?

2. The floor of the center room is a square with each side measuring 20 feet. What is the perimeter of this floor?

HANDS ON
Area of Plane Figures

▶ **Explore**

A **square unit** is a square with a side length of 1 unit. You use square units to measure area. **Area** is the number of square units needed to cover a flat surface.

1 square unit

1 unit

1 unit   1 unit

1 unit

VOCABULARY
square unit
area

MATERIALS
square tiles
grid paper

Use square tiles to find the area of your math book cover.

STEP 1   Estimate how many squares will cover your math book. Then place square tiles in rows on the front of your math book. Cover the whole surface.

STEP 2   Use grid paper. Draw a picture to show how you covered the math book.

STEP 3   Count and record the number of square tiles you used. This number is the book cover's area in square units.

How many rows of tiles do I need to cover an index card?

MATH IDEA You can find the area of a surface by counting the number of square units needed to cover the surface.

• Look at the picture you made. How could you use multiplication to find the area?

Try It

Use square tiles to find the area of each.

a. an index card
b. a sheet of notebook paper

CALIFORNIA STANDARDS O→ **MG 1.2** Estimate or determine the area and volume of solid figures by covering them with squares or by counting the number of cubes that would fill them. **MR 2.1** Use estimation to verify the reasonableness of calculated results. *also* **NS 1.1, NS 2.8, MR 2.3, MR 2.4**

To find the area of a rectangle, multiply the number of rows times the number in each row.

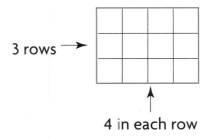

3 rows →

4 in each row

number of rows number in each row area

↓ ↓ ↓

3 × 4 = 12 square units

▶ **Practice**

TECHNOLOGY LINK

More Practice: Use E-Lab, *Finding Area*.

www.harcourtschool.com/ elab2002

Find the area of each figure. Write the area in square units.

1.

2.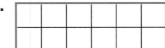

3.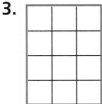

4.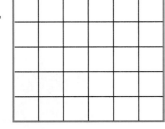

5.

6.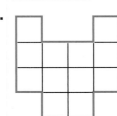

7. REASONING Look at 2 and 3. What do you notice about the area and perimeter of these figures?

8. **? What's the Question?** Rachel's blanket is 6 feet wide and 4 feet long. The answer is 24 square feet.

Mixed Review and Test Prep

Find each missing addend. (p. 68)

9. $5 + \blacksquare = 12$ **10.** $6 + \blacksquare = 14$

11. $\blacksquare + 9 = 15$ **12.** $\blacksquare + 4 = 11$

13. TEST PREP Erma had 45 beads. She put 9 beads on each key chain. How many key chains did she make? (p. 218)

A 10 **B** 9 **C** 5 **D** 3

Area of Solid Figures

▶ **Learn**

ABOUT FACE Find the total area that covers
a solid figure by finding the area of each face.

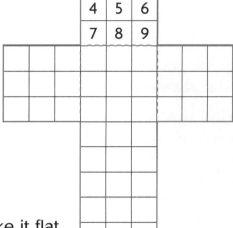

1	2	3
4	5	6
7	8	9

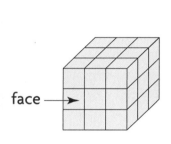

face ——▶

cube   Open the cube.   Make it flat.

To find the total area that covers this cube:

STEP 1

Count the number of square
units on each face to find the
area of each face.

STEP 2

Add the area of each face to
find the total area.

$9 + 9 + 9 + 9 + 9 + 9 = 54$

So, the total area that covers
this cube is 54 square units.

			1	2	3			
			4	5	6			
			7	8	9			
1	2	3	1	2	3	1	2	3
4	5	6	4	5	6	4	5	6
7	8	9	7	8	9	7	8	9
			1	2	3			
			4	5	6			
			7	8	9			
			1	2	3			
			4	5	6			
			7	8	9			

CALIFORNIA STANDARDS O—ᴨMG 1.2 Estimate or determine the area and volume of solid figures by covering
them with squares or by counting the number of cubes that would fill them. **MR 2.3** Use a variety of methods, such
as words, numbers, symbols, charts, graphs, tables, diagrams, and models, to explain mathematical reasoning.
also O—ᴨMG 2.3, MR 2.4

1. **Tell** the area of each face of the prism in Exercise 3.

Find the total area that covers each solid figure.

2.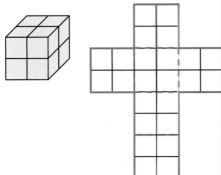

3.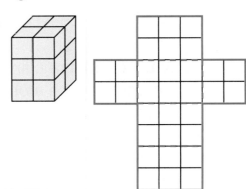

▶ **Practice and Problem Solving**

Find the total area that covers each solid figure.

4.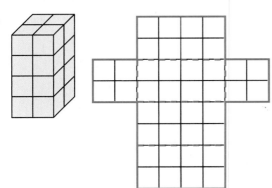

5.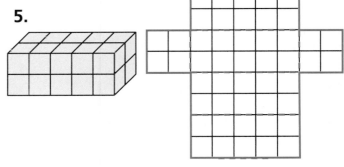

6.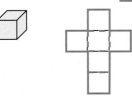

7.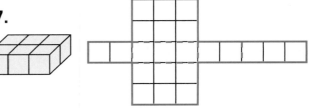

8. The product of two numbers is 24. The sum of the numbers is 11. What are the numbers?

9. **Write About It** How could you use multiplication to find the total area that covers the solid in Exercise 2?

Mixed Review and Test Prep

10. $4.49
    +$5.92

    (p. 88)

11. $6.00
    −$3.45

    (p. 88)

12. $7.07
    −$1.88

    (p. 88)

13. $72 \div 8 =$ ■ (p. 214)

14. **TEST PREP** There are 3 feet in 1 yard. How many feet are in 5 yards? (p. 430)

   A 2 feet   **C** 10 feet

   B 8 feet   **D** 15 feet

Problem Solving Skill
Make Generalizations

Understand → Plan → Solve → Check

DON'T FENCE ME IN Maura plans to plant a flower garden and put a fence around it. She has 12 feet of fencing to make a square or rectangular garden. If she wants to have the most area possible, should her fence be a square or a rectangle?

Maura draws a picture to show all the square and rectangular gardens she can make.

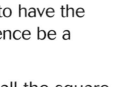

Perimeter:
1 + 5 + 1 + 5 = 12 feet
Area: 5 × 1 = 5 square feet

Perimeter:
2 + 4 + 2 + 4 = 12 feet
Area: 4 × 2 = 8 square feet

Perimeter:
3 + 3 + 3 + 3 = 12 feet
Area: 3 × 3 = 9 square feet

Order the areas: 9 > 8 > 5

9 square feet is the greatest area.

So, Maura should build a square fence.

Talk About It

* **Describe** how the area changes when rectangles with the same perimeter change from long and thin to square.

* **What if** Maura had 20 feet of fencing? To have the greatest area, should her fence be a square or a rectangle?

Quick Review
1. 4 + 2 + 4 + 2 = ■
2. 8 + 3 + 8 + 3 = ■
3. 3 × 4 = ■
4. 7 × 3 = ■
5. 8 × 4 = ■

CALIFORNIA STANDARDS MR 3.3 Develop generalizations of the results obtained and apply them in other circumstances. **MR 2.3** Use a variety of methods, such as words, numbers, symbols, charts, graphs, tables, diagrams, and models, to explain mathematical reasoning. *also* **NS 1.0, NS 2.8, O—n MG 1.3, MR 2.0, MR 2.4, MR 3.2**

466

1. **What if** Maura had 8 feet of fencing? She wants to make a rectangle or square with the greatest possible area. How long should it be? How wide should it be?

2. Kyle has 8 feet of fencing to make a play yard for his rabbit. What rectangle could he fence in that would have the least possible area?

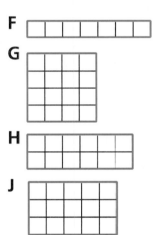

Jane used 16 inches of ribbon to make a rectangular frame.

3. Which drawing shows the frame Jane made?

A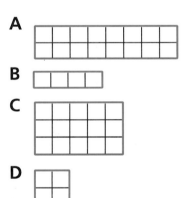

B

C

D

4. Which shows how Jane could increase the area of the frame?

F

G

H

J

Mixed Applications

5. Abe bought 3 muffins for $1 each and 2 cartons of milk for $0.50 each. How much did he spend?

6. Ted eats 1 sandwich and drinks 2 glasses of milk each day. How many glasses of milk does he drink in a week?

7. **REASONING** I am a 2-digit number less than 20. I can be divided evenly into groups of 4. I cannot be divided evenly into groups of 3. What number am I?

8. 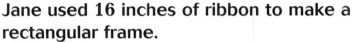 **Write a problem** about finding the perimeter of a pen for a guinea pig. Make sure you include how long and wide the pen is. Trade with a partner and solve.

6 Estimate and Find Volume

Quick Review

1. $1 \times 4 \times 2 = \blacksquare$
2. $2 \times 3 \times 2 = \blacksquare$
3. $5 \times 1 \times 2 = \blacksquare$
4. $3 \times 2 \times 8 = \blacksquare$
5. $4 \times 2 \times 3 = \blacksquare$

> ## Learn

FILL IT UP **Volume** is the amount of space a solid figure takes up.

A **cubic unit** is used to measure volume. A cubic unit is a cube with a side length of 1 unit. You can use color cubes to show cubic units.

1 cubic unit

VOCABULARY
volume
cubic unit

Activity

Use color cubes to find the volume of a box.

MATERIALS: color cubes, small box

STEP 1 Estimate how many cubes it will take to fill the box. Record your estimate.

STEP 2 Count the cubes you use. Place the cubes in rows along the bottom of the box. Then continue to make layers of cubes until the box is full.

STEP 3 Record how many cubes it took to fill the box. This is the volume of the box in cubic units.

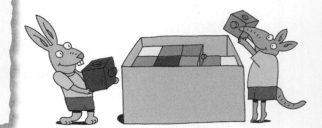

• How does your estimate compare with the actual volume?

MATH IDEA To measure the volume of a solid, find the number of cubic units needed to fill the solid.

CALIFORNIA STANDARDS ○━━MG 1.2 Estimate or determine the area and volume of solid figures by covering them with squares or by counting the number of cubes that would fill them. **MR 2.1** Use estimation to verify the reasonableness of calculated results. *also* **MG 2.5, MR 2.3, MR 2.4**

Find the Volume

When you cannot count each cube, you can think about layers to find the volume.

> **Example** **Find the volume of each solid.**
>
> Since you cannot see each cube, look at the top layer of cubes.
> Number of layers × cubes in each layer = Volume
>
>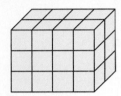
>
> 3 layers × 8 cubes per layer = 24 cubic units
> So, the volume of the cube is 24 cubic units.
>
> 2 layers × 6 cubes per layer = 12 cubic units
> So, the volume of the cube is 12 cubic units.

▶ Check

1. **Describe** how you could build another solid with 24 cubic units. You may use color cubes to help.

Use cubes to make each solid. Then write the volume in cubic units.

2.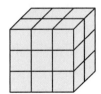

3.

4.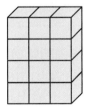

5.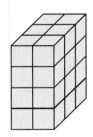

LESSON CONTINUES

▶ **Practice and Problem Solving**

Use cubes to make each solid. Then write the volume in cubic units.

6.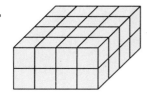

7.

8.

Find the volume of each solid. Write the volume in cubic units.

9.

10.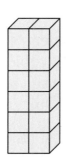

11.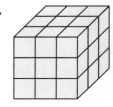

12.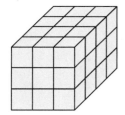

13.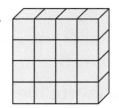

14.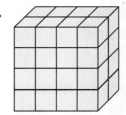

15. ⟨a+b/c **Algebra**⟩ Each layer of Andrew's prism is 6 cubic units. Its volume is 12 cubic units. How many layers are in the prism? You may use cubes to help.

16. ⟨? **What's the Error?**⟩ Justin found the volume of this solid. He said the volume was 16 cubic units. Describe his error. Give the correct volume.

17. Sally's box is 2 cubes long, 2 cubes wide, and 3 cubes high. What is the volume of her box?

18. Todd's box is 4 cubes long and 4 cubes wide. It has a volume of 32 cubic units. What is the height of the box?

19. ⟨**Write About It**⟩ How is finding the area different from finding the volume?

Mixed Review and Test Prep

Find each sum or missing addend.

(p. 68)

20. $45 + 12 + 5 = \blacksquare$

21. $35 + 15 + 20 = \blacksquare$

22. $102 + \blacksquare + 15 = 120$

23. $\blacksquare + 46 + 10 = 61$

24. $13 + 22 + \blacksquare = 42$

25. $200 + \blacksquare = 302$

26. **TEST PREP** Jane had 205 stickers. She gave 28 stickers to her sister. How many stickers does Jane have left? (p. 58)

A 175   **C** 180

B 177   **D** 182

27. **TEST PREP** Which is NOT true? (p. 132)

F $0 \times 4 = 0$ **H** $1 \times 3 = 3$

G $0 \times 0 = 1$ **J** $7 \times 1 = 7$

LINKUP

to Reading

STRATEGY • COMPARE When you **compare** information, you look for things that are alike. When you **contrast** information, you look for things that are different.

Read the information in the problem. Compare and contrast it with each answer choice.

A 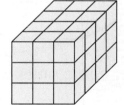   B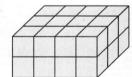

Craig made a prism with 12 cubes in each layer. Its volume was 24 cubic units. Which prism did Craig make?

PRISMS A AND B

Choice	Compare	Contrast
A	12 cubes in each layer	volume = 36 cubic units
B	12 cubes in each layer	volume = 24 cubic units

So, Craig made prism B.

1. Thelma's prism had 12 cubes in each layer. It had 3 layers. Which prism did she make?

2. Hikara made a prism. One of the faces had an area of 6 square units. Which prism did he make?

Review/Test

✔ CHECK VOCABULARY AND CONCEPTS

Choose the best term from the box.

> area
> cubic units
> perimeter

1. To measure volume, you use __?__ . (p. 468)

2. The distance around a figure is called its __?__ . (p. 456)

Find the perimeter of each figure.
(pp. 456–457)

3. 4. 5.

Write the area in square units.
(pp. 462–463)

6. 7. 8.

✔ CHECK SKILLS

Find the perimeter. (pp. 458–461)

9. 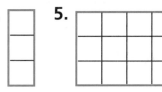 1 cm /\ 1 cm
1 cm

10. 1 cm
1 cm
1 cm
2 cm
1 cm
2 cm

Find the total area. (pp. 464–465)

11.

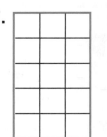

Write the volume in cubic units. (pp. 468–471)

12. 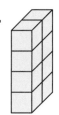 13.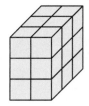

✔ CHECK PROBLEM SOLVING

Solve. (pp. 466–467)

14. Pedro has 20 inches of ribbon. He wants to make a rectangle or square with the greatest possible area. How wide should it be? How long should it be?

15. Nora has 14 feet of fencing. She wants to make a rectangle with an area less than 10 square feet. How long should it be? How wide should it be?

Cumulative Review

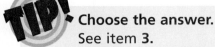

Choose the answer.
See item **3**.
Find the area for each answer choice. Compare and order the areas to find the greatest one.
Also see problem **6**, p. H65.

For 1–7, choose the best answer.

1. What is the area?

A 28 square units
B 35 square units
C 42 square units
D 45 square units

2. What is the perimeter?

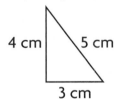

4 cm 5 cm
3 cm

F 5 cm   **H** 12 cm
G 7 cm   **J** 60 cm

3. Debbie has 24 feet of fencing. She wants to build a pen for her puppies. Which measurements would have the greatest area?

A 3 feet wide, 9 feet long
B 4 feet wide, 8 feet long
C 5 feet wide, 7 feet long
D 6 feet wide, 6 feet long

4. What is the perimeter?

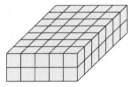

5 ft
2 ft

F 7 ft   **H** 21 ft
G 14 ft   **J** 28 ft

5. What is the volume?

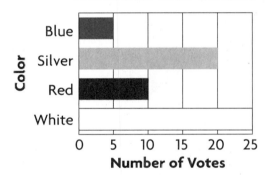

A 70 cubic units
B 35 cubic units
C 14 cubic units
D 10 cubic units

For 6–7, use the graph.

FAVORITE CAR COLOR

6. Which color received 20 votes?

F red   **H** blue
G white **J** silver

7. How many people voted for a favorite car color?

A 60   **C** 40
B 50   **D** NOT HERE

MATH DETECTIVE

Around and Around You Go

REASONING Use grid paper to copy the array. Write the letters in each square as shown. Then read each clue and follow directions to solve each case!

Clue 1: What is the perimeter of the rectangle?

A	B	C	D
E	F	G	H
I	J	K	L

Clue 2: Use the same sheet of grid paper. Draw the figure with squares A, D, I, and L missing. What is the perimeter of this figure?

Clue 3: How is the perimeter of the second figure related to the perimeter of the first figure?

Clue 4: Look at the figure you drew in Clue 2. Can you remove squares from this figure and keep the same perimeter? If so, draw the new figure.

Clue 5: Look at the figure you drew in Clue 1. How could you remove a square or squares to change the perimeter to 16 units? Draw the new figure.

STRETCH YOUR THINKING Use grid paper. Draw an array with a perimeter of 12 units. Then copy that array, but remove squares to make a different figure with a perimeter of 12 units.

Challenge

Find Circumference of a Circle

Suppose you want to make a bank out of an empty can. How could you find the distance around the can?

The distance around a circle is its **circumference.**

> **Remember**
> The distance around an object or plane figure is called its perimeter.

Activity

MATERIALS: can, string, centimeter ruler

STEP 1

Wrap the string around the can.

STEP 2

Use a ruler to measure the length of the string.

STEP 3

Copy the table. Write the circumference of the can to the nearest centimeter.

Object	Circumference
Can	■
Cup	■
Quarter	■
Doorknob	■

• How is finding the circumference of a circle like finding the perimeter of a rectangle? How is it different?

Try It

Find the circumference of each object. Record the measurement in your table.

1. drinking cup **2.** quarter **3.** doorknob

Study Guide and Review

VOCABULARY

Choose the best term from the box.

1. A square with a side length of 1 unit is a _?_. (p. 462)

square unit
perimeter

STUDY AND SOLVE

Chapter 24

Measure to the nearest inch.

The pin is 3 inches long to the nearest inch.

Measure to the nearest inch.
(pp. 420–423)

2.

3.

Change units.

Use a rule to find the number of quarts in 2 gallons.
Remember: 4 quarts = 1 gallon
Rule: Multiply the number of gallons by 4.
■ quarts = 2 gallons
■ = 2 × 4

8 = 2 × 4
So, there are 8 quarts in 2 gallons.

Use the rule to change the units (8 pints = 1 gallon). (pp. 430–433)

4. How many pints are in 5 gallons?
Rule: Multiply the number of gallons by 8.

■ = 5 × 8
■ pints = 5 gallons

Chapter 25

Measure to the nearest centimeter.

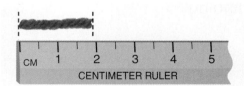

CM 1   2   3   4   5
CENTIMETER RULER

The string is 2 cm long to the nearest centimeter.

Measure to the nearest centimeter.
(pp. 440–443)

5.

6.

7.

8.

Estimate length and distance.

Choose the better estimate.

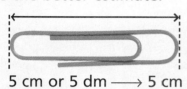

5 cm or 5 dm ⟶ 5 cm

Choose the better estimate.
(pp. 440–443)

9. Your thumb: 5 dm or 5 cm?

10. A lamp post: 3 dm or 3 m?

11. A pen: 15 cm or 15 dm?

Chapter 26

Find perimeter and area.

Add the lengths of the sides to find the perimeter.
$3 + 5 + 3 + 5 = 16$
So, the perimeter is 16 units.
Count the number of square units to find the area.
So, the area is 15 square units.

Find the perimeter. (pp. 456–461)

12.

3 cm

1 cm [   ] 1 cm

3 cm

Write the area in square units.
(pp. 462–463)

13.

14.

Find volume.

This figure has 2 layers of 4 cubes.
Find the volume by adding the number of cubes in each layer.
$4 + 4 = 8$
So, the volume is 8 cubic units.

Write the volume in cubic units.
(pp. 468–471)

15.

16.

PROBLEM SOLVING PRACTICE

Solve. (pp. 444–445, 466–467)

17. Linda's room is 400 cm wide. How many meters wide is it?

18. Jed makes a rectangle with the greatest possible area using 16 inches of string. What are the length and width of his rectangle?

California Connections

THE TOURNAMENT OF ROSES® PARADE

Each year on January 1 or 2, Pasadena is host to the famous Tournament of Roses Parade.

▲ In the parade there are floats covered with flowers, marching bands from all over the country, and trained horses and their riders.

Choose the unit you would use to measure each.
Write *inch, foot, yard,* or *mile.*

1.

the distance a band marches

2.

the width of a flower

3.

the length of a float

4. The thermometer shows the temperature at the start of the Rose Parade®. If the temperature is 63°F at the end of the parade, by how many degrees did the temperature rise?

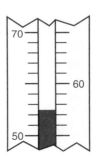

THE ROSE BOWL GAME®

The Rose Bowl college football game is played after the parade. The Rose Bowl stadium seats about 93,000 people.

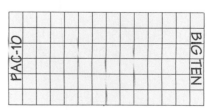

▲ The Rose Bowl Game has been a sellout every year since 1947.

Choose the unit you would use to measure each. Write *cm, m,* or *km.*

1.

the height of the stadium

2.

the width of a football helmet

3.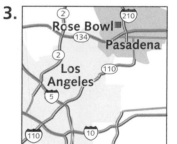

the distance between Los Angeles and the Rose Bowl stadium

USE DATA For 4–5, use the diagram of the football field.

4. What is the perimeter of the diagram of the football field?

5. What is the area of the diagram? Write the area in square units.

Understand Fractions

There are thousands of different kinds of insects in the world. Some are very large. Others are so small that we can't even see them. The table shows some insects and their sizes. Which insect is the smallest? Which is the largest?

SIZES OF INSECTS	
Insect	Length
Ant	$\frac{1}{4}$ inch
Mosquito	$\frac{1}{8}$ inch
Grasshopper	2 inches
Goliath beetle	$4\frac{1}{2}$ inches

African Goliath beetle

CHECK WHAT YOU KNOW ✓

Use this page to help you review and remember
important skills needed for Chapter 27.

✓ MODEL PARTS OF A WHOLE (See p. H28.)

Write how many equal parts make up the whole figure.
Then write how many parts are *not* shaded.

1.

2.

3.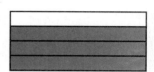

4.

5.

6.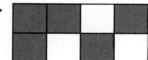

Write how many equal parts make up the whole
figure. Then write how many parts *are* shaded.

7.

8.

9.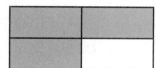

✓ MODEL PARTS OF A GROUP (See p. H29.)

Write the number in each group. Then write the
number in each group that is *not* green.

10.

11.

12.

Write the number in each group. Then write the
number in each group that *is* green.

13.

14.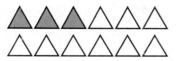

15.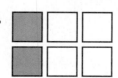

Count Parts of a Whole

Quick Review

Find the missing number in the pattern.

1. 2, 4, 6, 8, ■

2. 12, 11, 10, 9, ■

3. 3, 6, 9, 12, ■

4. 11, 9, 7, 5, ■

5. 4, 8, 12, 16, ■

▶ **Learn**

ALL TOGETHER

A number that names part of a whole or part of a set is called a **fraction**.

What fraction of this pizza has sausage?

1 part sausage   → $\frac{1}{6}$ ← numerator

6 equal parts in all →   ← denominator

Read: one sixth   **Write:** $\frac{1}{6}$

    one part out of six parts

    1 divided by 6

So, $\frac{1}{6}$ of the pizza has sausage.

The **numerator** tells how many parts are being counted.

The **denominator** tells how many equal parts are in the whole.

- What fraction of the pizza does *not* have sausage? Explain how you know.

VOCABULARY

fraction

numerator

denominator

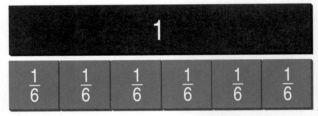

These fraction bars show how a whole can be divided into sixths, or six equal parts.

Examples A whole can be divided into equal parts.

A  $\frac{2}{5}$ two fifths

B 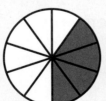 $\frac{4}{10}$ four tenths

C 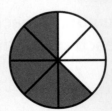 $\frac{5}{8}$ five eighths

CALIFORNIA STANDARDS NS 3.0 Students understand the relationship between whole numbers, simple fractions, and decimals. **MR 1.1** Analyze problems by identifying relationships, distinguishing relevant from irrelevant information, sequencing and prioritizing information, and observing patterns. *also* **MR 2.3**

Counting Equal Parts

You can count equal parts, such as sixths, to make one whole.

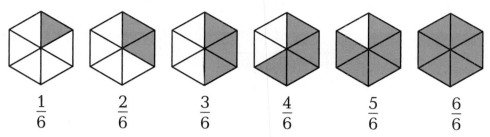

$\frac{1}{6}$ $\frac{2}{6}$ $\frac{3}{6}$ $\frac{4}{6}$ $\frac{5}{6}$ $\frac{6}{6}$

$\frac{6}{6}$ = one whole

A number line can show parts of one whole.

The part from 0 to 1 on these number lines shows one whole.
The line can be divided into any number of equal parts.

This number line is divided into sixths.

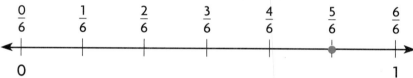

The point shows the location of $\frac{5}{6}$.

Examples

A This number line is divided into thirds.

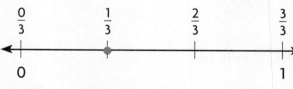

The point shows the location of $\frac{1}{3}$.

B This number line is divided into fourths.

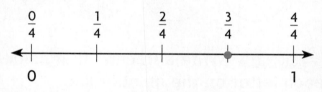

The point shows the location of $\frac{3}{4}$.

▶ Check

1. Write how to count by eighths to make one whole.

Write a fraction in numbers and words that names the shaded part.

2.

3.

4.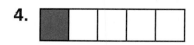

LESSON CONTINUES ▶

Write a fraction in numbers and words that names the shaded part.

5.

6.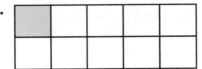

7.

8.

9.

10.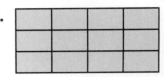

Write the fraction, using numbers.

11. one fourth

12. four out of nine

13. six sevenths

14. two divided by three

15. three fifths

16. five out of twelve

Write a fraction to describe each shaded part.

17.

18.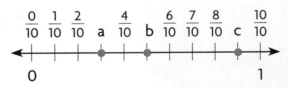

$\frac{a+b}{c}$ **Algebra** Write a fraction that names the point of each letter on the number line.

19.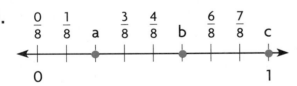

20.

21. REASONING There are two pizzas the same size. One is cut into 6 equal pieces. The other is cut into 8 equal pieces. Which pizza has smaller pieces?

22. **? What's the Error?** Lydia said the fraction that names this drawing is $\frac{6}{4}$. Explain Lydia's mistake.

23. Suppose you and 4 friends share equal pieces of a pie. What fraction describes your part?

24. Use crayons to draw a picture that shows each fraction.

    a. $\frac{1}{2}$ purple   **b.** $\frac{1}{4}$ green

Mixed Review and Test Prep

Find each difference. (p. 58)

25. $245 - 52 = \blacksquare$

26. $736 - 379 = \blacksquare$

27. $400 - 118 = \blacksquare$

Find each product.

28. $4 \times 8 = \blacksquare$ (p. 134)

29. $6 \times 6 = \blacksquare$ (p. 148)

30. $7 \times 9 = \blacksquare$ (p. 150)

31. $7 \times 0 = \blacksquare$ (p. 132)

32. $9 \times 9 = \blacksquare$ (p. 164)

33. $5 \times 1 = \blacksquare$ (p. 132)

34. **TEST PREP** Which polygon has 3 sides and 3 angles? (p. 382)

    A triangle   **C** pentagon
    B quadrilateral **D** hexagon

35. **TEST PREP** Theo measured the distance from his house to the hobby store across town. Which unit did he use to measure? (p. 424)

    F inch   **H** yard
    G foot   **J** mile

LINKUP to Social Studies

These alphabet flags are used on ships to send messages in code. For example, if a ship flies the flag for the letter P, that ship is about to sail out of the harbor. Ships carry books that explain the codes in nine different languages.

USE DATA For 1–4, use the flags.

1. Look at the flag for the letter G. What fraction names the part of the flag that is yellow?

2. Look at all of the flags. Which of them are divided into four equal parts, or fourths?

3. Look at the flag for the letter N. Into how many equal parts is the flag divided?

4. Write a fraction that names one part of each of these flags: L, O, and T.

Count Parts of a Group

Quick Review

Write the fraction that comes next in the counting pattern.

1. $\frac{0}{3}, \frac{1}{3}, \frac{2}{3}, \blacksquare$ 2. $\frac{1}{5}, \frac{2}{5}, \frac{3}{5}, \blacksquare$

3. $\frac{2}{6}, \frac{3}{6}, \frac{4}{6}, \blacksquare$ 4. $\frac{3}{8}, \frac{4}{8}, \frac{5}{8}, \blacksquare$

5. $\frac{3}{10}, \frac{4}{10}, \frac{5}{10}, \blacksquare$

▶ Learn

GO FISH Allison and her dad went to Pal's Pet Store. Allison chose 8 fish and her dad chose 8 fish. What fraction of the fish are black mollies?

Allison's Fish

part that is
black mollies → $\underset{4}{1}$ ← numerator
    total parts → $\overline{4}$ ← denominator

Read: one fourth, or one out of four

Write: $\frac{1}{4}$

So, $\frac{1}{4}$ of Allison's fish are black mollies.

Dad's Fish

part that is
black mollies → 1 ← numerator
    total parts → 2 ← denominator

Read: one half, or one out of two

Write: $\frac{1}{2}$

So, $\frac{1}{2}$ of Dad's fish are black mollies.

MATH IDEA Use fractions to show parts of a group.

▶ Check

1. **What if** Allison had 2 fish in each of 3 bags? What fraction names the fish in one bag?

Write the fraction that names the part of each group that is circled.

2.

3.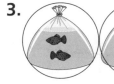

CALIFORNIA STANDARDS NS 3.0 Students understand the relationship between whole numbers, simple fractions, and decimals. **MR 1.1** Analyze problems by identifying relationships, distinguishing relevant from irrelevant information, sequencing and prioritizing information, and observing patterns. *also* **MR 2.3**

For 4–7, write the fraction that names the part of each group that is circled.

4.

5.

6.

7.

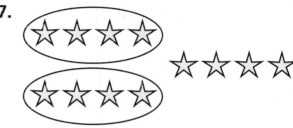

8. Draw 9 circles. Circle $\frac{1}{3}$ of them.

9. Draw 6 triangles. Circle $\frac{1}{2}$ of them.

10. Draw 16 stars. Circle $\frac{3}{4}$ of them.

Use a pattern to complete the table.

11.	Model	●●●●●	●●●●●	▦	●●●●●	●●●●●	●●●●●
12.	Total number of parts	5	▦	5	5	5	5
13.	Number of green parts	0	1	2	▦	4	5
14.	Fraction of green parts	$\frac{0}{5}$	$\frac{1}{5}$	$\frac{2}{5}$	$\frac{3}{5}$	$\frac{4}{5}$	▦

15. Esther has 12 ribbons. Of those ribbons, $\frac{1}{12}$ are red. The rest are yellow. How many yellow ribbons does she have?

16. **? What's the Question?** Jonas has 4 blue tiles, 3 green tiles, and 1 yellow tile. The answer is $\frac{7}{8}$.

17. ✎ **Write a problem** in which a fraction is used to name part of a group. Tell what the numerator and denominator mean.

Mixed Review and Test Prep

Find each quotient. (p. 204)

18. $4 \div 4 = $ ▦

19. $9 \div 1 = $ ▦

20. $0 \div 7 = $ ▦

21. $8 \div 8 = $ ▦

22. **TEST PREP** Find the missing factor.
$3 \times$ ▦ $\times 7 = 21$ (p. 170)

A 0

C 2

B 1

D 21

Equivalent Fractions

Quick Review

Name the fraction for the shaded part.

1.   2.

3.   4.

5.

▶ Learn

EQUAL PARTS   Two or more fractions that name the same amount are called **equivalent fractions**.

What other fractions name $\frac{1}{2}$?

VOCABULARY
equivalent fractions

Activity 1
MATERIALS: sheet of paper

STEP 1

Fold a sheet of paper in half. Shade one half of the paper blue.

1 out of 2 equal parts is blue.
$\frac{1}{2}$ of the paper is blue.

STEP 2

Fold the paper in half again.

2 out of 4 equal parts are blue.
$\frac{2}{4}$ of the paper is blue.

TECHNOLOGY LINK

More Practice: Use E-Lab, *Equivalent Fractions*.
www.harcourtschool.com/elab2002

STEP 3

Fold the paper in half a third time.

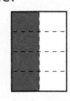

4 out of 8 equal parts are blue.
$\frac{4}{8}$ of the paper is blue.

$$\frac{1}{2} = \frac{2}{4} = \frac{4}{8}$$

So, $\frac{1}{2}$, $\frac{2}{4}$, and $\frac{4}{8}$ are all names for $\frac{1}{2}$.

They are equivalent fractions.

 CALIFORNIA STANDARDS NS 3.1 Compare fractions represented by drawings or concrete materials to show equivalency and to add and subtract simple fractions in context. **MR 1.1** Analyze problems by identifying relationships, distinguishing relevant from irrelevant information, sequencing and prioritizing information, and observing patterns. *also* **NS 3.0, MR 2.3**

Activity 2

Use fraction bars to find equivalent fractions.
What fraction is equivalent to $\frac{2}{3}$?

STEP 1

Start with the bar for 1 whole. Line up the bars for $\frac{2}{3}$.

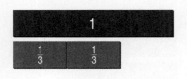

STEP 2

Use $\frac{1}{6}$ bars to match the length of the two $\frac{1}{3}$ bars.

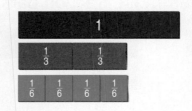

STEP 3

Count the number of $\frac{1}{6}$ bars that make up $\frac{2}{3}$. Write the equivalent fraction.

Count: $\frac{1}{6}$ $\frac{2}{6}$ $\frac{3}{6}$ $\frac{4}{6}$

Write: $\frac{2}{3} = \frac{4}{6}$

- How can you tell that the fraction bars show equivalent fractions?

- Use sixths, eighths, and tenths fraction bars. Find fractions that are equivalent to $\frac{1}{2}$.

Examples

A

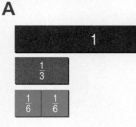

$$\frac{1}{3} = \frac{2}{6}$$

B

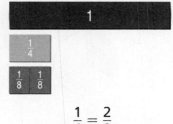

$$\frac{1}{4} = \frac{2}{8}$$

C

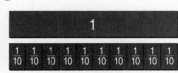

$$1 = \frac{10}{10}$$

▶ Check

1. **Explain** how to use fraction bars to decide if $\frac{2}{4}$ and $\frac{2}{3}$ are equivalent.

Find an equivalent fraction. Use fraction bars.

2.

3.

LESSON CONTINUES

Find an equivalent fraction. Use fraction bars.

4.

5.

6.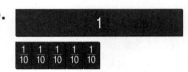

Find the missing numerator. Use fraction bars.

7.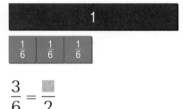

$$\frac{3}{6} = \frac{\blacksquare}{2}$$

8.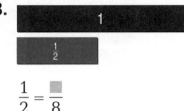

$$\frac{1}{2} = \frac{\blacksquare}{8}$$

9.

$$\frac{3}{5} = \frac{\blacksquare}{10}$$

10. $\frac{3}{9} = \frac{\blacksquare}{3}$

11. $\frac{1}{2} = \frac{\blacksquare}{10}$

12. $\frac{2}{4} = \frac{\blacksquare}{12}$

13. $\frac{1}{4} = \frac{\blacksquare}{8}$

14. $\frac{2}{6} = \frac{\blacksquare}{3}$

15. $\frac{3}{5} = \frac{\blacksquare}{10}$

16. Write the fraction that names the shaded part of each. Then tell which fractions are equivalent.

   a. 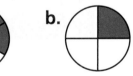   b.   c.   d.   e.

17. Molly's age is an even number between 20 and 29. Her age can be evenly divided by 3. How old is Molly?

18. Pablo likes oranges and plums, but not apples or pears. What fraction of the fruit does Pablo like?

19. **a+b/c Algebra** In his yard, Mr. York has a 42 foot tall pine tree. The pine tree is 15 feet taller than his oak tree. How tall is his oak tree?

20. **Write About It** Explain how to use fraction bars to find equivalent fractions for $\frac{4}{8}$.

21. Jason ate $\frac{2}{6}$ of the pie, and Wesley ate $\frac{1}{3}$ of it. Who ate more pie?

22. **REASONING** Sumi used 6 of one kind of fraction bar to show $\frac{1}{2}$. What kind of fraction bars did she use?

Mixed Review and Test Prep

Round to the nearest thousand. (p. 30)

23. 833   **24.** 2,497   **25.** 5,555   **26.** 39,670

Find the quotient. (p. 220)

27. 90 ÷ 9   **28.** 56 ÷ 8   **29.** 49 ÷ 7   **30.** 72 ÷ 9

31. (**TEST PREP**) Emma left at 11:15 A.M. and returned home at 12:37 P.M. How long was she gone? (p. 98)
- **A** 1 hour 12 minutes
- **B** 1 hour 22 minutes
- **C** 1 hour 42 minutes
- **D** 1 hour 52 minutes

32. (**TEST PREP**) Which set is equivalent to 3 quarters? (p. 80)
- **F** 7 nickels, 25 pennies
- **G** 6 dimes, 5 pennies
- **H** 1 quarter, 9 nickels
- **J** 5 dimes, 5 nickels

LINKUP to Science

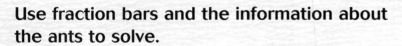

Did you know that there may be as many as 14,000 different kinds of ants in the world?

Not all ants are alike. Different species of ants live in different places, eat different foods, and are different colors and sizes. Here are some interesting facts about three kinds of ants.

Use fraction bars and the information about the ants to solve.

1. Find an equivalent fraction for the length of
   a. the fire ant   b. the bulldog ant.

2. Find how many odorous house ants it would take to equal the length of one bulldog ant. Write the equivalent fractions.

Fire ants
- are red in color
- will sting
- are about $\frac{1}{4}$ inch long

Bulldog ants
- eat meat
- are found in Australia
- are about $\frac{4}{5}$ inch long

Odorous house ants
- are brown or black in color
- smell bad when crushed
- are about $\frac{1}{10}$ inch long

Compare and Order Fractions

Quick Review

Compare. Write $<$, $>$, or $=$ for each ●.

1. 27 ● 37

2. 398 ● 399

3. 241 ● 214

4. 4,012 ● 4,012

5. 5,431 ● 5,341

▶ **Learn**

SIZE IT UP Fraction bars can help you compare parts of a whole.

Examples

A Compare $\frac{1}{4}$ and $\frac{2}{4}$.

The bar for $\frac{1}{4}$ is shorter than the bars for $\frac{2}{4}$.

So, $\frac{1}{4} < \frac{2}{4}$, or $\frac{2}{4} > \frac{1}{4}$.

B Compare $\frac{1}{3}$ and $\frac{1}{4}$.

The bar for $\frac{1}{4}$ is shorter than the bar for $\frac{1}{3}$.

So, $\frac{1}{4} < \frac{1}{3}$, or $\frac{1}{3} > \frac{1}{4}$.

• When the denominator is larger, is the fraction bar larger or smaller? Why?

Tiles can help you compare parts of a group.

Examples

A Compare $\frac{2}{5}$ and $\frac{3}{5}$.

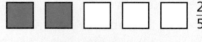 $\frac{2}{5}$

$\frac{3}{5}$

3 tiles are more than 2 tiles.

So, $\frac{3}{5} > \frac{2}{5}$, or $\frac{2}{5} < \frac{3}{5}$.

B Compare $\frac{4}{6}$ and $\frac{2}{3}$.

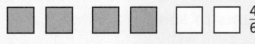 $\frac{4}{6}$

 $\frac{2}{3}$

4 tiles is the same as 4 tiles.

So, $\frac{4}{6} = \frac{2}{3}$.

• How do you compare fractions when the denominators are the same? when the denominators are different?

CALIFORNIA STANDARDS NS 3.1 Compare fractions represented by drawings or concrete materials to show equivalency and to add and subtract simple fractions in context. **MR 1.1** Analyze problems by identifying relationships, distinguishing relevant from irrelevant information, sequencing and prioritizing information, and observing patterns. *also* **NS 3.0, MR 2.3**

Ordering Fractions

You can order three or more fractions from least to greatest or from greatest to least.

Cassie needs $\frac{3}{8}$ cup raisins, $\frac{1}{4}$ cup chocolate chips, and $\frac{2}{3}$ cup peanuts to make a trail mix. She wants to know which ingredient she needs the most of. Use fraction bars to order the fractions from greatest to least.

STEP 1 Compare the fractions.

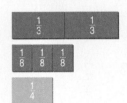

$$\frac{2}{3} > \frac{3}{8}$$

$$\frac{3}{8} > \frac{1}{4}$$

STEP 2 Order the fractions from greatest to least.

Think: $\frac{2}{3} > \frac{3}{8} > \frac{1}{4}$

Write: $\frac{2}{3}, \frac{3}{8}, \frac{1}{4}$

So, the fractions in order from greatest to least are $\frac{2}{3}, \frac{3}{8},$ and $\frac{1}{4}$.

• Order the fractions from least to greatest.

▶ Check

1. **Describe** what happens to the size of fraction bars when the denominators become greater, such as $\frac{1}{2}, \frac{1}{3},$ and $\frac{1}{4}$.

2. **Describe** what happens to the size of fraction bars when the denominators become smaller, such as $\frac{1}{8}, \frac{1}{6},$ and $\frac{1}{4}$.

Compare. Write <, >, or = for each ●.

3.
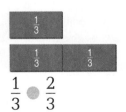

$$\frac{1}{3} ● \frac{2}{3}$$

4.
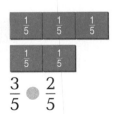

$$\frac{3}{5} ● \frac{2}{5}$$

5.
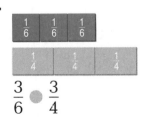

$$\frac{3}{6} ● \frac{3}{4}$$

LESSON CONTINUES ▶

Compare. Write <, >, or = for each ⬤.

6.

$\dfrac{1}{2}$ ⬤ $\dfrac{1}{4}$

7.

$\dfrac{9}{10}$ ⬤ $\dfrac{9}{10}$

8.

$\dfrac{3}{12}$ ⬤ $\dfrac{5}{8}$

9.

$\dfrac{1}{4}$ ⬤ $\dfrac{2}{4}$

10.

$\dfrac{3}{6}$ ⬤ $\dfrac{1}{3}$

11.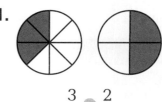

$\dfrac{3}{8}$ ⬤ $\dfrac{2}{4}$

Compare the part of each group that is green.
Write <, >, or = for each ⬤.

12.

$\dfrac{2}{7}$ ⬤ $\dfrac{3}{7}$

13.

$\dfrac{4}{4}$ ⬤ $\dfrac{3}{4}$

14.

$\dfrac{4}{9}$ ⬤ $\dfrac{4}{9}$

15. Order $\dfrac{1}{3}$, $\dfrac{1}{6}$, and $\dfrac{4}{6}$ from greatest to least.

16. Order $\dfrac{1}{2}$, $\dfrac{3}{4}$, and $\dfrac{2}{5}$ from least to greatest.

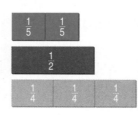

Use Data For 17–19, use the bar graph.

17. How many leaves did Kevin collect in all?

18. What fraction of the leaves are red? yellow? orange?

19. Order the fractions of leaves from the greatest to the least amount.

KEVIN'S LEAVES

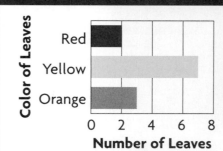

20. REASONING Terry says that because 10 is greater than 6, $\frac{5}{10}$ is greater than $\frac{5}{6}$. Do you agree or disagree? Explain.

21. Write the next two fractions in the pattern. Explain the rule.
$\frac{1}{8}, \frac{2}{8}, \frac{3}{8}, \frac{4}{8}, \frac{5}{8}, \blacksquare, \blacksquare$

Mixed Review and Test Prep

Order each set of numbers from least to greatest. (p. 24)

22. 45, 54, 50   **23.** 110, 111, 101   **24.** 199, 89, 98   **25.** 455, 555, 545

Find the missing factor. (p. 142)

26. $4 \times \blacksquare = 36$ **27.** $\blacksquare \times 9 = 45$ **28.** $7 \times \blacksquare = 56$ **29.** $\blacksquare \times 9 = 81$

30. **TEST PREP** Cheryl bought two markers for $0.55 each and a notebook for $1.79. How much did she spend in all? (p. 88)

   A $1.89   **C** $2.79
   B $2.34   **D** $2.89

31. **TEST PREP** How many feet are in 4 yards? (p. 430)

   F 20 feet   **H** 9 feet
   G 12 feet   **J** 6 feet

Thinker's Corner

MIXED NUMBERS Use whole numbers and fractions to name amounts that are greater than 1. A **mixed number** is made up of a whole number and a fraction.

Randy ate 1 whole sandwich and $\frac{1}{2}$ of another sandwich for lunch. So, Randy ate 3 halves, or $1\frac{1}{2}$ sandwiches, for lunch.

Write: $1\frac{1}{2}$
Read: one and one half

For 1–4, write a mixed number for the parts that are shaded.

1.

2.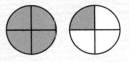

3.

4.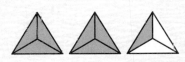

Problem Solving Strategy
Make a Model

Understand ➡ Plan ➡ Solve ➡ Check

Quick Review

Compare. Write <, >, or = for each ⬤.

1. $\frac{2}{3}$ ⬤ $\frac{2}{3}$ 2. $\frac{4}{6}$ ⬤ $\frac{3}{6}$

3. $\frac{1}{4}$ ⬤ $\frac{1}{2}$ 4. $\frac{4}{8}$ ⬤ $\frac{5}{10}$

5. $\frac{2}{3}$ ⬤ $\frac{1}{6}$

PROBLEM   Three classmates ran a relay in the Spring Sports Day race. George ran $\frac{2}{3}$ mile, Rosa ran $\frac{7}{8}$ mile, and Ben ran $\frac{5}{6}$ mile. Who ran the farthest?

- What are you asked to find?

- What information will you use?

- Is there information you will not use? If so, what?

- What strategy can you use to solve the problem?

  Make a model to show what part of a mile each person ran.

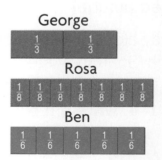

Solve

- How can you use the strategy to solve the problem?

  Model the problem by using fraction bars.

  Line up the fraction bars for $\frac{2}{3}$, $\frac{7}{8}$, and $\frac{5}{6}$.

  Compare the lengths of the fraction bars.

  Since $\frac{7}{8} > \frac{5}{6} > \frac{2}{3}$, Rosa ran the farthest.

Check

- What other strategy could you use to solve the problem?

CALIFORNIA STANDARDS NS 3.1 Compare fractions represented by drawings or concrete materials to show equivalency and to add and subtract simple fractions in context. **MR 1.1** Analyze problems by identifying relationships, distinguishing relevant from irrelevant information, sequencing and prioritizing information, and observing patterns. *also* **NS 3.0, MR 2.0, MR 2.3**

▶ Problem Solving Practice

Use *make a model* to solve.

1. **What if** George had run $\frac{9}{10}$ mile? Who would have run the farthest?

2. Tina used $\frac{2}{3}$ cup of milk, $\frac{1}{4}$ cup of sugar, and $\frac{1}{2}$ cup of nuts in a recipe. Of which ingredient did she use the most?

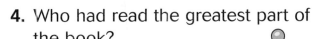
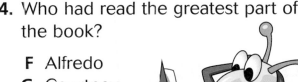

PROBLEM SOLVING STRATEGIES

Draw a Diagram or Picture
Make a Model or Act It Out
Make an Organized List
Find a Pattern
Make a Table or Graph
Predict and Test
Work Backward
Solve a Simpler Problem
Write an Equation
Use Logical Reasoning

Mr. Collins asked his science class to read a book on insects. By Friday, Alfredo had read $\frac{1}{2}$ of the book. Courtney had read $\frac{2}{5}$ of the book. Sandy had read $\frac{3}{8}$ of the book.

3. Which statement is true?

    A $\frac{3}{8} > \frac{2}{5} > \frac{1}{2}$

    B $\frac{1}{2} > \frac{2}{5} > \frac{3}{8}$

    C $\frac{1}{2} > \frac{3}{8} > \frac{2}{5}$

    D $\frac{2}{5} > \frac{3}{8} > \frac{1}{2}$

4. Who had read the greatest part of the book?

    F Alfredo
    G Courtney
    H Sandy
    J Peter

Mixed Strategy Practice

5. **REASONING** Mohammed folded a sheet of notebook paper in half and then folded it in half again. He unfolded the paper and shaded one of the equal parts red. What fraction shows how many parts are red?

6. Eric had 74 baseball cards. He traded 26 of his cards for 12 of Paul's cards. How many cards does he have now?

7. Arlo has basketball practice from 3:30 to 5 o'clock. It takes him a half hour to get home and one hour to do his homework before dinner. At what time does he eat dinner?

8. **REASONING** Kathy is cutting an apple into 8 equal slices. Julie is cutting an apple into 6 equal slices. Kathy says her slices will be larger because 8 is greater than 6. Do you agree? Explain.

Review/Test

✓ CHECK VOCABULARY AND CONCEPTS

Choose the best term from the box.

1. In the fraction $\frac{3}{8}$, the 3 is called the __?__ . (p. 482)

2. A number that names part of a whole or part of a set is a __?__ . (p. 482)

> fraction
> numerator
> denominator

Find an equivalent fraction. Use fraction bars. (pp. 488–491)

3.
4.
5.

✓ CHECK SKILLS

Write a fraction in numbers and in words that names the shaded part. (pp. 482–485)

6.
7.
8.
9.

Compare. Write <, >, or = for each ●. (pp. 492–495)

10.

$$\frac{1}{3} \bullet \frac{1}{4}$$

11.

$$\frac{3}{8} \bullet \frac{3}{5}$$

12.

$$\frac{4}{12} \bullet \frac{3}{6}$$

13.

$$\frac{2}{4} \bullet \frac{4}{8}$$

✓ CHECK PROBLEM SOLVING

Use *make a model* to solve. (pp. 496–497)

14. Luke walks $\frac{1}{2}$ mile to the park, Karen walks $\frac{4}{5}$ mile to the park, and Darius walks $\frac{1}{3}$ mile. Who walks the farthest?

15. Kumiko has 3 yellow candles, 4 blue candles, and 2 red candles. What fraction of the candles are red?

Cumulative Review

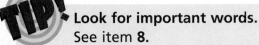

TIP! Look for important words. See item **8**.

Important words are *order from least to greatest*. Compare and order the fractions in each answer choice. Be sure you start with the least.

Also see problem **2**, p. H62.

For 1–8, choose the best answer.

1. Which fraction names the shaded part of the group?

A $\frac{1}{9}$ **C** $\frac{3}{9}$

B $\frac{1}{5}$ **D** $\frac{5}{9}$

2. Which fraction names the shaded part?

F $\frac{1}{5}$ **H** $\frac{3}{5}$

G $\frac{2}{5}$ **J** $\frac{2}{3}$

3. Which number makes this number sentence true?

$$213 + \blacksquare = 701$$

A 488   **C** 588

B 498   **D** 914

4. Ray and 7 friends share a melon cut into equal parts. What part of the melon will each person get?

F $\frac{1}{8}$ **G** $\frac{1}{7}$ **H** $\frac{1}{6}$ **J** $\frac{7}{8}$

5.
$$\begin{array}{r} 416 \\ \times\ \ 9 \\ \hline \end{array}$$

A 36,954   **C** 3,694

B 3,744   **D** NOT HERE

6. Which symbol makes this a true number sentence?

$\frac{1}{2}$ ● $\frac{2}{3}$ 

F >   **H** <

G =   **J** ×

7. Which number makes this number sentence true?

$\frac{1}{3} = \frac{\blacksquare}{6}$

A 1   **C** 3

B 2   **D** 6

8. Which set of fractions is in order from least to greatest?

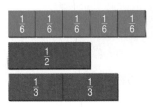

F $\frac{5}{6}, \frac{1}{2}, \frac{2}{3}$ **H** $\frac{1}{2}, \frac{5}{6}, \frac{2}{3}$

G $\frac{1}{2}, \frac{2}{3}, \frac{5}{6}$ **J** $\frac{2}{3}, \frac{1}{2}, \frac{5}{6}$

Add and Subtract Like Fractions

Corn bread, a bread made with cornmeal, was a favorite food of early pioneers. It is also a favorite food of many people today. It is usually served warm with butter. This recipe serves 4 people. How can you change the recipe to serve 8 people? Rewrite the recipe.

Corn Bread

1 cup cornmeal

$\frac{1}{4}$ cup flour

$\frac{1}{2}$ teaspoon baking powder

$\frac{1}{8}$ teaspoon baking soda

$\frac{3}{4}$ teaspoon salt

1 egg

1 tablespoon corn oil

$\frac{3}{4}$ cup buttermilk

Mix dry ingredients. Add egg, oil, and buttermilk. Mix well. Pour batter into a hot iron skillet. Bake at 425° for 20 to 25 minutes.

CHECK WHAT YOU KNOW

Use this page to help you review and remember important skills needed for Chapter 28.

✔ VOCABULARY

Choose the best term from the box.

1. Two or more fractions that name the same amount, such as $\frac{2}{4}$ and $\frac{1}{2}$, are called __?__ .

> numerators
> equivalent
> fractions

✔ NAME THE FRACTION (See p. H29.)

Name the fraction for the part that is shaded.

2.

3.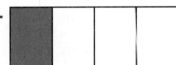

4.

5.

6.

7.

Write the fraction for the part that is green.

8.

9.

10.

✔ COMPARE FRACTIONS (See p. H30.)

Compare. Write $<$, $>$, or $=$ for each ●.

11.
$\frac{1}{3}$

$\frac{1}{3}$ $\frac{1}{3}$

$\frac{1}{3}$ ● $\frac{2}{3}$

12.
$\frac{1}{5}$ $\frac{1}{5}$ $\frac{1}{5}$ $\frac{1}{5}$

$\frac{1}{5}$ $\frac{1}{5}$

$\frac{4}{5}$ ● $\frac{2}{5}$

13.
$\frac{1}{6}$ $\frac{1}{6}$ $\frac{1}{6}$

$\frac{1}{6}$ $\frac{1}{6}$ $\frac{1}{6}$ $\frac{1}{6}$ $\frac{1}{6}$

$\frac{3}{6}$ ● $\frac{5}{6}$

14.
$\frac{1}{2}$

$\frac{1}{6}$ $\frac{1}{6}$ $\frac{1}{6}$

$\frac{1}{2}$ ● $\frac{3}{6}$

15.
$\frac{1}{5}$ $\frac{1}{5}$ $\frac{1}{5}$ $\frac{1}{5}$

$\frac{1}{10}$ $\frac{1}{10}$ $\frac{1}{10}$ $\frac{1}{10}$ $\frac{1}{10}$ $\frac{1}{10}$

$\frac{4}{5}$ ● $\frac{6}{10}$

16.
$\frac{1}{4}$ $\frac{1}{4}$

$\frac{1}{8}$ $\frac{1}{8}$ $\frac{1}{8}$ $\frac{1}{8}$

$\frac{2}{4}$ ● $\frac{4}{8}$

HANDS ON
Add Fractions

Quick Review

Name an equivalent fraction.

1. $\frac{4}{6}$ 2. $\frac{2}{8}$ 3. $\frac{2}{4}$

4. $\frac{2}{10}$ 5. $\frac{1}{2}$

VOCABULARY
like fractions

MATERIALS
fraction bars

▶ Explore

Fractions that have the same denominator are called **like fractions**.

Ryan and his mother are making fruit punch. The recipe says to add $\frac{2}{4}$ cup orange juice and $\frac{1}{4}$ cup pineapple juice. How much juice is needed altogether?

Use fraction bars to find $\frac{2}{4} + \frac{1}{4}$.

Remember
$\frac{1}{2}$ → numerator
$\quad$ → denominator

 STEP 1 Model $\frac{2}{4}$ with fraction bars.

$\frac{2}{4}$

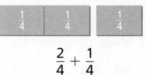 **STEP 2** Add one more $\frac{1}{4}$ fraction bar.

$\frac{2}{4} + \frac{1}{4}$

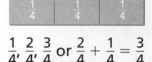 **STEP 3** Count the number of $\frac{1}{4}$ bars.

$\frac{1}{4}, \frac{2}{4}, \frac{3}{4}$ or $\frac{2}{4} + \frac{1}{4} = \frac{3}{4}$

So, the recipe calls for $\frac{3}{4}$ cup juice altogether.

- Why doesn't the denominator change when you find the sum?

- Explain how you could use fraction bars to find $\frac{1}{3} + \frac{1}{3}$.

I'm counting the fraction bars to find the sum: $\frac{1}{6}, \frac{2}{6}, \frac{3}{6}$... What comes next?

Try It

Use fraction bars to find the sum.

a. $\frac{3}{6} + \frac{1}{6}$ b. $\frac{3}{5} + \frac{2}{5}$

CALIFORNIA STANDARDS NS 3.1 Compare fractions represented by drawings or concrete materials to show equivalency and to add and subtract simple fractions in context. **○┑ NS 3.2** Add and subtract simple fractions. *also* **MR 1.1, MR 2.0, MR 2.3**

▶ Connect

Add like fractions by adding the numerators.

Tom peeled an orange. It had 8 wedges. He ate 1 wedge, or $\frac{1}{8}$ of it. Then he ate 4 more wedges, or $\frac{4}{8}$ of it. What fraction of the orange did he eat?

Model

Add the number of $\frac{1}{8}$ wedges that Tom ate.

$$\frac{1}{8} \quad + \quad \frac{4}{8}$$

Record

1 wedge + 4 wedges = 5 wedges
↓   ↓   ↓

$$\frac{1}{8} \quad + \quad \frac{4}{8} \quad = \quad \frac{5}{8}$$

So, Tom ate $\frac{5}{8}$ of the orange.

MATH IDEA When you add like fractions, you add the numerators, and the denominators stay the same.

▶ Practice

Find the sum.

1. [fraction bars: $\frac{1}{5}$ $\frac{1}{5}$ $\frac{1}{5}$]

$$\frac{1}{5} + \frac{2}{5} = \blacksquare$$

2. [fraction bars: $\frac{1}{8}$ $\frac{1}{8}$ $\frac{1}{8}$ $\frac{1}{8}$ $\frac{1}{8}$ $\frac{1}{8}$]

$$\frac{3}{8} + \frac{3}{8} = \blacksquare$$

3. [fraction bars: $\frac{1}{3}$ $\frac{1}{3}$]

$$\frac{1}{3} + \frac{1}{3} = \blacksquare$$

Use fraction bars to find the sum.

4. $\frac{2}{5} + \frac{2}{5} = \blacksquare$ **5.** $\frac{1}{6} + \frac{4}{6} = \blacksquare$ **6.** $\frac{4}{10} + \frac{3}{10} = \blacksquare$ **7.** $\frac{3}{8} + \frac{2}{8} = \blacksquare$

8. Kris has 6 plums and 2 apples. She buys 2 more apples. What fraction of the fruit are apples?

9. A recipe calls for $\frac{1}{4}$ cup sugar. Celia wants to double the recipe. How much sugar will she need?

Mixed Review and Test Prep

10. $7 \times 9 = \blacksquare$ (p. 150)

11. $8 \times 6 = \blacksquare$ (p. 152)

12. $72 \div 9 = \blacksquare$ (p. 218)

13. 1 foot = $\blacksquare$ inches (p. 424)

14. **TEST PREP** What is the quotient and remainder for $73 \div 4$? (p. 328)

A 20 r3   C 18 r1

B 19 r1   D 17 r3

Add Fractions

Quick Review
Compare. Write $<$, $>$, or $=$ for each ●.

1. $\frac{5}{8}$ ● $\frac{7}{8}$ 2. $\frac{3}{4}$ ● $\frac{1}{4}$

3. $\frac{3}{6}$ ● $\frac{1}{2}$ 4. $\frac{6}{10}$ ● $\frac{2}{5}$

5. $\frac{5}{10}$ ● $\frac{4}{8}$

▶ Learn

PIECES PLUS When you add fractions, you can show the sum in simplest form. A fraction is in **simplest form** when it uses the largest fraction bars possible.

Abby and Chris are sharing a sub sandwich for lunch. The whole sandwich has 8 pieces. Abby ate 2 pieces. Chris ate 2 pieces. How much of the sandwich did they eat?

VOCABULARY
simplest form

$\frac{2}{8}$ $+$ $\frac{2}{8}$ $=$?

Find $\frac{2}{8} + \frac{2}{8}$ in simplest form.

Activity

Materials: fraction bars

STEP 1 Model $\frac{2}{8}$ with fraction bars.

| $\frac{1}{8}$ | $\frac{1}{8}$ |

$\frac{2}{8}$

STEP 2 Add $\frac{2}{8}$.

| $\frac{1}{8}$ | $\frac{1}{8}$ | $\frac{1}{8}$ | $\frac{1}{8}$ |

$\frac{2}{8} + \frac{2}{8} = \frac{4}{8}$

STEP 3 Find the largest fraction bar that is the same length.

| $\frac{1}{8}$ | $\frac{1}{8}$ | $\frac{1}{8}$ | $\frac{1}{8}$ |
| $\frac{1}{2}$ | | | |

$\frac{2}{8} + \frac{2}{8} = \frac{4}{8}$

$\frac{4}{8}$ in simplest form is $\frac{1}{2}$.

So, Abby and Chris ate $\frac{1}{2}$ of the sub sandwich.

• How do you know if a fraction is in simplest form?

CALIFORNIA STANDARDS NS 3.1 Compare fractions represented by drawings or concrete materials to show equivalency and to add and subtract simple fractions in context. **O⊓ NS 3.2** Add and subtract simple fractions. *also* **MR 1.1, MR 2.3**

Adding Fractions

What if Chris ate 2 more pieces of the sandwich? How much of the sandwich was eaten altogether?

Use fraction bars to find $\frac{4}{8} + \frac{2}{8}$ in simplest form.

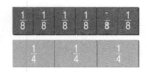

$$\frac{4}{8} \quad + \quad \frac{2}{8} \quad = \quad \frac{6}{8}, \text{ or } \frac{3}{4}$$

So, Abby and Chris ate $\frac{3}{4}$ of the sub sandwich.

• How do you know that $\frac{6}{8} = \frac{3}{4}$?

• Is $\frac{3}{8}$ in simplest form? Explain.

MATH IDEA You can use fraction bars to add fractions and to find the simplest form.

▶ Check

1. Explain how to find $\frac{3}{12} + \frac{1}{12}$ in simplest form.

Find the sum. Write the answer in simplest form.

2.

$$\frac{1}{4} + \frac{1}{4} = \blacksquare, \text{ or } \blacksquare$$

3.

$$\frac{2}{6} + \frac{2}{6} = \blacksquare, \text{ or } \blacksquare$$

4.

$$\frac{2}{10} + \frac{3}{10} = \blacksquare, \text{ or } \blacksquare$$

5.

$$\frac{2}{8} + \frac{4}{8} = \blacksquare, \text{ or } \blacksquare$$

6.

$$\frac{1}{12} + \frac{3}{12} = \blacksquare, \text{ or } \blacksquare$$

7.

$$\frac{2}{10} + \frac{2}{10} = \blacksquare, \text{ or } \blacksquare$$

LESSON CONTINUES

Find the sum. Write the answer in simplest form.

8.

$\frac{4}{12} + \frac{2}{12} = $ ■, or ■

9.

$\frac{1}{8} + \frac{1}{8} = $ ■, or ■

Find the sum. Write the answer in simplest form.
Use fraction bars if you wish.

10.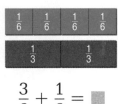

$\frac{3}{6} + \frac{1}{6} = $ ■

11.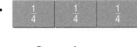

$\frac{2}{4} + \frac{1}{4} = $ ■

12.

$\frac{4}{8} + \frac{3}{8} = $ ■

13. $\frac{1}{5} + \frac{1}{5} = $ ■

14. $\frac{4}{6} + \frac{1}{6} = $ ■

15. $\frac{5}{10} + \frac{4}{10} = $ ■

16. $\frac{2}{6} + \frac{2}{6} = $ ■

17. $\frac{4}{10} + \frac{4}{10} = $ ■

18. $\frac{1}{3} + \frac{1}{3} = $ ■

19. $\frac{3}{12} + \frac{6}{12} = $ ■

20. $\frac{3}{5} + \frac{1}{5} = $ ■

21. $\frac{5}{8} + \frac{1}{8} = $ ■

22. $\frac{3}{10} + \frac{6}{10} = $ ■

23. $\frac{2}{8} + \frac{5}{8} = $ ■

24. $\frac{3}{8} + \frac{2}{8} = $ ■

25. $\frac{3}{12} + \frac{4}{12} = $ ■

26. $\frac{5}{10} + \frac{3}{10} = $ ■

27. $\frac{4}{8} + \frac{2}{8} = $ ■

28. $\frac{3}{5} + \frac{2}{5} = $ ■

29. **REASONING** Lamar's mother cut a cake into 12 equal pieces. Lamar's family ate 4 pieces. What fraction, in simplest form, tells how much of the cake was *not* eaten?

30. At camp, the boys hiked $\frac{2}{4}$ of the trail in the morning and $\frac{1}{4}$ of the trail in the afternoon. What part of the trail did they hike?

31. ✏️ **Write a problem** about a sub sandwich. Use fractions in your problem.

32. ❓ **What's the Question?** Sam folded his paper into 8 equal sections. He drew pictures on 3 of the sections. The answer is $\frac{5}{8}$.

33. A basketball game is divided into four quarters. Why is the break period after the first two quarters called *halftime*?

34. James bought a model car for $9.49. He paid with a $20 bill. How much change did he receive?

Mixed Review and Test Prep

Find the difference.

35. 724 (p. 58)
-182

36. 400 (p. 58)
$- 63$

37. 888 (p. 58)
-491

38. 1,355 (p. 62)
$- 628$

39. 3,742 (p. 62)
$-1,281$

Find the missing addend. (p. 42)

40. $35 + \blacksquare = 74$

41. $\blacksquare + 658 = 845$

42. $198 + \blacksquare = 451$

43. **TEST PREP** Which fraction is equivalent to $\frac{4}{6}$? (p. 488)

A $\frac{3}{4}$ **B** $\frac{2}{3}$ **C** $\frac{6}{12}$ **D** $\frac{1}{2}$

44. **TEST PREP** A box is 4 cubes long, 3 cubes wide, and 2 cubes high. What is the volume of this box? (p. 468)

F 9 cubic units **H** 20 cubic units
G 16 cubic units **J** 24 cubic units

Thinker's Corner

1,2,3, 4,5,?

Solve the riddle!

On which side does a chicken have more feathers?

Find the sum. Write the answer in simplest form. Then find the fraction below that matches. Record the letter from that box.

$\frac{3}{9} + \frac{1}{9}$	**T**	$\frac{2}{6} + \frac{2}{6}$	**O**	$\frac{5}{12} + \frac{5}{12}$	**S**	$\frac{1}{8} + \frac{1}{8}$	**D**
$\frac{5}{8} + \frac{2}{8}$	**U**	$\frac{2}{5} + \frac{2}{5}$	**E**	$\frac{5}{12} + \frac{6}{12}$	**N**	$\frac{3}{8} + \frac{3}{8}$	**E**
$\frac{6}{12} + \frac{1}{12}$	**O**	$\frac{5}{10} + \frac{1}{10}$	**I**	$\frac{1}{6} + \frac{2}{6}$	**T**	$\frac{1}{10} + \frac{2}{10}$	**H**

$\dfrac{?}{\frac{7}{12}}$ $\dfrac{?}{\frac{11}{12}}$ $\dfrac{?}{\frac{4}{9}}$ $\dfrac{?}{\frac{3}{10}}$ $\dfrac{?}{\frac{4}{5}}$ $\dfrac{?}{\frac{2}{3}}$ $\dfrac{?}{\frac{7}{8}}$ $\dfrac{?}{\frac{1}{2}}$ $\dfrac{?}{\frac{5}{6}}$ $\dfrac{?}{\frac{3}{5}}$ $\dfrac{?}{\frac{1}{4}}$ $\dfrac{?}{\frac{3}{4}}$!

HANDS ON
Subtract Fractions

Quick Review

Find the difference.

1. 12 – 8   2. 15 – 6

3. 18 – 5   4. 21 – 8

5. 25 – 11

▶ **Explore**

Rebecca's dad has $\frac{7}{8}$ of his pan of corn bread to share with the family. If Rebecca eats $\frac{2}{8}$ of the corn bread, how much of the corn bread is left?

Use fraction bars to find $\frac{7}{8} - \frac{2}{8}$.

MATERIALS
fraction bars

STEP 1

Model $\frac{7}{8}$ with fraction bars.

STEP 2

Take away 2 of the fraction bars, or $\frac{2}{8}$.

STEP 3

Count the number of $\frac{1}{8}$ fraction bars left.

There are five $\frac{1}{8}$ fraction bars left.

So, $\frac{5}{8}$ of the corn bread is left.

- Why doesn't the denominator change when you find the difference?

If I take away one of the $\frac{1}{6}$ bars, how many sixths will be left?

Try It

Use fraction bars to find the difference.

a. $\frac{3}{6} - \frac{1}{6} = $ ▇

b. $\frac{5}{8} - \frac{2}{8} = $ ▇

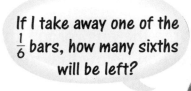

CALIFORNIA STANDARDS NS 3.1 Compare fractions represented by drawings or concrete materials to show equivalency and to add and subtract simple fractions in context. ⊶ **NS 3.2** Add and subtract simple fractions. *also* **MR 1.1, MR 2.0, MR 2.3**

▶ Connect

You subtract like fractions by subtracting the numerators. $\frac{7}{8} - \frac{2}{8} = \blacksquare$

Model

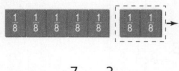

$\frac{7}{8} - \frac{2}{8}$

Record

bars you start with ↓	bars you take away ↓	bars that are left ↓
$\frac{7}{8}$ −	$\frac{2}{8}$ =	$\frac{5}{8}$

MATH IDEA When you subtract like fractions, you subtract the numerators, and the denominators stay the same.

▶ Practice

Find the difference.

1. $\frac{7}{8} - \frac{4}{8} = \blacksquare$

2. $\frac{8}{10} - \frac{3}{10} = \blacksquare$

3. $\frac{6}{12} - \frac{2}{12} = \blacksquare$

Use fraction bars to find the difference.

4. $\frac{4}{5} - \frac{1}{5} = \blacksquare$

5. $\frac{6}{8} - \frac{3}{8} = \blacksquare$

6. $\frac{9}{10} - \frac{7}{10} = \blacksquare$

7. $\frac{2}{3} - \frac{1}{3} = \blacksquare$

8. $\frac{4}{6} - \frac{2}{6} = \blacksquare$

9. $\frac{11}{12} - \frac{6}{12} = \blacksquare$

10. $\frac{7}{8} - \frac{4}{8} = \blacksquare$

11. $\frac{8}{10} - \frac{7}{10} = \blacksquare$

12. Malcolm began practicing his drums at 11:40 A.M. He stopped at 12:55 P.M. How long did he practice?

Mixed Review and Test Prep

Find the sum. (p. 36)

13.
```
  73
  86
+95
```

14.
```
  45
  19
+71
```

15.
```
  67
  24
+36
```

16. Write the value of the 2 in 56,299. (p. 10)

17. **TEST PREP** Wanda earned $30 mowing lawns, $25 raking leaves, and $24 baby-sitting. How much more does Wanda need to have a total of $100? (p. 172)

   A $100 more   C $45 more

   B $79 more   D $21 more

Subtract Fractions

Quick Review

Find the sum. Write the answer in simplest form.

1. $\frac{3}{12} + \frac{1}{12} = \blacksquare$ **2.** $\frac{1}{4} + \frac{2}{4} = \blacksquare$

3. $\frac{5}{8} + \frac{1}{8} = \blacksquare$ **4.** $\frac{3}{10} + \frac{6}{10} = \blacksquare$

5. $\frac{2}{6} + \frac{3}{6} = \blacksquare$

▶ **Learn**

MANY MORE   Kara and Eli shared a pack of 10 graham crackers. Kara ate 5 crackers, or $\frac{5}{10}$ of them. Eli ate 3 crackers, or $\frac{3}{10}$ of them. What fraction tells how many more of the crackers Kara ate than Eli?

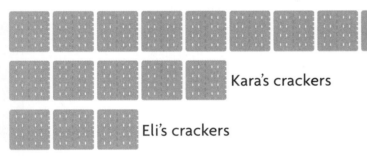

Total number of crackers

Kara's crackers

Eli's crackers

Use fraction bars to find $\frac{5}{10} - \frac{3}{10}$ in simplest form.

Materials: fraction bars

STEP 1   Model $\frac{5}{10}$ and $\frac{3}{10}$ with fraction bars.

STEP 2   Compare the bars to find the difference.

STEP 3   Find the largest fraction bar that is the same length.

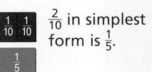

$\frac{2}{10}$ in simplest form is $\frac{1}{5}$.

$$\frac{5}{10} - \frac{3}{10} = \frac{2}{10}, \text{ or } \frac{1}{5}$$

So, Kara ate $\frac{1}{5}$ more of the graham crackers than Eli.

• How much of the pack did Kara and Eli eat altogether? How much is left?

CALIFORNIA STANDARDS NS 3.1 Compare fractions represented by drawings or concrete materials to show equivalency and to add and subtract simple fractions in context. **○━ NS 3.2** Add and subtract simple fractions. *also* **MR 1.1, MR 2.3**

Subtracting Fractions

What if Kara and Eli made 8 peanut butter crackers and ate 4 of the crackers. What fraction of the crackers are left?

Use fraction bars to find $\frac{8}{8} - \frac{4}{8}$ in simplest form.

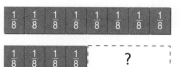

$\frac{8}{8}$ = the total number of crackers

$\frac{4}{8}$ = the 4 crackers they ate

$\frac{4}{8}$ = the crackers that are left

$\frac{4}{8}$ in simplest form is $\frac{1}{2}$

$$\frac{8}{8} - \frac{4}{8} = \frac{4}{8}, \text{ or } \frac{1}{2}$$

So, $\frac{1}{2}$ of the crackers are left.

MATH IDEA You can compare fraction bars to subtract fractions and to find the simplest form.

▶ Check

1. **Explain** how you can use fraction bars to find $\frac{3}{4} - \frac{1}{4}$ in simplest form.

Compare. Find the difference. Write the answer in simplest form.

2.

$$\frac{5}{6} - \frac{2}{6} = \blacksquare$$

3.

$$\frac{6}{12} - \frac{2}{12} = \blacksquare$$

4.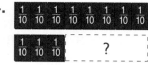

$$\frac{8}{10} - \frac{3}{10} = \blacksquare$$

5.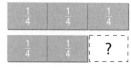

$$\frac{3}{4} - \frac{2}{4} = \blacksquare$$

6.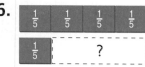

$$\frac{4}{5} - \frac{1}{5} = \blacksquare$$

7.

$$\frac{5}{8} - \frac{4}{8} = \blacksquare$$

LESSON CONTINUES

▶ Practice and Problem Solving

Compare. Find the difference. Write the answer in simplest form.

8.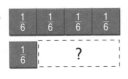

$$\frac{4}{6} - \frac{1}{6} = \blacksquare$$

9.

$$\frac{6}{12} - \frac{3}{12} = \blacksquare$$

10.

$$\frac{3}{8} - \frac{2}{8} = \blacksquare$$

11.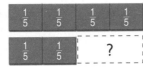

$$\frac{4}{5} - \frac{2}{5} = \blacksquare$$

12.

$$\frac{6}{8} - \frac{4}{8} = \blacksquare$$

13.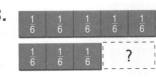

$$\frac{5}{6} - \frac{3}{6} = \blacksquare$$

Find the difference. Write the answer in simplest form. Use fraction bars.

14. $\dfrac{2}{3} - \dfrac{1}{3} = \blacksquare$

15. $\dfrac{4}{5} - \dfrac{1}{5} = \blacksquare$

16. $\dfrac{4}{6} - \dfrac{2}{6} = \blacksquare$

17. $\dfrac{5}{6} - \dfrac{1}{6} = \blacksquare$

18. $\dfrac{7}{10} - \dfrac{2}{10} = \blacksquare$

19. $\dfrac{7}{8} - \dfrac{5}{8} = \blacksquare$

20. $\dfrac{10}{12} - \dfrac{7}{12} = \blacksquare$

21. $\dfrac{6}{8} - \dfrac{1}{8} = \blacksquare$

22. $\dfrac{11}{12} - \dfrac{9}{12} = \blacksquare$

23. $\dfrac{8}{10} - \dfrac{5}{10} = \blacksquare$

24. $\dfrac{5}{6} - \dfrac{3}{6} = \blacksquare$

25. $\dfrac{6}{8} - \dfrac{2}{8} = \blacksquare$

26. REASONING There are 10 letters in the word *California*. The letter *a* is $\frac{2}{10}$ of the word, and the letter *C* is $\frac{1}{10}$ of the word. What fraction of the word is the letter *i*? What fraction of the word are the vowels?

27. Dana cut an apple pie into 8 equal pieces. She shared the pie with 4 of her friends. Dana and each of her friends ate 1 piece of pie. What fraction of the pie is left?

28. MEASUREMENT A pancake recipe calls for $\frac{2}{3}$ cup pancake mix and $\frac{1}{3}$ cup milk. How much more mix than milk is needed?

29. **? What's the Error?** Haley wrote $\frac{5}{12} - \frac{3}{12} = 2$. What was her error?

30. Viola has $\frac{7}{8}$ of a cake to share with friends. If she and her friends eat $\frac{4}{8}$ of the cake, how much of the cake will Viola have left?

31. REASONING An apple was cut into equal-size slices. Kinji ate 3 of them. If 6 slices of apple are left, what fraction of the apple did Kinji eat?

Mixed Review and Test Prep

Find the quotient. (p. 218)

32. $36 \div 9$ **33.** $40 \div 10$

34. $63 \div 9$ **35.** $100 \div 10$

36. $70 \div 10$ **37.** $45 \div 9$

38. $72 \div 9$ **39.** $10 \div 10$

40. $20 \div 10$ **41.** $54 \div 9$

42. $81 \div 9$ **43.** $60 \div 10$

44. TEST PREP How many centimeters are in 3 meters? (p. 444)

A 3   **C** 300
B 30   **D** 3,000

45. TEST PREP Roller-coaster tickets cost $1.50 each. Samantha bought 2 tickets. How much change should she receive from $5.00? (p. 88)

F $2.00   **H** $3.50
G $3.00   **J** $4.00

LINKUP
to Geography

You can use what you know about adding and subtracting fractions to find distances on a map. The scale tells you how distances are measured. On this map each unit represents $\frac{1}{10}$ of a mile.

What if you are visiting your friend Carl at his house after school? Use the map to answer the questions.

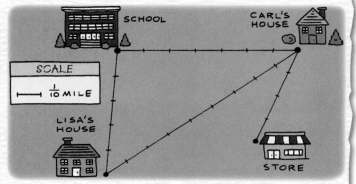

1. How much farther is the school than the store from Carl's house?

2. How far is Lisa's house from the school?

3. Which is farther from Carl's house, Lisa's house or the school? How much farther?

4. REASONING Carl says the store is $\frac{1}{2}$ mile from his house. Is he right? Explain your reasoning.

Problem Solving Skill
Reasonable Answers

Understand → Plan → Solve → Check

Quick Review

Find the difference.

1. $\frac{8}{12} - \frac{3}{12} = \blacksquare$

2. $\frac{9}{10} - \frac{2}{10} = \blacksquare$ 3. $\frac{5}{6} - \frac{4}{6} = \blacksquare$

4. $\frac{7}{8} - \frac{4}{8} = \blacksquare$ 5. $\frac{2}{3} - \frac{1}{3} = \blacksquare$

SOUNDS GOOD Whenever you solve a problem, always check to see that your answer is reasonable and makes sense.

Tanya's mother made a very large chocolate chip cookie. Tanya and Allie each ate $\frac{3}{8}$ of the cookie. What fraction of the cookie was left?

1 whole cookie , or $\frac{8}{8}$

STEP 1 Find out what the problem asks.

How much of the cookie was left after Tanya and Allie each ate $\frac{3}{8}$ of it?

STEP 2 Add to find how much of the cookie was eaten.

$\frac{3}{8} + \frac{3}{8} = \frac{6}{8}$

STEP 3 Subtract the amount eaten from the whole cookie to find the amount left.

$\frac{8}{8} - \frac{6}{8} = \frac{2}{8}$

So, $\frac{2}{8}$, or $\frac{1}{4}$, of the cookie was left.

STEP 4 Check to see that your answer is reasonable and makes sense.

Think: The girls ate more than $\frac{1}{2}$ of the cookie. So, less than $\frac{1}{2}$ will be left. Since $\frac{2}{8}$ is less than $\frac{1}{2}$, it is a reasonable answer.

Talk About It

• Why is $\frac{8}{8}$ used for the whole cookie?

• Why wouldn't it be reasonable to decide that $\frac{5}{8}$ of the cookie was left?

CALIFORNIA STANDARDS O—π **NS 3.2** Add and subtract simple fractions. **MR 3.1** Evaluate the reasonableness of the solution in the context of the original situation. *also* **NS 3.1, MR 2.0, MR 2.3, MR 2.4**

Solve. Tell how you know your answer is reasonable.

1. Gil hiked $\frac{2}{5}$ of the mountain trail and rested. Then he hiked another $\frac{1}{5}$ of the trail. How much of the trail does he have left to hike?

2. Clyde opened a new box of cereal. He ate $\frac{1}{3}$ of the cereal in the box. How much cereal was left?

Amy planted $\frac{1}{4}$ of her garden on Monday and $\frac{1}{4}$ of her garden on Tuesday. She planted the rest on Wednesday. How much of the garden did she plant on Wednesday?

3. How can you solve the problem?

   A Compare $\frac{1}{4}$ and $\frac{1}{4}$.

   B Find $\frac{1}{4} + \frac{1}{4}$.

   C Find $\frac{1}{4} + \frac{1}{4}$, and then subtract from $\frac{4}{4}$.

   D Find $\frac{1}{4} - \frac{1}{4}$.

4. What is the answer to the question?

   F $\frac{1}{4}$ G $\frac{1}{2}$ H $\frac{3}{4}$ J $\frac{4}{4}$

Mixed Applications

5. **REASONING** Selma cut a pan of brownies into 3 equal pieces. Then she cut each of those pieces in half. She ate 2 of the pieces. What fraction of the brownies were left? Explain.

6. **Algebra** Satoko had a total of 12 apples and pears. She traded each pear for 2 apples. Then she had 15 apples in all. How many of each fruit did Satoko have to begin with?

7. **Write About It** Describe the pattern below. What will the fourteenth shape in the pattern be?

Review/Test

✓ CHECK VOCABULARY AND CONCEPTS

Choose the best term from the box.

| like fractions |
| simplest form |
| denominator |

1. Fractions that have the same denominator are __?__. (p. 502)

2. When a fraction uses the largest fraction bar or bars possible, it is in __?__. (p. 504)

✓ CHECK SKILLS

Find the sum. Write the answer in simplest form. Use fraction bars. (p. 504–507)

3. $\frac{2}{4} + \frac{1}{4} = \blacksquare$

4. $\frac{3}{5} + \frac{1}{5} = \blacksquare$

5. $\frac{2}{8} + \frac{4}{8} = \blacksquare$

6. $\frac{3}{12} + \frac{2}{12} = \blacksquare$

7. $\frac{2}{6} + \frac{1}{6} = \blacksquare$

8. $\frac{3}{8} + \frac{2}{8} = \blacksquare$

9. $\frac{2}{10} + \frac{7}{10} = \blacksquare$

10. $\frac{1}{3} + \frac{2}{3} = \blacksquare$

Find the difference. Write the answer in simplest form. Use fraction bars. (pp. 510–513)

11. $\frac{4}{6} - \frac{2}{6} = \blacksquare$

12. $\frac{7}{10} - \frac{3}{10} = \blacksquare$

13. $\frac{6}{8} - \frac{4}{8} = \blacksquare$

14. $\frac{5}{5} - \frac{3}{5} = \blacksquare$

15. $\frac{10}{12} - \frac{7}{12} = \blacksquare$

16. $\frac{2}{4} - \frac{1}{4} = \blacksquare$

17. $\frac{8}{10} - \frac{5}{10} = \blacksquare$

18. $\frac{8}{12} - \frac{4}{12} = \blacksquare$

✓ CHECK PROBLEM SOLVING

Solve. Tell how you know your answer is reasonable.

(pp. 514–515)

19. Joe gave $\frac{1}{8}$ of his football cards to Pete and $\frac{3}{8}$ of his cards to Ron. What fraction of his cards did he keep for himself?

20. On Monday Raquel read $\frac{1}{10}$ of her library book. On Tuesday she read $\frac{1}{10}$ of her book. What fraction of her book does she have left to read?

Cumulative Review

Choose the answer.
See item **10.**

Subtract the fractions and look for an answer that matches yours. If your answer doesn't match, look for another form of the number, such as simplest form.

Also see problem **6,** p. H64.

For 1–10, choose the best answer.

1. $\frac{7}{10} + \frac{2}{10} = \blacksquare$

A $\frac{9}{20}$ **C** $\frac{9}{10}$

B $\frac{5}{20}$ **D** $\frac{5}{20}$

2. Trisha had $\frac{6}{8}$ of a pizza to share with her friend. She and her friend ate $\frac{3}{8}$ of the pizza. What fraction was left?

F $\frac{9}{8}$ **H** $\frac{9}{16}$

G $\frac{3}{8}$ **J** $\frac{3}{4}$

3. $\frac{2}{6} + \frac{3}{6} = \blacksquare$

A $\frac{5}{6}$ **C** $\frac{5}{12}$

B $\frac{2}{6}$ **D** 1

4. Which number is even?

F 2,227   **H** 2,685

G 2,243   **J** 3,330

5. $7 \times (2 \times 4) = \blacksquare$

A 13   **C** 56

B 14   **D** NOT HERE

6. $\frac{5}{6} - \frac{3}{6} = \blacksquare$

F $\frac{1}{6}$ **H** $\frac{2}{3}$

G $\frac{1}{3}$ **J** $\frac{8}{6}$

7. Libby ordered 4 books from her book club. Each book cost $3.98. What was the total cost of the books Libby ordered?

A $15.92   **C** $7.98

B $15.62   **D** NOT HERE

8. Rudy read $\frac{2}{5}$ of his new book last night. How much is left to read?

F $\frac{2}{5}$ **H** $\frac{4}{5}$

G $\frac{3}{5}$ **J** $\frac{5}{2}$

9. The third graders want to measure the length of a hallway. Which unit would be best to measure the length of a hall?

A cups   **C** degrees

B pounds   **D** yards

10. $\frac{7}{9} - \frac{4}{9} = \blacksquare$

F $\frac{1}{3}$ **H** $\frac{5}{9}$

G $\frac{1}{9}$ **J** $\frac{11}{9}$

Decimals and Fractions

Swimming is a sport in which amounts of time are used to keep score. The times have to be very exact, so decimals are used to show 100 parts of one second. The table shows some of the world's fastest freestyle swimmers and their recorded times. In an Olympic-size pool, this distance of 100 meters is one lap of the pool. Use the table to find which swimmer had the fastest time.

FINAL TIMES IN 100 M FREESTYLE	
Swimmer	Time
Alexander Popov	48.21 seconds
Matt Biondi	48.42 seconds
Michael Klim	48.98 seconds
Fernando Scherer	48.69 seconds

Use this page to help you review and remember important skills needed for Chapter 29.

✓ NAME THE FRACTION (See p. H29.)

Write a fraction for the shaded part.

1.

2.

3.

Write a fraction that names the part of each group that is shaded.

4.

5.

6.

✓ COMPARE FRACTIONS (See p. H30.)

Compare. Write <, >, or = for each ●.

7.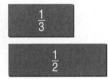
$\frac{1}{3}$ ● $\frac{1}{2}$

8.
$\frac{1}{4}$ ● $\frac{2}{8}$

9.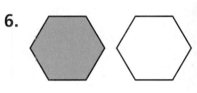
$\frac{2}{10}$ ● $\frac{5}{10}$

Compare the part of each group that is shaded.
Write <, >, or = for each ●.

10.
$\frac{5}{5}$ ● $\frac{2}{5}$

11.
$\frac{1}{3}$ ● $\frac{2}{3}$

12.
$\frac{3}{4}$ ● $\frac{0}{4}$

13.
$\frac{3}{6}$ ● $\frac{4}{6}$

14.
$\frac{4}{4}$ ● $\frac{4}{4}$

15.
$\frac{6}{7}$ ● $\frac{5}{7}$

Relate Fractions and Decimals

▶ **Learn**

FAIR SHARE A **decimal** is a number with one or more digits to the right of the decimal point. A decimal uses place value to show values less than one, such as tenths.

VOCABULARY
decimal
tenth

This square has 10 equal parts. Each equal part is one **tenth**.

TECHNOLOGY LINK

More Practice:
Use Mighty Math
Number Heroes,
Fraction Fireworks,
Levels Q and Y.

Fraction
Write: $\frac{4}{10}$

Decimal
Write: 0.4
↑decimal point

Read: four tenths   **Read:** four tenths

The fraction $\frac{4}{10}$ and the decimal 0.4 name the same amount.

MATH IDEA You can use a fraction or a decimal to show values in tenths.

Examples

A

Fraction: $\frac{9}{10}$
Decimal: 0.9
Read: nine tenths

B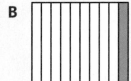

Fraction: $\frac{1}{10}$
Decimal: 0.1
Read: one tenth

C

Fraction: $\frac{10}{10}$
Decimal: 1.0
Read: ten tenths or one

• How many parts on the square would you shade to show 0.3?

CALIFORNIA STANDARDS NS 3.4 Know and understand that fractions and decimals are two different representations of the same concept. **MR 2.3** Use a variety of methods, such as words, numbers, symbols, charts, graphs, tables, diagrams, and models, to explain mathematical reasoning. *also* **MR 2.4, MR 3.2**

► Check

1. **Write** a fraction that shows the amount of crispy treats that are left. Then write the same amount as a decimal.

Write the fraction and decimal for the shaded part.

2. 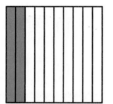   3. 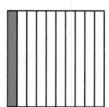   4.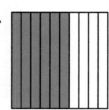

► Practice and Problem Solving

Write the fraction and decimal for the shaded part.

5. 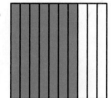   6.    7. 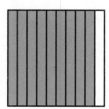   8.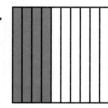

9. 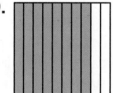   10.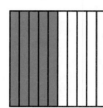

11. **REASONING** Look at Exercise 8. How could you use subtraction to find the decimal amount that is NOT shaded?

12. **Write About It** Look at the fraction bars at the right. Explain why 0.5 is the same as $\frac{1}{2}$.

$\frac{1}{2}$

| $\frac{1}{10}$ | $\frac{1}{10}$ | $\frac{1}{10}$ | $\frac{1}{10}$ | $\frac{1}{10}$ |

─ Mixed Review and Test Prep ─

Find the quotient. (p. 214)

13. $49 \div 7 = \blacksquare$ 14. $63 \div 7 = \blacksquare$

15. $42 \div 6 = \blacksquare$ 16. $32 \div 8 = \blacksquare$

17. **TEST PREP** Frieda puts 5 beads on each of 9 necklaces. How many beads does she use? (p. 118)

A 45   **B** 54   **C** 72   **D** 81

Extra Practice page H60, Set A

HANDS ON
Tenths

Quick Review

1. $\frac{1}{10} + \frac{2}{10} = \blacksquare$

2. $\frac{2}{10} + \frac{5}{10} = \blacksquare$

3. $\frac{2}{10} - \frac{1}{10} = \blacksquare$

4. $\frac{4}{10} - \frac{4}{10} = \blacksquare$

5. $\frac{7}{10} + \frac{2}{10} = \blacksquare$

▶ Explore

This decimal model shows six tenths.

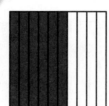

Write: $\frac{6}{10}$ or 0.6

Use a decimal model to show 0.2, or $\frac{2}{10}$.

MATERIALS
decimal models
markers

STEP 1 Shade the decimal model to show two tenths.

Remember
A mixed number is made up of a whole number and a fraction. $1\frac{1}{4}$

STEP 2 Below your decimal model, write the fraction and decimal amount you have shown.

• Use a new decimal model. Shade and label the decimal model to show 0.8, or $\frac{8}{10}$.

We shaded 9 out of 10 equal parts on a decimal model. What fraction does this show?

Try It

Shade and label decimal models to show each amount.

   a. $\frac{9}{10}$, or 0.9　　　**b.** $\frac{5}{10}$, or 0.5

   c. $\frac{7}{10}$, or 0.7　　　**d.** $\frac{1}{10}$, or 0.1

CALIFORNIA STANDARDS **NS 3.4** Know and understand that fractions and decimals are two different representations of the same concept. **MR 2.3** Use a variety of methods, such as words, numbers, symbols, charts, graphs, tables, diagrams, and models, to explain mathematical reasoning. *also* **NS 2.8, NS 3.0, MR 2.4, MR 3.2**

You can write a fraction or a decimal to show tenths.

Write: $\frac{7}{10}$ or 0.7

Read: seven tenths

Write: $\frac{1}{10}$ or 0.1

Read: one tenth

Write: $1\frac{5}{10}$ or 1.5

Read: one and five tenths

▶ **Practice**

Use decimal models to show each amount. Then write the decimal.

1. $\frac{3}{10}$ 2. $\frac{8}{10}$ 3. $\frac{4}{10}$ 4. $1\frac{6}{10}$

Write each fraction or mixed number as a decimal.

5. $\frac{2}{10}$ 6. $\frac{9}{10}$ 7. $1\frac{1}{10}$ 8. $\frac{7}{10}$

Write each decimal as a fraction or mixed number.

9. 0.5   10. 1.4   11. 0.3   12. 0.1

13. **? What's the Question?** Greg has a total of 10 blueberry and pumpkin muffins. Two tenths of the muffins are blueberry. The answer is 8.

14. Hidori had 64 party favors. She put an equal number in each of 8 bags. How many party favors were in 2 bags?

Mixed Review and Test Prep

Write <, >, or = for each ●. (p. 138)

15. 4×9 ● 36

16. 9×7 ● 8×8

17. 3×6 ● 9×2

18. 8×7 ● 9×6

19. **TEST PREP** Sara had $4.08. She bought a sticker for $0.29. How much money did she have left?
(p. 88)

A $3.89   C $3.79
B $3.81   D $3.70

HANDS ON
Hundredths

▶ **Explore**

Each of these decimal models has 100 equal parts. Each equal part is one **hundredth**.

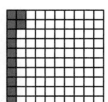

Write: $\frac{12}{100}$ or 0.12

Read: twelve hundredths

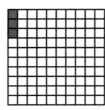

Write: $\frac{3}{100}$ or 0.03

Read: three hundredths

Use a decimal model to show 0.05, or $\frac{5}{100}$.

STEP 1   Shade the decimal model to show five hundredths.

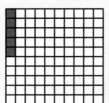

STEP 2   Below your model, write the fraction and decimal you have shown.

How many equal parts should I shade to show 0.06?

• Use a new decimal model. Shade and label the decimal model to show 0.13, or $\frac{13}{100}$.

Try It

Use decimal models to show:

a. 0.06

b. 0.60

CALIFORNIA STANDARDS NS 3.4 Know and understand that fractions and decimals are two different representations of the same concept. **MR 2.3** Use a variety of methods, such as words, numbers, symbols, charts, graphs, tables, diagrams, and models, to explain mathematical reasoning. *also* **MR 2.4, MR 3.2**

▶ Connect

You can write a fraction or a decimal to show hundredths.

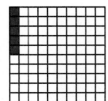 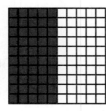  

Write: $\frac{5}{100}$ or 0.05
Read: five hundredths

Write: $\frac{50}{100}$ or 0.50
Read: fifty hundredths

Write: $1\frac{55}{100}$ or 1.55
Read: one and fifty-five hundredths

▶ Practice

TECHNOLOGY LINK

More Practice:
Use E-Lab, *Hundredths*.
www.harcourtschool.com/elab2002

Use decimal models to show each amount. Then write the decimal.

1. $\frac{8}{100}$ **2.** $\frac{10}{100}$ **3.** $\frac{24}{100}$ **4.** $1\frac{22}{100}$

Write each fraction or mixed number as a decimal.

5. $\frac{6}{100}$ **6.** $\frac{15}{100}$ **7.** $\frac{70}{100}$ **8.** $1\frac{57}{100}$

Write each decimal as a fraction or mixed number.

9. 0.01 **10.** 0.56 **11.** 1.20 **12.** 0.02 **13.** 1.05

14. **? What's the Error?**
Sean said that this model shows 0.90. Describe his error. Write the correct decimal.

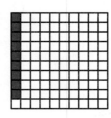

15. **Write About It** Describe how 0.35 and 0.53 are different. You may use decimal models to help.

Mixed Review and Test Prep

16. $3{,}921 + 765 = \blacksquare$ (p. 46)

17. $593 + 1{,}421 = \blacksquare$ (p. 46)

18. $843 - 258 = \blacksquare$ (p. 58)

19. $706 - 55 = \blacksquare$ (p. 58)

20. **TEST PREP** Jacob's photo is 5 in. wide and 7 in. long. What is its perimeter? (p. 458)

A 12 in.   **C** 26 in.
B 24 in.   **D** 35 in.

Read and Write Decimals

Quick Review

Write each number in standard form.

1. $400 + 50 + 3$

2. $100 + 70 + 6$

3. $300 + 0 + 3$

4. $900 + 20 + 1$

5. $200 + 40 + 9$

▶ **Learn**

CUT THE CAKE   Richard went to a party. There was a large cake, cut into 100 equal pieces. If 35 pieces were eaten, what decimal shows how much of the cake was eaten?

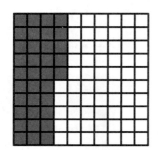

ONES	•	TENTHS	HUNDREDTHS
0	•	3	5

Write: 0.35   **Read:** thirty-five hundredths

So, 0.35 of the cake was eaten.

You can also show a decimal in expanded form.

expanded form: $0.3 + 0.05$

0.35 is the same as 3 tenths 5 hundredths.

Example

ONES	•	TENTHS	HUNDREDTHS
2	•	4	6

standard form: 2.46

word form: two and forty-six hundredths

expanded form: $2.0 + 0.4 + 0.06$

2.46 is the same as 2 ones 4 tenths 6 hundredths.

- What is the standard form for $0.5 + 0.06$, or 5 tenths 6 hundredths?

CALIFORNIA STANDARDS NS 3.4 Know and understand that fractions and decimals are two different representations of the same concept. **O▬ NS 1.5** Use expanded notation to represent numbers. *also* **MR 2.3, MR 2.4, MR 3.2**

1. Write the expanded form for 0.32.

Write the word form and expanded form for each decimal.

2.

ONES	•	TENTHS	HUNDREDTHS
0	•	6	2

3.

ONES	•	TENTHS	HUNDREDTHS
3	•	5	5

► **Practice and Problem Solving**

Write the word form and expanded form for each decimal.

4.

ONES	•	TENTHS	HUNDREDTHS
0	•	7	4

5.

ONES	•	TENTHS	HUNDREDTHS
0	•	1	3

6.

ONES	•	TENTHS	HUNDREDTHS
4	•	6	1

7.

ONES	•	TENTHS	HUNDREDTHS
0	•	2	9

Write _tenths_ or _hundredths_.

8. $0.16 = 1$ tenth 6 _?_

9. $0.20 = 2$ _?_ 0 hundredths

Write the missing number.

10. $0.90 = $ ■ tenths 0 hundredths

11. $0.09 = $ ■ tenths 9 hundredths

12. **? What's the Error?** Beth wrote 0.03 to show 3 tenths. Describe her error. Write the correct decimal to show 3 tenths.

13. Raul says that 4 tenths is the same as 40 hundredths. Do you agree? Explain. Use this decimal square to help.

Mixed Review and Test Prep

Find the product. (p. 122)

14. $3 \times 4 = $ ■   **15.** ■ $= 7 \times 3$

16. $8 \times 3 = $ ■   **17.** ■ $= 3 \times 3$

18. **TEST PREP** Pilar put 6 olives on each pizza. If she made 8 pizzas, how many olives did she use? (p. 148)

A 14 **B** 24 **C** 40 **D** 48

5 Compare and Order Decimals

Quick Review

Compare. Use <, >, or = for each ⬤.

1. 35 ⬤ 53

2. 901 ⬤ 1,093

3. 1,243 ⬤ 1,423

4. 30 + 5 ⬤ 25 + 10

5. 45 + 6 ⬤ 55 − 5

▶ **Learn**

TAKE YOUR PLACE A place-value chart can help you compare tenths.

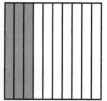 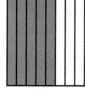

ONES	•	TENTHS
0	•	3
0	•	5

0.3   <   0.5

* Compare ones.
* Compare tenths. 0.3 is less than 0.5.

You can also compare decimals with hundredths.

Swimmers and other athletes compare their speeds using times written in hundredths of seconds.

ONES	•	TENTHS	HUNDREDTHS
4	•	7	9
4	•	5	1

4.79   >   4.51

* Begin with the digit in the greatest place value.
* Compare digits in each place.
* 7 tenths > 5 tenths. So, 4.79 is greater than 4.51.

You can use a number line to order decimals.

Example Write 0.7, 0.3, and 0.9 in order from least to greatest.

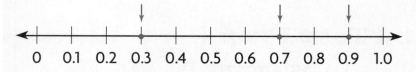

0   0.1   0.2   0.3   0.4   0.5   0.6   0.7   0.8   0.9   1.0

The numbers from least to greatest are 0.3, 0.7, 0.9.

* How could you use the number line to compare 0.5 and 0.8?

CALIFORNIA STANDARDS NS 3.4 Know and understand that fractions and decimals are two different representations of the same concept. **MR 2.3** Use a variety of methods, such as words, numbers, symbols, charts, graphs, tables, diagrams, and models, to explain mathematical reasoning. *also* **AF 1.0, MR 2.4, MR 3.2**

▶ Check

1. Explain how to compare 3.54 and 3.52.

Compare. Write < or > for each ⬤.

2.

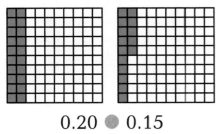

0.20 ⬤ 0.15

3.

ONES	•	TENTHS
4	•	1
4	•	6

4.1 ⬤ 4.6

▶ Practice and Problem Solving

Compare. Write < or > for each ⬤.

4.

ONES	•	TENTHS
1	•	5
1	•	6

1.5 ⬤ 1.6

5.

ONES	•	TENTHS	HUNDREDTHS
2	•	5	1
2	•	3	9

2.51 ⬤ 2.39

6. 2.4 ⬤ 2.1 **7.** 3.5 ⬤ 2.6 **8.** 0.22 ⬤ 0.25 **9.** 1.67 ⬤ 1.76

For 10–14, order the decimals from least to greatest. For 10–12, use the number line.

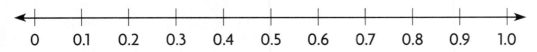

10. 0.5, 0.2, 0.1 **11.** 0.6, 0.5, 0.8 **12.** 0.1, 0.3, 0.2

13. 0.13, 0.09, 0.2 **14.** 0.11, 0.3, 0.04

15. ✎ **Write a problem** comparing decimals with ones, tenths, and hundredths. Trade with a partner and solve.

Mixed Review and Test Prep

Find the sum. (p. 42)

16. $132 + 65 = $ ■

17. $209 + 371 = $ ■

18. $587 + 439 = $ ■

19. $76 + 366 = $ ■

20. **TEST PREP** $72 \div 9 = $ ■ (p. 218)

    A 8 **B** 7 **C** 6 **D** 5

Extra Practice page H60, Set C

Problem Solving Skill
Reasonable Answers

Understand → Plan → Solve → Check

LIGHT WEIGHT   Brody weighed a grapefruit at the grocery store. The grapefruit weighed 0.5 pound. He took 2 grapefruits to the checkout counter. The checker said that he bought 3.3 pounds of grapefruit. Is that amount reasonable?

MATH IDEA When you check if an answer is reasonable, decide if the answer makes sense compared with the facts in the problem.

Read the problem again. What information do you have?

One grapefruit weighs 0.5 pound.
Brody bought 2 grapefruits.

Find about how much 2 grapefruits weigh.

Think: 0.5 is the same as $\frac{5}{10}$.

$\frac{5}{10}$ is the same as $\frac{1}{2}$.

Each grapefruit weighs about a half pound. That means that two grapefruits weigh about 1 pound.

Compare 1 pound and 3.3 pounds. Is 3.3 pounds reasonable?

No, 3.3 pounds is not reasonable.

CALIFORNIA STANDARDS NS 3.4 Know and understand that fractions and decimals are two different representations of the same concept. **MR 1.1** Analyze problems by identifying relationships, distinguishing relevant from irrelevant information, sequencing and prioritizing information, and observing patterns. *also* **NS 1.0, NS 2.8,** ⊶ **NS 3.2,** ⊶ **NS 3.3, MR 1.0, MR 2.0, MR 2.1, MR 2.3, MR 2.4, MR 3.1, MR 3.2**

Solve.

1. Brody found that a plum weighs $\frac{1}{4}$ pound. He took 2 plums to the grocery clerk. The clerk said that 2 plums weigh $\frac{1}{2}$ pound. Is the clerk's total reasonable? Explain.

2. Zach had 2.5 pounds of hamburger to cook for dinner. After his family ate dinner, he said he had about 3 pounds of hamburger left. Is his estimate reasonable? Explain.

Nieta spent $4.15 on lunch. She paid with a $5 bill.

3. Which is a reasonable answer for how much change Nieta got?

   A No change
   B $0.85
   C $1.85
   D $3.85

4. **What if** Nieta gave the clerk a $5 bill to pay for her lunch, which cost $3.15? Which is a reasonable answer for how much change she got?

   F No change   **H** $1.85
   G $0.85   **J** $3.85

Mixed Applications

5. Ms. Cortez built a fence around her garden. The length of the garden is 4 feet and the width is 3 feet. What is the perimeter of her fence?

6. Mia baked 30 cookies. She gave 6 cookies to each of 4 teachers. Then she gave half of the rest of the cookies to her sister. How many cookies did she keep?

7. I am a number less than 40 that can be divided equally into groups of 7. The sum of my digits is 8. What number am I?

8. Neil bought 2 sandwiches. He paid with a $10 bill. He received $4 in change. How much did each sandwich cost?

USE DATA For 9-10, use the graph.

9. How many more students voted for otters and seals than voted for whales and seals?

10. How many students did NOT vote for dolphins?

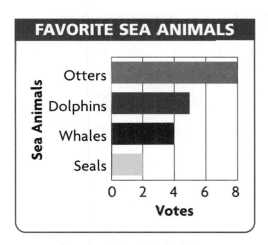

FAVORITE SEA ANIMALS

Review/Test

✔ CHECK VOCABULARY AND CONCEPTS

Choose the best term from the box.

> decimal
> product

1. A number with one or more digits to the right of the decimal point is a __?__ . (p. 520)

Write each fraction or mixed number as a decimal. (pp. 522–525)

2. $\frac{3}{10}$
3. $\frac{9}{10}$
4. $1\frac{2}{10}$
5. $\frac{3}{100}$
6. $\frac{80}{100}$

✔ CHECK SKILLS

Write the fraction and decimal for the shaded part. (pp. 520–521)

7.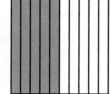
8.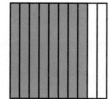
9.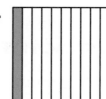
10.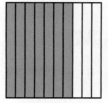

Write the missing number. (pp. 526–527)

11. 0.11 = ■ tenth 1 hundredth
12. 0.07 = 0 tenths ■ hundredths
13. 0.02 ■ tenths 2 hundredths
14. 0.35 = ■ tenths 5 hundredths

Compare. Write < or > for each ●. (pp. 528–529)

15. 1.9 ● 1.1
16. 3.42 ● 3.41
17. 0.13 ● 0.31

Order from least to greatest. (pp. 528–529)

18. 0.9, 0.3, 0.4
19. 0.1, 0.8, 0.7

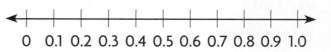

✔ CHECK PROBLEM SOLVING

Solve. (pp. 530–531)

20. Quinn is being weighed. He asks that 2 pounds be subtracted from his weight because he has his shoes on. Each shoe weighs 0.4 pound. Is his request reasonable? Explain.

Cumulative Review

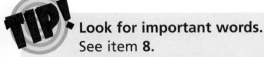

Look for important words.
See item **8.**

An important word is *reasonable*. All answer choices include the word *about* so an exact amount is not needed.

Also see problem **2**, p. H62.

For 1–10, choose the best answer.

1. Which is another name for 0.43?

- **A** forty-three ones
- **B** forty-three tenths
- **C** forty-three hundredths
- **D** forty-three tens

2. Which symbol makes this true?

0.3 ● 0.07

F <   **G** >   **H** =

3. Which decimal names the shaded part?

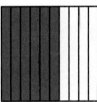

- **A** 0.5
- **B** 0.6
- **C** 0.7
- **D** 6.0

4. What time does the clock show?

- **F** 7:17
- **G** 6:17
- **H** 4:37
- **J** 3:37

5. 6,147
      −3,218

- **A** 2,929
- **B** 2,829
- **C** 2,819
- **D** NOT HERE

6. Which is another name for $\frac{33}{100}$?

- **F** 33
- **G** 3.3
- **H** 0.33
- **J** 0.033

7. Which point on the number line represents 0.5?

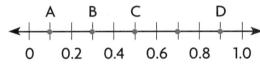

- **A** point A
- **B** point B
- **C** point C
- **D** point D

8. Jan spent $5.29 on school supplies. She paid with a $10 bill. Which is a reasonable amount for Jan's change?

- **F** about $0.70
- **G** about $5.00
- **H** about $6.00
- **J** about $7.00

9. Which is another name for $\frac{5}{100}$?

- **A** 50.0
- **B** 5.0
- **C** 0.50
- **D** 0.05

10. Jo's mother has 24 tomato plants. She wants to plant them in 4 equal rows. Which expression tells how many plants will be in each row?

- **F** 24 ÷ 4
- **G** 24 × 4
- **H** 24 + 4
- **J** 24 − 4

Decimals and Money

Coin banks have been made out of wood, tin, glass, cast iron, and pottery. Banks with moving parts were first made in the 1870s. These banks move when a coin is inserted. Look at the table. Using the fewest coins, list the dimes and pennies that could make up each amount.

MONEY IN COIN BANKS	
Erik	$0.70
Letticia	$0.87
Amal	$0.56
Pilar	$0.94

CHECK WHAT YOU KNOW

Use this page to help you review and remember important skills needed for Chapter 30.

✓ USE MONEY NOTATION (See p. H30.)

Count and write the amount.

1.

2.

3.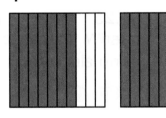

4.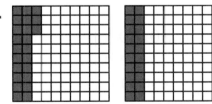

✓ NAME THE DECIMAL (See p. H31.)

Write each fraction or mixed number as a decimal.

5. $\frac{1}{10}$ 6. $\frac{7}{10}$ 7. $2\frac{6}{10}$ 8. $\frac{2}{10}$ 9. $\frac{5}{10}$

10. $1\frac{65}{100}$ 11. $\frac{4}{100}$ 12. $\frac{30}{100}$ 13. $\frac{3}{100}$ 14. $\frac{10}{100}$

✓ COMPARE AND ORDER DECIMALS (See p. H31.)

Compare. Write < or > for each ●.

15.

0.7 ● 0.5

16.

0.23 ● 0.20

Use the number line to order the decimals from least to greatest.

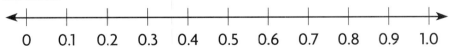

| 0 | 0.1 | 0.2 | 0.3 | 0.4 | 0.5 | 0.6 | 0.7 | 0.8 | 0.9 | 1.0 |

17. 0.7, 0.5, 0.9   18. 0.3, 0.2, 0.5   19. 0.1, 0, 0.6   20. 0.8, 0.1, 1.0

HANDS ON
Relate Decimals and Money

▶ **Explore**

 =

100 pennies = 1 dollar

 =

10 dimes = 1 dollar

 = $\frac{1}{100}$ or 0.01 of a dollar

 = $\frac{1}{10}$ or 0.1 of a dollar

Use play money to connect money and decimals.

Copy Tables A and B.

STEP 1

Use pennies to show $\frac{31}{100}$ or 0.31 of a dollar. Record the number of pennies you used.

STEP 3

Use dimes and pennies to show $\frac{31}{100}$ or 0.31 of a dollar. Use as few coins as possible. Record the coins you used.

STEP 4

STEP 2

Repeat Step 1 for 0.02 of a dollar.

Repeat Step 3 for 0.02 of a dollar.

TABLE A	
Decimal	Number of Pennies
0.31	▪
0.02	▪

TABLE B		
Decimal	Number of Dimes	Number of Pennies
0.31	▪	▪
0.02	▪	▪

Try It

a. Write 53¢ or $\frac{53}{100}$ as a decimal.

b. What coins would you use to show 0.12 of a dollar?

► Connect

You can think of dimes as tenths and pennies as hundredths.

0.49			
Ones	.	Tenths	Hundredths
0	.	4	9

0.49 = 49 hundredths

0.49 = 4 tenths 9 hundredths

$0.49			
Dollars	.	Dimes	Pennies
0	.	4	9

$0.49 = 49 pennies =
49 hundredths of a dollar

$0.49 = 4 dimes 9 pennies =
4 tenths 9 hundredths of a dollar

► Practice

Write the money amount for each fraction of a dollar.

1. $\frac{59}{100}$
2. $\frac{3}{100}$
3. $\frac{13}{100}$
4. $\frac{20}{100}$
5. $\frac{10}{100}$

Write the money amount.

6. 3 hundredths of a dollar

7. 24 hundredths of a dollar

8. 19 hundredths of a dollar

Write the missing numbers. Use the fewest coins possible.

9. $0.52 = ■ dimes ■ pennies $0.52 = ■ tenths ■ hundredths of a dollar

10. $0.80 = ■ dimes ■ pennies $0.80 = ■ tenths ■ hundredths of a dollar

11. $0.06 = ■ dimes ■ pennies $0.06 = ■ tenths ■ hundredths of a dollar

12. **Write About It** You know that 1 dime is 0.1 or 1 tenth of a dollar. Is 1 dime also 0.10 or 10 hundredths of a dollar? Explain.

13. **? What's the Question?** Juan has $1.25 to spend at the school fair. He spends $\frac{3}{4}$ of a dollar on a hot dog. The answer is $0.50.

Mixed Review and Test Prep

Subtract 352 from each number. (p. 58)

14. 556

15. 398

16. 409

17. 500

18. **TEST PREP** How many sides does a rectangle have? (p. 382)

A 1

B 3

C 2

D 4

Add and Subtract Decimals and Money

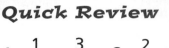

Quick Review

1. $\frac{1}{10} + \frac{3}{10}$ 2. $\frac{2}{10} + \frac{5}{10}$

3. $\frac{5}{10} - \frac{1}{10}$ 4. $\frac{7}{10} - \frac{4}{10}$

5. $\frac{9}{10} - \frac{3}{10}$

▶ **Learn**

MOVING RIGHT ALONG   Gary walked to Dan's house and then went to the playground. How far did Gary walk in all?

One Way   Use decimal models to add.

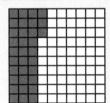 

Shade 0.33 and 0.42. Add the shaded squares.

$$0.33 + 0.42 = 0.75$$

So, Gary walked 0.75 kilometer.

GARY'S HOUSE

0.33 KM

DAN'S HOUSE

0.42 KM

PLAYGROUND

Another Way   Add decimals by using pencil and paper. Add 0.33 and 0.42.

STEP 1 Line up the decimal points.	**STEP 2** Add as you would add whole numbers. Regroup if necessary.	**STEP 3** Write the decimal point in the sum.
0.33 +0.42 ↑ decimal point	0.33 +0.42 0 75	0.33 +0.42 0.75

Examples

A 0.5 +0.3 0.8	B 1.33 +0.06 1.39	C $\overset{1}{}$ $0.38 +$0.45 $0.83	D $\overset{1}{}$ $2.17 +$3.58 $5.75

CALIFORNIA STANDARDS O━ⁿ **NS 3.3** Solve problems involving addition, subtraction, multiplication, and division of money amounts in decimal notation and multiply and divide money amounts in decimal notation by using whole-number multipliers and divisors. *also* **MR 2.3, MR 2.4**

Subtract Decimals

One Way Use decimal models to subtract.

Find 0.73 − 0.50.

Shade 0.73 of a decimal model.

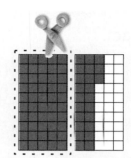

Take away 50 shaded squares. Count the shaded squares that are left to find the difference. 23 shaded squares are left.

So, 0.73 − 0.50 = 0.23.

Another Way Subtract decimals by using pencil and paper. Subtract 0.38 from 0.71.

STEP 1 Line up the decimal points.

$$\begin{array}{r} 0.71 \\ -0.38 \\ \hline \end{array}$$

↑

decimal point

STEP 2 Subtract as you would with whole numbers.

$$\begin{array}{r} {\scriptstyle 6\ 11} \\ 0.\,7\,\cancel{1} \\ -0.\,3\,8 \\ \hline 0\ 3\ 3 \end{array}$$

STEP 3 Write the decimal point in the difference.

$$\begin{array}{r} {\scriptstyle 6\ 11} \\ 0.\,7\,\cancel{1} \\ -0.\,3\,8 \\ \hline 0.\,3\,3 \end{array}$$

Examples

A
$$\begin{array}{r} 0.9 \\ -0.5 \\ \hline 0.4 \end{array}$$

B
$$\begin{array}{r} 2.75 \\ -1.33 \\ \hline 1.42 \end{array}$$

C
$$\begin{array}{r} {\scriptstyle 4\ 14} \\ \$3.\,\cancel{5}\,\cancel{4} \\ -\$0.\,0\,5 \\ \hline \$3.\,4\,9 \end{array}$$

D
$$\begin{array}{r} {\scriptstyle 5\ 13} \\ 0.\,\cancel{6}\,\cancel{3} \\ -0.\,3\,7 \\ \hline 0.\,2\,6 \end{array}$$

• How is subtracting money like subtracting hundredths?

▶ Check

1. Tell why you write a 5 above the 6 tenths in Example D.

Add or subtract.

2.
$$\begin{array}{r} 0.46 \\ +0.19 \\ \hline \end{array}$$

3.
$$\begin{array}{r} 0.75 \\ -0.56 \\ \hline \end{array}$$

4.
$$\begin{array}{r} 0.29 \\ +0.33 \\ \hline \end{array}$$

LESSON CONTINUES ▶

Practice and Problem Solving

Add or subtract.

5. 0.47
    +0.39

6. 0.60
    −0.45

7. 0.51
    +0.18

TECHNOLOGY LINK

To learn more about adding and subtracting decimals and money, watch the **Harcourt Math Newsroom Video**, *Rare Coins*.

8. 0.42
    +0.12

9. 4.57
    +0.09

10. 0.8
    +0.1

11. $7.73
    −$4.55

12. $0.39
    +$0.46

13. 5.5
    −1.3

14. 0.80
    −0.53

15. 2.3
    −1.1

16. 2.35
    +6.19

17. $0.13
    +$2.41

18. 0.73
    −0.32

19. $3.33
    −$0.25

20. 0.5
    +0.4

21. 1.6
    −1.2

22. 0.79
    +0.12

23. **? What's the Error?** Gina wrote 0.20 + 0.07 = 0.90. Describe her error. Give the correct sum.

24. **REASONING** Do you have to regroup to find 0.64 − 0.27? Explain.

25. Kendra bikes 0.55 kilometer to school and 0.34 kilometer to a friend's house. How many kilometers does she bike in all?

26. Rhea lives 0.92 kilometer from school. She has walked 0.38 kilometer so far this morning. How much farther must she walk to get to school?

27. **$\frac{a+b}{c}$ Algebra** Use the price list. Conner bought popcorn and one other item with a $5 bill. His change was $4.05. What was the other item?

28. ✎ **Write a problem** about adding or subtracting money amounts. Use the price list. Trade with a partner and solve.

PRICE LIST	
Item	**Cost**
Popcorn	$0.50
Juice	$0.45
Trail mix	$0.39

29. David went to the store with $10.00. He bought milk, apples, and eggs. He had $4.78 when he left the store. The milk was $2.89, and the eggs were $0.89. How much were the apples?

Mixed Review and Test Prep

Find each missing factor. (p. 142)

30. ■ × 8 = 32   **31.** 8 × ■ = 48

32. ■ × 4 = 28   **33.** 7 × ■ = 21

Find each quotient. (p. 214)

34. 12 ÷ 6 = ■   **35.** 54 ÷ 6 = ■

36. 24 ÷ 8 = ■   **37.** 49 ÷ 7 = ■

38. **TEST PREP** Find 309 − 87. (p. 58)

   A 396   **C** 241
   B 322   **D** 222

39. **TEST PREP** Lori had 5 carrots. She cut each carrot into 3 pieces. Then she ate 2 pieces. How many pieces of carrot did she have left? (p. 172)

   F 10   **G** 12   **H** 13   **J** 15

--LINKUP--- to Reading

STRATEGY • MAKE PREDICTIONS You can use a graph to help you make a prediction. A prediction is a guess that is based on information.

USE DATA Emmett's sunflower should grow to be about 1 meter tall. Using the graph, predict whether the flower will be taller than 0.65 m at the end of the fifth week.

The sunflower has grown at least 0.15 m each week. It was 0.60 m tall at the end of the fourth week.

$$0.60 \text{ m} + 0.15 \text{ m} = 0.75 \text{ m}$$

So, the sunflower will probably be taller than 0.65 m at the end of the fifth week.

GROWTH OF SUNFLOWER

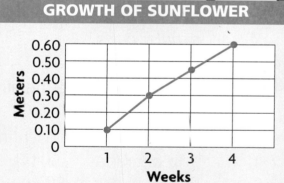

- Between week 1 and week 2 the flower grew 0.20 meter.
- Between week 2 and week 3 it grew 0.15 meter.
- Between week 3 and week 4 it grew 0.15 meter.

1. Predict whether the sunflower will be taller than 0.75 m at the end of the sixth week. Explain.

2. Between which two weeks was the sunflower 0.35 m tall?

Problem Solving Strategy

Break Problems into Simpler Parts

Understand ➡ Plan ➡ Solve ➡ Check

Quick Review

1. $0.24 + 0.24$
2. $0.3 + 0.3$
3. $0.44 + 0.44$
4. $0.2 + 0.2$
5. $0.39 + 0.39$

PROBLEM Alma has $5 to buy fruit. If she buys 1 pound of grapes and 2 pounds of bananas, how much money will she have left?

 Understand

- What are you asked to find?
- What information will you use?
- Is there information you will not use? If so, what?

 Plan

- What strategy can you use to solve the problem?

  You can *break the problem into simpler parts*.

 Solve

- How can you use the strategy to solve the problem?

Find the price for each fruit.
1 lb of grapes: $0.98
2 lb of bananas: $0.45 + $0.45

Add to find the total cost.
grapes: $0.98
bananas: +$0.90
              $1.88

Subtract the total cost of the fruit from $5.

$5.00
−$1.88
$3.12

So, Alma will have $3.12 left.

 Check

- Look at the Problem. Does your answer make sense for the problem? Explain.

CALIFORNIA STANDARDS MR 1.2 Determine when and how to break a problem into simpler parts. **O–n NS 3.3** Solve problems involving addition, subtraction, multiplication, and division of money amounts in decimal notation and multiply and divide money amounts in decimal notation by using whole-number multipliers and divisors. *also* **O–n AF 2.1, MR 1.0, MR 2.2, MR 2.3, MR 2.4, MR 2.6, MR 3.1, MR 3.2**

Problem Solving Practice

PROBLEM SOLVING STRATEGIES

Draw a Diagram or Picture
Make a Model or Act It Out
Make an Organized List
Find a Pattern
Make a Table or Graph
Predict and Test
Work Backward
▶ **Solve a Simpler Problem**
Write an Equation
Use Logical Reasoning

Use the prices on page 544. Break the problem into simpler parts to solve.

1. **What if** Alma buys 2 pounds of grapes and 1 pound of bananas? How much change will she receive from $5?

2. Brandon has $3. Does he have enough money to buy 2 apples and 2 pounds of grapes? Explain.

Maddie bought 3 apples and 1 pound of grapes. She gave the clerk $2. How much change did she get? Use the prices on page 544.

3. Which number sentence could help you break the problem into simpler parts?

   A $0.24 + $4 = $4.24
   B $0.24 + $0.61 = $0.85
   C $0.24 + $0.24 + $0.24 = $0.72
   D $0.98 + $0.98 = $1.96

4. Which answers the question?

   F $0.30
   G $0.75
   H $0.99
   J $1.55

Mixed Strategy Practice

5. On Wednesday, Emily had $24.75. She earned $5.50 on Thursday. Then she spent $6.25. How much money does she have now?

6. Norm biked $\frac{7}{10}$ of a mile one Saturday. Joel biked $\frac{9}{10}$ of a mile. How much farther did Joel bike?

7. Mark is 1 year younger than Rick. Jaime is twice as old as Rick. Jaime is 12 years old. How old are Mark and Rick?

8. **? What's the Question?** The perimeter of Kiyo's rectangular rabbit pen is 20 ft. The length is 6 ft. The answer is 4 ft.

9. Shelly worked on her science project for 45 minutes on Monday, and for 1 hour and 20 minutes on Tuesday. How much time did she work on her project in the two days?

10. Toni baby-sat from 4:30 P.M. until 8:30 P.M. She earned $3 per hour. How long did she baby-sit? How much did she earn?

Review/Test

✓ CHECK CONCEPTS

Write the money amount for each fraction of a dollar. (pp. 538–539)

1. $\frac{20}{100}$

2. $\frac{7}{100}$

3. $\frac{44}{100}$

4. $\frac{75}{100}$

Write the money amount.

5. 14 hundredths of a dollar

6. 10 hundredths of a dollar

7. 54 hundredths of a dollar

✓ CHECK SKILLS

Write the amount of money shown. Then write the amount as a fraction of a dollar. (pp. 536–537)

8.

9.

10.

Add or subtract. (pp. 540–543)

11. 0.6
  +0.2

12. 2.5
  +7.4

13. 0.06
  +0.19

14. $3.52
  +$4.16

15. 0.9
  −0.6

16. 2.8
  −0.1

17. $1.70
  −$0.05

18. 0.61
  −0.59

✓ PROBLEM SOLVING

Solve. (pp. 544–545)

19. Felix bought a notebook for $1.29 and two pens for $0.41 each. He gave the clerk $5.00. How much change did he receive?

20. Ginger wants to buy 2 pears for $0.35 each and 1 cookie for $1.05. She has $2.00. Will she have enough money? Explain.

Cumulative Review

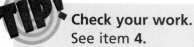

Check your work.
See item **4**.

Be sure you used the correct information from the table. You need to subtract to find how much change.

Also see problem **7**, p. H65.

For 1–9, choose the best answer.

1. 0.5
+ 0.3

    A 0.15   **C** 8
    B 0.2   **D** NOT HERE

2. $0.46
+$0.38

    F $0.08   **H** $0.84
    G $0.46   **J** NOT HERE

3. Which fraction names the shaded part?

    A $\frac{3}{8}$ **C** $\frac{3}{5}$
    B $\frac{1}{2}$ **D** $\frac{5}{8}$

4. Tim buys 4 pencils and 3 erasers. He gives the clerk $4.00. How much change should he receive?

ITEM	COST
pencil	$0.39
eraser	$0.69

    F $0.37   **H** $2.44
    G $1.93   **J** $3.63

5. Which fraction names the part of a dollar that is shown?

    A $\frac{1}{9}$ **C** $\frac{1}{4}$
    B $\frac{1}{5}$ **D** $\frac{1}{2}$

6. How many dimes equal one dollar?

    F 5   **H** 20
    G 10   **J** 50

7. Nancy gave $\frac{1}{8}$ of her stickers to Jill and $\frac{3}{8}$ of her stickers to Laurie. What fraction of her stickers did she give away?

    A $\frac{1}{8}$ **C** $\frac{1}{2}$
    B $\frac{2}{8}$ **D** $\frac{5}{8}$

8. Gerald ran 0.4 mile to the library and then 0.3 mile to school. How far did he run in all?

    F 1.2 mi   **H** 0.1 mi
    G 0.7 mi   **J** 0.07 mi

9. Jerry had $0.61. He spent $0.25. Which fraction names the part of a dollar that Jerry has left?

    A $\frac{36}{100}$ **C** $\frac{86}{100}$
    B $\frac{46}{100}$ **D** $\frac{96}{100}$

MATH DETECTIVE

I Found the Whole Thing

If you know what a part of a whole looks like, can you figure out what the whole looks like? Use the clues below to solve each case!

Draw a picture of each clue to help you find the answer to the question.

Case 1

If $\frac{1}{8} = \square$, what does one whole equal?

Case 2

If $\frac{1}{4} = $ 🌼🌼🌼, what does one whole equal?

Case 3

If $\frac{2}{3} = \square\ \square$, what does one whole equal?

Case 4

If $2 = \hexagon$, what does one whole equal?

Case 5

If $\frac{3}{2} = \frac{\triangledown\triangledown\triangledown}{\triangledown\triangledown\triangledown}$, what does one whole equal?

Case 6

If $\frac{3}{4} = $ 🍎🍎🍎🍎🍎, what does one whole equal?

STRETCH YOUR THINKING Draw a fraction as a picture. Have a classmate use your clues and draw a picture showing what the whole looks like.

Challenge

Fractions and Decimals on a Number Line

You can use decimal models and a number line to help you understand fractions and decimals.

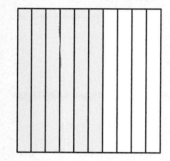

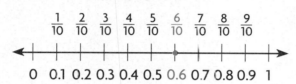

$$\frac{1}{10} \quad \frac{2}{10} \quad \frac{3}{10} \quad \frac{4}{10} \quad \frac{5}{10} \quad \frac{6}{10} \quad \frac{7}{10} \quad \frac{8}{10} \quad \frac{9}{10}$$

0   0.1 0.2 0.3 0.4 0.5 0.6 0.7 0.8 0.9 1

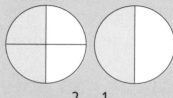

$$\frac{2}{4} = \frac{1}{2}$$

Since 0.6 and $\frac{6}{10}$ name 6 parts out of 10, they name the same amount, or are equivalent.

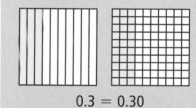

0.3 = 0.30

One whole can be divided into 100 equal parts on a number line.

0      $\frac{25}{100}$ $\frac{50}{100}$ $\frac{75}{100}$ 1

0   0.10   0.20   0.30   0.40   0.50   0.60   0.70   0.80   0.90   1

Practice

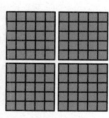

$$\frac{25}{100} = \frac{1}{4}$$

1. Find the point on the number line that is halfway between 0 and 1. Write a fraction name and a decimal name for that point.

2. Find the point on the number line that is one fourth of the way from 0 to 1. Write a fraction name and decimal name for that point.

Use the number line. Write each as a fraction.

3. 0.30   **4.** 0.45   **5.** 0.75   **6.** 0.99

Study Guide and Review

VOCABULARY

Choose the best term from the box.

1. A number with one or more digits to the right of the decimal point is a __?__. (p. 520)

2. A fraction is in __?__ when a model of it uses the largest fraction bars possible. (p. 504)

> decimal
> fraction
> equivalent fractions
> simplest form

STUDY AND SOLVE

Chapter 27

Compare fractions.

Compare $\frac{1}{2}$ and $\frac{3}{5}$.

The bar for $\frac{1}{2}$ is shorter than the bars for $\frac{3}{5}$.

So, $\frac{1}{2} < \frac{3}{5}$ or $\frac{3}{5} > \frac{1}{2}$.

Compare. Write <, >, or = for each ●. (pp. 492–495)

3.

$\frac{3}{8}$ ● $\frac{1}{3}$

4.

$\frac{2}{6}$ ● $\frac{5}{12}$

Chapter 28

Add and subtract fractions.

Write the sum or difference in simplest form.

$\frac{3}{8} + \frac{1}{8} = \frac{4}{8}$ $\frac{4}{6} - \frac{2}{6} = \frac{2}{6}$

So, the sum, $\frac{4}{8}$, is $\frac{1}{2}$ in simplest form.

So, the difference, $\frac{2}{6}$, is $\frac{1}{3}$ in simplest form.

Use fraction bars to find the sum or difference in simplest form. (pp. 502–513)

5. $\frac{1}{8} + \frac{1}{8} = $ ■

6. $\frac{2}{4} + \frac{1}{4} = $ ■

7. $\frac{1}{6} + \frac{3}{6} = $ ■

8. $\frac{9}{12} - \frac{5}{12} = $ ■

9. $\frac{2}{5} - \frac{1}{5} = $ ■

10. $\frac{4}{10} - \frac{2}{10} = $ ■

11. $\frac{1}{12} + \frac{4}{12} = $ ■

12. $\frac{4}{4} - \frac{2}{4} = $ ■

Chapter 29

Relate fractions and decimals.

Write the fraction and the decimal for the shaded part.

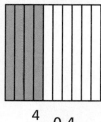

$\frac{4}{10}$, 0.4 $\frac{71}{100}$, 0.71

Write the fraction and the decimal for the shaded part. (pp. 524–525)

13.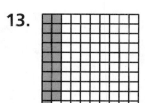

Write the decimal for each fraction.
(pp. 520–525)

14. $\frac{6}{10}$ 15. $\frac{12}{100}$ 16. $\frac{55}{100}$

Chapter 30

Add and subtract decimals.

Subtract. $0.83 − $0.56

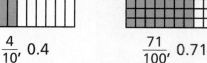

$\begin{array}{r} \overset{7\ 13}{\$0.\cancel{8}\cancel{3}} \\ -\$0.56 \\ \hline \$0.27 \end{array}$

• Line up the decimal points.
• Subtract. Regroup if necessary.
• Write the decimal point in the difference.

Remember to write the dollar sign in your answer when you are adding or subtracting money amounts.

Find the sum or difference. (pp. 540–543)

17. $0.23
    +$0.44

18. 0.78
    +0.04

19. $0.64
    +$0.29

20. 0.5
    −0.4

21. $0.72
    −$0.19

22. $0.19
    −$0.12

PROBLEM SOLVING PRACTICE

Solve. (pp. 514–515, 544–545)

23. Cole and Jon are sharing a pizza. If they each eat $\frac{1}{3}$ of the pizza, is it reasonable to say that $\frac{1}{2}$ of the pizza will be left? Explain.

24. Mr. Jackson is training for a race. Last week, he ran 1.25 miles on each of two days. He ran 1.1 miles on each of two other days. How far did Mr. Jackson run last week?

California Connections

HUMMINGBIRDS

Several species of hummingbirds are found in California. Hummingbirds are the smallest birds in the world.

▲ The nest of the Anna's hummingbird may be as small as a thimble and is lined with feathers and spider webs.

CALIFORNIA HUMMINGBIRDS	AVERAGE WEIGHT OF MALE (IN OUNCES)
Allen's	0.11
Anna's	0.15
Calliope	0.09
Black-Chinned	0.11

For 1–5, use the table.

1. Write the weight of the male Anna's hummingbird as a fraction.

2. Use a decimal square to show the weight of the male Black-Chinned hummingbird.

3. Order the weights of the male Allen's, Anna's, and Calliope hummingbirds from least to greatest.

4. How much more does the Anna's hummingbird weigh than the Calliope hummingbird?

5. The average weight of a female Black-Chinned hummingbird is 0.12 ounces. Does the female weigh more or less than the male? Explain how you know.

6. A hummingbird's egg weighs less than 0.02 ounce. Do three eggs weigh more than or less than $\frac{1}{2}$ ounce? Explain how you know.

TUCKER WILDLIFE SANCTUARY

Tucker Wildlife Sanctuary, in Mission Viejo, is one of the best places in California to view hummingbirds. This 12-acre wilderness is located between Los Angeles and San Diego.

◀ Signs on a porch at the Tucker Wildlife Sanctuary help visitors identify plants and birds.

▲ Black-Chinned hummingbird

The Black-Chinned hummingbird can be seen in the sanctuary during the months of April to September.

1. Write a fraction for the part of the year this hummingbird can be seen in the sanctuary.

2. Write an equivalent fraction for the fraction you wrote for Problem 1.

The Anna's hummingbird can be seen in the sanctuary during the months of January to December.

3. Write a fraction for the part of the year this hummingbird can be seen in the sanctuary.

4. Explain what the fraction you wrote for Problem 3 means.

5. Visitors to the sanctuary are asked to make a $1.50 donation. How much change would you receive from $5.00 if two people visit the sanctuary? Explain.

STUDENT HANDBOOK

Troubleshooting H2

PREREQUISITE SKILLS REVIEW Do you have the math skills needed to start a new chapter? Use this list of skills to review and remember your skills from last year.

Skill	Page	Skill	Page
Ordinal Numbers	H2	Read a Chart	H17
Read and Write Ones, Tens, Hundreds	H2	Tally Data	H17
Place Value with Ones, Tens, Hundreds	H3	Read Pictographs	H18
Compare 2-Digit Numbers	H3	Identify Parts of a Whole	H18
Order Numbers	H4	Compare Parts of a Whole	H19
Addition Facts	H4	Use a Tally Table	H19
2-Digit Addition	H5	Regroup Ones and Tens	H20
Mental Math: Add 2-Digit Numbers	H5	Multiplication Facts	H20
Subtraction Facts	H6	Multiply 2-Digit Numbers	H21
2-Digit Subtraction	H6	Division Facts	H21
Mental Math: Subtract 2-Digit Numbers	H7	Regroup Hundreds and Tens	H22
Money: Count Bills and Coins	H7	Multiplication and Division Facts	H22
Tell Time	H8	Identify Solid Figures	H23
Calendar	H8	Identify Plane Figures	H23
Skip-Count	H9	Sides and Corners	H24
Equal Groups	H9	Geometric Patterns	H24
Column Addition	H10	Congruent Figures	H25
Addition	H10	Types of Angles	H25
Order Property of Addition	H11	Use a Customary Ruler	H26
Multiplication Facts	H11	Measure to the Nearest Inch	H26
Add Equal Groups	H12	Find a Rule	H27
Multiplication Facts Through 5	H12	Use a Metric Ruler	H27
Skip-Count by Tens	H13	Measure to the Nearest Centimeter	H28
Multiplication Facts Through 8	H13	Model Parts of a Whole	H28
Meaning of Multiplication	H14	Model Parts of a Group	H29
Multiplication Facts Through 10	H14	Name the Fraction	H29
Subtraction	H15	Compare Fractions	H30
Division Facts Through 5	H15	Use Money Notation	H30
Order Property of Multiplication	H16	Name the Decimal	H31
Missing Factors	H16	Compare and Order Decimals	H31

SILLY CIRCLES

Extra Practice H32

When learning new skills, it may help you to practice a little more. On these pages you'll find extra practice exercises to do so you can be sure of your new skills!

Sharpen Your Test-Taking Skills H62

Before a test, sharpen your test-taking skills by reviewing these pages. Here you can find tips such as how to get ready for a test, how to understand the directions, and how to keep track of time.

Basic Facts Tests H66

Review addition, subtraction, multiplication, and division facts by taking the basic facts tests throughout the year to improve your memorization skills.

Table of Measures H70

All the important measures used in this book are in this table. If you've forgotten exactly how many feet are in a mile, this table will help you.

California Standards H71

This section will help you learn about the standards that describe what you must know and be able to do.

Glossary H78

This glossary will help you speak and write the language of mathematics. Use the glossary to check the definitions of important terms.

Index H88

Use the index when you want to review a topic. It lists the page numbers where the topic is taught.

✓ ORDINAL NUMBERS

Ordinal numbers tell order or position.

first   second   third   fourth   fifth   sixth   seventh   eighth   ninth   tenth

Example

Write the position of the yellow marble.

The yellow marble is second.

▶ Practice

Write the position of each marble.

1. purple   **2.** white   **3.** red   **4.** green

✓ READ AND WRITE ONES, TENS, HUNDREDS

You can use base-ten blocks to show numbers.

Example

Write the number.

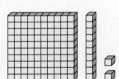

3 tens 4 ones = 34   1 hundred 1 ten 2 ones = 112

▶ Practice

Write the number.

1.

2.
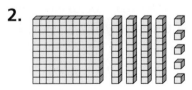

3.

4.

PLACE VALUE WITH ONES, TENS, HUNDREDS

You can use a place-value chart to understand numbers.

Hundreds	Tens	Ones
5	8	2

582 = 5 hundreds 8 tens 2 ones

▶ Practice

Write the number.

1. 3 hundreds 7 tens 9 ones

2. 8 tens 1 one

3. 6 tens 5 ones

4. 1 hundred 8 ones

5. 7 hundreds 5 tens 1 one

6. 4 hundreds 2 tens

COMPARE 2-DIGIT NUMBERS

You can use base-ten blocks to compare numbers.

13 is **greater than** 11.

31 is **less than** 32.

Examples

Write the words *greater than* or *less than*.

A 95 is _?_ 72.

Think: 95 has more tens than 72.

So, 95 is *greater than* 72.

B 34 is _?_ 36.

Think: 34 and 36 have the same number of tens. 34 has fewer ones than 36.

So, 34 is *less than* 36.

▶ Practice

Write the words *greater than* or *less than*.

1. 15 is _?_ 51.

2. 77 is _?_ 68.

3. 84 is _?_ 83.

4. 99 is _?_ 97.

5. 55 is _?_ 65.

6. 12 is _?_ 21.

7. 23 is _?_ 33.

8. 14 is _?_ 40.

9. 90 is _?_ 80.

✔ ORDER NUMBERS

You can write numbers in order.

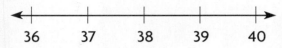

36   37   38   39   40

37 is just **after** 36.
38 is just **before** 39.
39 is **between** 38 and 40.

▶ Practice

Write the number that is just after, just before, or between.

1. 29, ▪

2. ▪, 179

3. 60, ▪

4. 506, ▪, 508

5. 552, ▪

6. ▪, 13

7. 27, ▪, 29

8. ▪, 60

9. 439, ▪, 441

✔ ADDITION FACTS

The answer to an addition problem is the **sum**. If you don't remember a fact, you can think about how to **make a ten** to help find the sum.

> **Example** Find 7 + 5.
>
>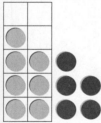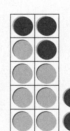
>
> Move 3 counters to fill the ten frame. 7 + 3 = 10
>
> There are 2 more counters to add. 10 + 2 = 12
>
> Show 7 + 5.   So, 7 + 5 = 12.

▶ Practice

Add. You may wish to make a ten.

1.   5
       + 6

2.   8
       + 5

3.   6
       + 4

4.   7
       + 7

5.   9
       + 6

✓ 2-DIGIT ADDITION

Sometimes you must regroup to add 2-digit numbers.

Find 14 + 17.

STEP 1	STEP 2	STEP 3
Add the ones.	Regroup.	Add the tens.
$4 + 7 = 11$	11 ones = 1 ten 1 one	$1 + 1 + 1 = 3$

tens	ones
1	4
+ 1	7

tens	ones
1	
1	4
+ 1	7
	1

tens	ones
1	
1	4
+ 1	7
3	1

▶ Practice

Add.

1.　62
　　+ 23

2.　31
　　+ 49

3.　58
　　+ 18

4.　73
　　+ 17

✓ MENTAL MATH: ADD 2-DIGIT NUMBERS

You can use mental math to add.

Examples

A Find $32 + 46 = \blacksquare$.
- Add the tens. $\quad 30 + 40 = 70$
- Add the ones. $\quad 2 + 6 = 8$
- Add the sums. $\quad 70 + 8 = 78$

So, $32 + 46 = 78$.

B Find $19 + 21 = \blacksquare$.
- Add the tens. $\quad 10 + 20 = 30$
- Add the ones. $\quad 9 + 1 = 10$
- Add the sums. $\quad 30 + 10 = 40$

So, $19 + 21 = 40$.

▶ Practice

Use mental math to add.

1. $12 + 11 = \blacksquare$　　**2.** $45 + 19 = \blacksquare$　　**3.** $59 + 31 = \blacksquare$　　**4.** $46 + 47 = \blacksquare$

5. $27 + 13 = \blacksquare$　　**6.** $32 + 29 = \blacksquare$　　**7.** $64 + 28 = \blacksquare$　　**8.** $39 + 43 = \blacksquare$

9. $76 + 25 = \blacksquare$　　**10.** $89 + 32 = \blacksquare$　　**11.** $94 + 46 = \blacksquare$　　**12.** $49 + 83 = \blacksquare$

TROUBLESHOOTING

✔ SUBTRACTION FACTS

The answer to a subtraction problem is the **difference**.
If you can't remember a fact, you can **count back** to help
find the difference.

Example

Find 7 − 2.

Say 7. Count back 2.

6, 5

The difference is 5.
So, 7 − 2 = 5.

▶ **Practice**

Subtract.

1. 11 − 2	**2.** 11 − 5	**3.** 13 − 7	**4.** 15 − 8	**5.** 18 − 9

✔ 2-DIGIT SUBTRACTION

Check whether you must regroup to subtract 2-digit numbers.

Find 32 − 15.

STEP 1

Since 5 > 2, you
must regroup.

tens	ones
3	2
− 1	5

STEP 2

Regroup.
3 tens 2 ones = 2 tens 12 ones.
Subtract the ones.

tens	ones
2	12
3̸	2̸
− 1	5
	7

STEP 3

Subtract the tens.

tens	ones
2	12
3̸	2̸
− 1	5
1	7

▶ **Practice**

Subtract.

1. 57 − 22	**2.** 73 − 34	**3.** 66 − 18	**4.** 31 − 17	**5.** 80 − 27

✅ MENTAL MATH: SUBTRACT 2-DIGIT NUMBERS

You can use mental math to subtract.

Use mental math to find 95 − 19.

STEP 1	STEP 2	STEP 3
Add more to make the smaller number a ten.	Add the same number to the larger number.	Subtract your answers.
95 − 19 ▇ Think: 19 + 1 = 20	95 + 1 = 96 19 + 1 = 20	96 − 20 76 So, 95 − 19 = 76.

▶ Practice

Use mental math to subtract.

1. 41 − 19 = ▇   **2.** 78 − 29 = ▇   **3.** 37 − 18 = ▇   **4.** 56 − 28 = ▇

✅ MONEY: COUNT BILLS AND COINS

1 dollar, or $1.00   1 quarter = 25¢, or $0.25   1 dime = 10¢, or $0.10   1 nickel = 5¢, or $0.05   1 penny = 1¢, or $0.01

Count on to find the value of a group of coins.

Example

Count and write the amount.

$1.00 $1.25 $1.50 $1.60 $1.65 $1.66

So, the value of the bill and coins is $1.66.

▶ Practice

Count and write the amount.

1.

2.

TROUBLESHOOTING

✓ TELL TIME

The short hand on a clock is the **hour hand**.
The long hand on a clock is the **minute hand**.

You can **count by fives** to find the minutes after the hour.

This clock shows 9:35.

Examples

Write the time.

A

3:00

B

3:30

C

3:45

▶ Practice

Write the time.

1.

2.

3.

✓ CALENDAR

You can use a calendar to keep track of days, weeks, and months.

There are 5 Wednesdays in October. The third Friday in this month is October 18.

▶ Practice
Use the calendar.

1. The second Tuesday in October is ?.

2. The last day in October is a ?.

3. There are ? Mondays in October.

4. The fifth Wednesday in October is ?.

OCTOBER						
Sun	Mon	Tue	Wed	Thu	Fri	Sat
		1	2	3	4	5
6	7	8	9	10	11	12
13	14	15	16	17	18	19
20	21	22	23	24	25	26
27	28	29	30	31		

✅ SKIP-COUNT

To skip-count, start with any number and add or subtract the same number.

> **Example**
>
> Skip-count to find the missing numbers. 6, 9, 12, ■, ■
>
> Think: 6 + 3 = 9   9 + 3 = 12
>
> So, skip-count by threes to find the missing numbers.
>
> 12 + 3 = 15   15 + 3 = 18
>
> So, the numbers are 6, 9, 12, 15, 18.

▶ Practice

Skip-count to find the missing numbers.

1. 8, 10, 12, ■

2. 25, 30, 35, ■

3. 100, 90, 80, ■, ■

4. 36, 38, 40, ■

5. 12, 15, 18, ■

6. 35, 40, 45, ■, ■

✅ EQUAL GROUPS

When you have **equal groups**, you can skip-count to find how many in all.

> **Example**
>
> Write how many there are in all.
>
>
>
> Think: Since there are 2 in each group, you can skip-count by twos: 2, 4, 6, 8.
>
> So, 4 groups of 2 = 8.

▶ Practice

Write how many there are in all.

1.

3 groups of 2 = ■

2.

5 groups of 3 = ■

3.

2 groups of 2 = ■

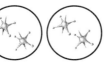

TROUBLESHOOTING

✓ COLUMN ADDITION

To add more than two numbers, choose two numbers
to add first. Look for facts you know.

> **Example**
>
> $$\begin{array}{r} 2 \\ 5 \\ +\ 8 \\ \hline 15 \end{array}$$
> $2 + 8 = 10$
> $10 + 5 = 15$
>
> $$\begin{array}{r} 2 \\ 5 \\ +\ 8 \\ \hline 15 \end{array}$$
> $2 + 5 = 7$
> $7 + 8 = 15$
>
> $$\begin{array}{r} 2 \\ 5 \\ +\ 8 \\ \hline 15 \end{array}$$
> $5 + 8 = 13$
> $13 + 2 = 15$

▶ Practice

Find the sum.

1.	**2.**	**3.**	**4.**	**5.**	**6.**
4	6	9	7	2	1
4	2	1	3	7	8
+ 2	+ 3	+ 2	+ 5	+ 4	+ 1

✓ ADDITION

When addends are the same, you can skip-count to
find the sum. Any number plus 0 equals that same number.

> **Examples**
>
> **A** Find the sum.
>
> $3 + 3 + 3 + 3 + 3 = $ ■
>
> **Think:** Since each addend is 3,
> skip-count by threes.
> 3, 6, 9, 12, 15
>
> So, $3 + 3 + 3 + 3 + 3 = 15$.
>
> **B** Find the missing number.
>
> $7 + $ ■ $ = 7$
>
> **Think:** What number plus 7
> equals 7?
>
> So, $7 + 0 = 7$.

▶ Practice

Find the sum or missing number.

1. $5 + 5 + 5 = $ ■   **2.** ■ $ + 9 = 9$ **3.** $4 + 4 + 4 + 4 = $ ■

4. $2 + 2 + 2 = $ ■   **5.** $0 + $ ■ $ = 8$ **6.** ■ $ + 3 = 3$

7. $3 + 3 + 3 + 3 = $ ■   **8.** ■ $ + 1 = 1$ **9.** $7 + 7 + 7 = $ ■

✓ ORDER PROPERTY OF ADDITION

Changing the order of addends does not change the sum.

Example

Find the missing number.

$2 + \blacksquare = 3 + 2$ Think: $3 + 2 = 5$
So, I must solve $2 + \blacksquare = 5$.
The missing number is 3.

So, $2 + 3 = 3 + 2$.

▶ **Practice**

Find the missing number.

1. $4 + 1 = 1 + \blacksquare$ **2.** $7 + 6 = 6 + \blacksquare$ **3.** $\blacksquare + 5 = 5 + 8$

4. $9 + \blacksquare = 5 + 9$ **5.** $\blacksquare + 2 = 2 + 4$ **6.** $3 + \blacksquare = 0 + 3$

✓ MULTIPLICATION FACTS

The answer to a multiplication problem is the **product**.

Examples

A Find 2×6.

2 rows of $6 = 12$
So, $2 \times 6 = 12$.

B Find 3×5.

3 rows of $5 = 15$
So, $3 \times 5 = 15$.

▶ **Practice**

Find the product.

1. $2 \times 2 = \blacksquare$ **2.** $4 \times 3 = \blacksquare$ **3.** $5 \times 2 = \blacksquare$

4. $\blacksquare = 5 \times 5$ **5.** $3 \times 2 = \blacksquare$ **6.** $\blacksquare = 5 \times 9$

7. $8 \times 3 = \blacksquare$ **8.** $\blacksquare = 7 \times 5$ **9.** $3 \times 7 = \blacksquare$

10. $3 \times 9 = \blacksquare$ **11.** $2 \times 8 = \blacksquare$ **12.** $5 \times 1 = \blacksquare$

TROUBLESHOOTING

✓ ADD EQUAL GROUPS

When you multiply, you add equal groups.

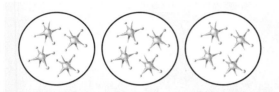

$4 + 4 + 4 = 12$
So, 3 groups of 4 = 12.

► Practice

Write how many there are in all.

1.

2 groups of 3 = ■

2.

5 groups of 3 = ■

3.

2 groups of 1 = ■

4.

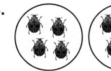

4 groups of 4 = ■

✓ MULTIPLICATION FACTS THROUGH 5

When one of the factors of a problem is even, you can double a fact you already know.

> **Example**
>
> Find 4×7.
>
> **Think:** To find a 4's fact, you can double a 2's fact.
>
> • First find the 2's fact. **$2 \times 7 = 14$**
>
> • Double the product. **$14 + 14 = 28$**
>
> So, $4 \times 7 = 28$.

► Practice

Find the product.

1. $8 \times 2 = $ ■   **2.** $4 \times 4 = $ ■   **3.** $4 \times 6 = $ ■   **4.** $6 \times 5 = $ ■

5. $3 \times 6 = $ ■   **6.** ■ $= 4 \times 7$ **7.** $9 \times 4 = $ ■   **8.** $7 \times 6 = $ ■

9. $8 \times 3 = $ ■   **10.** $4 \times 5 = $ ■   **11.** ■ $= 6 \times 2$ **12.** $8 \times 4 = $ ■

✓ SKIP-COUNT BY TENS

You can use a pattern to skip-count by tens.

Example

Skip-count by tens to find the missing numbers.

18, 28, 38, ■, ■, 68

1	2	3	4	5	6	7	8	9	10
11	12	13	14	15	16	17	18	19	20
21	22	23	24	25	26	27	28	29	30
31	32	33	34	35	36	37	38	39	40
41	42	43	44	45	46	47	48	49	50
51	52	53	54	55	56	57	58	59	60
61	62	63	64	65	66	67	68	69	70

$38 + 10 = 48$
$48 + 10 = 58$

So, the numbers are 18, 28, 38, 48, 58, 68.

▶ **Practice**

Skip-count to find the missing numbers.

1. 15, 25, 35, 45, ■, 65

2. 11, 21, 31, ■, 51, 61

3. 54, 44, 34, ■, ■

4. 16, 26, 36, ■, 56, ■

✓ MULTIPLICATION FACTS THROUGH 8

Break apart an array to find a multiplication fact.

Find 4×8.

STEP 1

Show 4 rows of 8.

$4 \times 8 = $ ■

STEP 2

Break apart the array.

$2 \times 8 = 16$ $2 \times 8 = 16$

STEP 3

Add the products.

$$\begin{array}{r} 16 \\ + 16 \\ \hline 32 \end{array}$$

So, $4 \times 8 = 32$.

▶ **Practice**

Find the product.

1. $7 \times 6 = $ ■   **2.** $7 \times 4 = $ ■   **3.** $8 \times 6 = $ ■   **4.** $8 \times 9 = $ ■

5. $9 \times 6 = $ ■   **6.** $9 \times 4 = $ ■   **7.** $7 \times 7 = $ ■   **8.** $8 \times 8 = $ ■

9. $4 \times 6 = $ ■   **10.** $3 \times 8 = $ ■   **11.** $8 \times 4 = $ ■   **12.** $6 \times 3 = $ ■

TROUBLESHOOTING

✔ MEANING OF MULTIPLICATION

When you multiply, you add equal groups.

5 groups of 4
$4 + 4 + 4 + 4 + 4 = 20$
So, $5 \times 4 = 20$.

▶ Practice
Copy and complete.

1.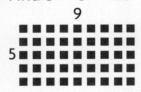

   a. ■ groups of ■
   b. ■ + ■ + ■ = ■
   c. ■ × ■ = ■

2.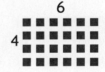

   a. ■ groups of ■
   b. ■ + ■ = ■
   c. ■ × ■ = ■

✔ MULTIPLICATION FACTS THROUGH 10

You can use an array to find a multiplication fact.

Examples

A Find $5 \times 9 = $ ■.

9
5 ▪▪▪▪▪▪▪▪▪
   ▪▪▪▪▪▪▪▪▪
   ▪▪▪▪▪▪▪▪▪
   ▪▪▪▪▪▪▪▪▪
   ▪▪▪▪▪▪▪▪▪

So, $5 \times 9 = 45$.

B Find $4 \times 6 = $ ■.

6
4 ▪▪▪▪▪▪
   ▪▪▪▪▪▪
   ▪▪▪▪▪▪
   ▪▪▪▪▪▪

So, $4 \times 6 = 24$.

▶ Practice
Find the product.

1. $9 \times 7 = $ ■   **2.** $8 \times 10 = $ ■   **3.** $6 \times 9 = $ ■   **4.** $5 \times 9 = $ ■

5. $10 \times 7 = $ ■   **6.** $9 \times 9 = $ ■   **7.** $7 \times 7 = $ ■   **8.** $10 \times 6 = $ ■

9. $9 \times 6 = $ ■   **10.** $3 \times 8 = $ ■   **11.** $3 \times 10 = $ ■   **12.** $6 \times 8 = $ ■

13. $9 \times 8 = $ ■   **14.** $9 \times 4 = $ ■   **15.** $9 \times 10 = $ ■   **16.** $8 \times 8 = $ ■

✔ SUBTRACTION

To find a missing number in a subtraction sentence, think about fact families. Remember how numbers are related.

$5 + 7 = 12$ $7 + 5 = 12$ $12 - 5 = 7$ $12 - 7 = 5$

Examples

Find the missing number.

A $12 - \blacksquare = 7$

To solve $12 - \blacksquare = 7$,
find $12 - 7 = \blacksquare$.
$12 - 7 = 5$
So, $12 - 5 = 7$.

B $32 - \blacksquare = 28$

To solve $32 - \blacksquare = 28$,
find $32 - 28 = \blacksquare$.
$32 - 28 = 4$
So, $32 - 4 = 28$.

▶ Practice

Find the missing number.

1. $21 - \blacksquare = 15$ **2.** $32 - \blacksquare = 29$ **3.** $43 = 51 - \blacksquare$ **4.** $98 - \blacksquare = 92$

5. $31 - \blacksquare = 17$ **6.** $40 - \blacksquare = 25$ **7.** $50 = 70 - \blacksquare$ **8.** $67 - \blacksquare = 11$

✔ DIVISION FACTS THROUGH 5

The answer in a division problem is a **quotient**. You can think about a related multiplication fact to find a quotient.

Example

Find $3\overline{)15}$.

Read: 15 divided by 3
Write: $15 \div 3 = \blacksquare$

Think: 15 counters in 3 groups
How many in each group?

5 in each group

So, $15 \div 3 = 5$, or $3\overline{)15}$.

▶ Practice

Find the quotient.

1. $25 \div 5 = \blacksquare$ **2.** $30 \div 5 = \blacksquare$ **3.** $24 \div 3 = \blacksquare$ **4.** $14 \div 2 = \blacksquare$

5. $40 \div 5 = \blacksquare$ **6.** $36 \div 4 = \blacksquare$ **7.** $28 \div 4 = \blacksquare$ **8.** $45 \div 5 = \blacksquare$

✔ ORDER PROPERTY OF MULTIPLICATION

Two numbers can be multiplied in any order. The product will be the same.

Example

3 groups of 5 = 15
So, 3 × 5 = 15.

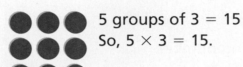

5 groups of 3 = 15
So, 5 × 3 = 15.

If you can't remember 3 × 5, try thinking of 5 × 3.

So, 3 × 5 = 5 × 3.

▶ **Practice**

Use the Order Property of Multiplication to help you find each product.

1. 4 × 5 = ■   **2.** 8 × 4 = ■   **3.** 7 × 8 = ■

4. 9 × 6 = ■   **5.** 7 × 3 = ■   **6.** 6 × 3 = ■

✔ MISSING FACTORS

You can use division to find a missing factor.

Examples

Find the missing factor.

A 4 × ■ = 36

Think: 36 ÷ 4 = 9

So, the missing factor is 9.
4 × 9 = 36

B 5 × ■ = 40

Think: 40 ÷ 5 = 8

So, the missing factor is 8.
5 × 8 = 40

▶ **Practice**

Find the missing factor.

1. ■ × 3 = 18   **2.** 4 × ■ = 20   **3.** 3 × ■ = 24

4. ■ × 4 = 32   **5.** ■ × 5 = 15   **6.** 4 × ■ = 16

7. 4 × ■ = 40   **8.** ■ × 5 = 35   **9.** ■ × 3 = 27

✔ READ A CHART

A chart is used to organize information.

FAVORITE BREAKFAST	
Type	**Students**
Cereal	12
Pancakes	6
Eggs	7
Fruit	3

← Title

The data shows the number of students who chose each type of breakfast food.

This chart shows that
- Cereal had the most votes.
- 7 students chose eggs.
- 28 students in all voted.

▶ Practice

For 1–3, use the information in the chart above.

1. Which breakfast food had the fewest votes?

2. How many students chose pancakes?

3. How many students in all chose cereal or eggs?

✔ TALLY DATA

A **tally table** has tally marks to record data.
Tally marks are grouped by fives.   (5 = 卌)

FAVORITE SPORT	
Sport	**Tally**
Baseball	卌 I
Basketball	卌 卌
Hockey	IIII
Soccer	卌 IIII

← 6 students chose baseball.

← 9 students chose soccer.

▶ Practice

For 1–2, use the tally table.

1. How many students chose basketball as their favorite sport?

2. How many students did NOT choose soccer as their favorite sport?

TROUBLESHOOTING

✔ READ PICTOGRAPHS

A **pictograph** shows data by using pictures.

The **key** in this pictograph shows that each ◆ equals 2 votes.

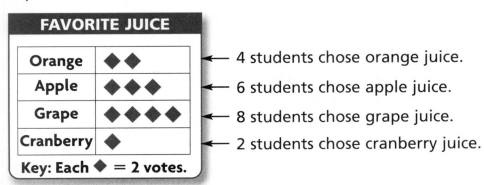

FAVORITE JUICE

Orange	◆ ◆	← 4 students chose orange juice.
Apple	◆ ◆ ◆	← 6 students chose apple juice.
Grape	◆ ◆ ◆ ◆	← 8 students chose grape juice.
Cranberry	◆	← 2 students chose cranberry juice.

Key: Each ◆ = 2 votes.

▶ Practice
Use the value of the symbol to find the missing number.

1. If ✱ = 2, then
✱ + ✱ + ✱ = ▪.

2. If ✱ = 4, then
✱ + ✱ = ▪.

3. If ★ = 10, then
★ + ★ + ★ + ★ = ▪.

4. If ✿ = 5, then
✿ + ✿ + ✿ = ▪.

✔ IDENTIFY PARTS OF A WHOLE

A **fraction** names a part of a whole.

The top number tells how many parts are being used.

The bottom number tells how many equal parts are in the whole.

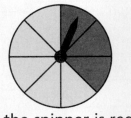

$\frac{3}{8}$ of the spinner is red.

▶ Practice
Write a fraction that names the red part of the spinner.

1.

2.
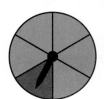

3.

4.

✅ COMPARE PARTS OF A WHOLE

You compare fractions by comparing parts of a whole.

$\frac{1}{6}$ of the spinner is blue.

$\frac{5}{6}$ of the spinner is yellow.

The largest part of the spinner is yellow.

▶ Practice

Write the color shown by the largest part of each spinner.

1. 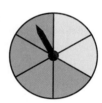 **2.**  **3.**  **4.**

✅ USE A TALLY TABLE

You can record the number of times a pointer stops on each color by using a tally table.

SPINNER RESULTS	
Color	**Tally**
Yellow	III
Red	ЖH
Blue	IIII

This spinner was used 12 times.
The pointer stopped on yellow 3 times.

▶ Practice

Use the tally table.

1. How many times did the pointer stop on red?

2. Which color was landed on the fewest times?

3. How many more times was red landed on than blue?

TROUBLESHOOTING

✔ REGROUP ONES AND TENS

You can regroup 10 ones as 1 ten.

10 ones = 1 ten

Example

Regroup 2 tens 16 ones.

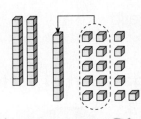

So, 2 tens 16 ones = 3 tens 6 ones, or 36.

▶ Practice
Regroup.

1. 6 tens 12 ones = ■ tens 2 ones

2. 3 tens 15 ones = ■ tens 5 ones

3. 4 tens 25 ones = ■ tens 5 ones

4. 4 tens ■ ones = 5 tens 8 ones

✔ MULTIPLICATION FACTS

There are many ways you have learned to remember multiplication facts.

Example

Find $4 \times 9 = $ ■.

- Skip-count by equal groups.
- Use an array.
- Break apart an array.
- If one of the factors is even, use doubling.
- If you can't remember a fact, try changing the order of the factors.

So, $4 \times 9 = 36$.

▶ Practice
Find the product.

1. $6 \times 8 = $ ■

2. $3 \times 9 = $ ■

3. $5 \times 7 = $ ■

4. $6 \times 7 = $ ■

5. $3 \times 5 = $ ■

6. $7 \times 7 = $ ■

7. $5 \times 8 = $ ■

8. $2 \times 9 = $ ■

✓ MULTIPLY 2-DIGIT NUMBERS

You can use base-ten blocks to multiply 2-digit numbers.

Find $3 \times 16 = \blacksquare$.

STEP 1

Model 3 groups of 16.
Multiply the ones.

$$\begin{array}{r} 16 \\ \times\ 3 \\ \hline 18 \end{array} \quad (3 \times 6 \text{ ones})$$

STEP 2

Multiply the tens.

$$\begin{array}{r} 16 \\ \times\ 3 \\ \hline 18 \\ 30 \end{array} \quad (3 \times 1 \text{ ten})$$

STEP 3

Add to find the product.

$$\begin{array}{r} 16 \\ \times\ 3 \\ \hline 18 \\ +\ 30 \\ \hline 48 \end{array}$$

So, $3 \times 16 = 48$.

▶ Practice

Find the product.

1. $18 \times 8 = \blacksquare$ **2.** $27 \times 4 = \blacksquare$ **3.** $23 \times 6 = \blacksquare$ **4.** $81 \times 9 = \blacksquare$

5. $64 \times 4 = \blacksquare$ **6.** $25 \times 3 = \blacksquare$ **7.** $62 \times 5 = \blacksquare$ **8.** $85 \times 2 = \blacksquare$

✓ DIVISION FACTS

You have learned many ways to remember division facts.

Example Find $18 \div 3 = \blacksquare$.

A Use counters.

B Think about an array.

C Use repeated subtraction.

$$\begin{array}{r} 18 \\ -\ 3 \\ \hline 15 \end{array} \nearrow \begin{array}{r} 15 \\ -\ 3 \\ \hline 12 \end{array} \nearrow \begin{array}{r} 12 \\ -\ 3 \\ \hline 9 \end{array} \nearrow \begin{array}{r} 9 \\ -\ 3 \\ \hline 6 \end{array} \nearrow \begin{array}{r} 6 \\ -\ 3 \\ \hline 3 \end{array} \nearrow \begin{array}{r} 3 \\ -\ 3 \\ \hline 0 \end{array}$$

So, $18 \div 3 = 6$.

D Think about fact families.

$3 \times 6 = 18 \quad 18 \div 6 = 3$

$6 \times 3 = 18 \quad 18 \div 3 = 6$

▶ Practice

Divide.

1. $20 \div 5 = \blacksquare$ **2.** $36 \div 6 = \blacksquare$ **3.** $27 \div 3 = \blacksquare$ **4.** $42 \div 6 = \blacksquare$

5. $72 \div 9 = \blacksquare$ **6.** $24 \div 3 = \blacksquare$ **7.** $56 \div 8 = \blacksquare$ **8.** $63 \div 7 = \blacksquare$

TROUBLESHOOTING

✔ REGROUP HUNDREDS AND TENS

You can regroup 10 tens as 1 hundred and 10 ones
as 1 ten.

10 ones = 1 ten   10 tens = 1 hundred

Example

Regroup 3 hundreds 12 tens.

So, 3 hundreds 12 tens = 4 hundreds 2 tens, or 420.

▶ Practice
Regroup.

1. 2 hundreds ■ tens = 350   **2.** 910 = ■ hundreds 11 tens

3. 1 hundred 17 tens = 2 hundreds ■ tens

✔ MULTIPLICATION AND DIVISION FACTS

Fact families can help you remember multiplication and
division facts.

Examples

A Find $21 \div 3 = $ ■.

Think of a related multiplication fact.

If you know $3 \times 7 = 21$,
then you know $21 \div 3 = 7$.

B Find $4 \times 6 = $ ■.

Think of a related division fact.

If you know $24 \div 4 = 6$,
then you know $4 \times 6 = 24$.

▶ Practice
Find the product or quotient.

1. $6 \times 7 = $ ■   **2.** $9 \times 3 = $ ■   **3.** $56 \div 7 = $ ■   **4.** $6 \times 6 = $ ■

5. $45 \div 5 = $ ■   **6.** $3 \times 8 = $ ■   **7.** $42 \div 6 = $ ■   **8.** $72 \div 9 = $ ■

✔ IDENTIFY SOLID FIGURES

These are solid figures.

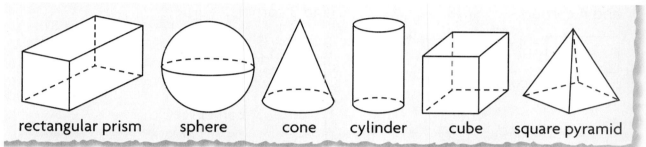

rectangular prism   sphere   cone   cylinder   cube   square pyramid

▶ Practice

Write the name of each solid figure.

1.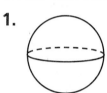

2.

3.

4.

✔ IDENTIFY PLANE FIGURES

These are plane figures.

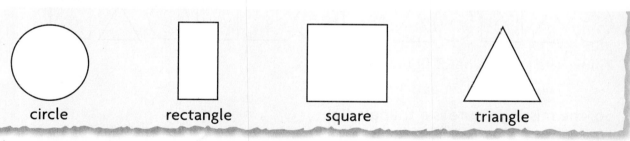

circle   rectangle   square   triangle

▶ Practice

Write the name of each plane figure.

1.

2.

3.

4.

✔ SIDES AND CORNERS

A rectangle has 4 sides
and 4 corners.

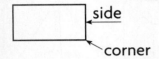

A triangle has 3 sides
and 3 corners.

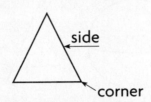

▶ Practice

Write the number of sides and corners in each figure.

1.   2.   3.   4.

✔ GEOMETRIC PATTERNS

You can find a pattern by looking for the place where
a figure, or group of figures, repeats.

Example

Name the missing figure in the pattern. Describe the pattern.

○△□ ○△□ ○ _?_ □ ○△□

Think: The pattern has 3 figures: ○△□.

These figures repeat.

So, the missing figure is a triangle.

▶ Practice

Describe the pattern. Name the missing figure in the pattern.

1. ○△○△○△ _?_ △○△

2. ▯△△▯△△▯ _?_ △▯△△

✅ CONGRUENT FIGURES

Figures that are the same size and shape are **congruent**.

These figures have the same size and shape. They are **congruent**.

These figures have the same shape, but they are not the same size. They are **not congruent**.

▶ Practice

Tell whether the two figures are congruent.
Write *yes* or *no*.

1.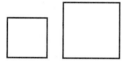

2. ☆ ☆

3.

4.

5.

6.

✅ TYPES OF ANGLES

A **right angle** is an angle that forms a square corner.

This is a **right angle**.

This angle is **less than** a right angle.

This angle is **greater than** a right angle.

▶ Practice

Write if each angle is a *right angle, greater than* a right angle, or *less than* a right angle.

1.

2.

3.

4.

5.

6.

TROUBLESHOOTING

✓ USE A CUSTOMARY RULER

An **inch (in.)** is used to measure length.
Use an inch ruler to measure.

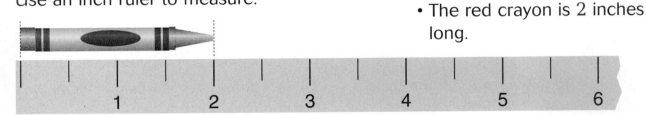

• The red crayon is 2 inches long.

▶ Practice

For 1–2, use the drawing below.

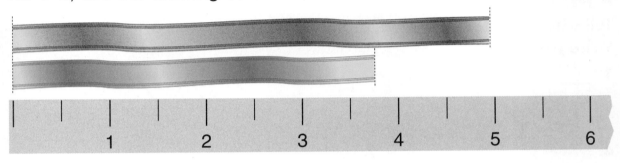

1. Which ribbon is about 5 inches long?

2. Which ribbon is between 3 and 4 inches long?

✓ MEASURE TO THE NEAREST INCH

Example

Write the length to the nearest inch.

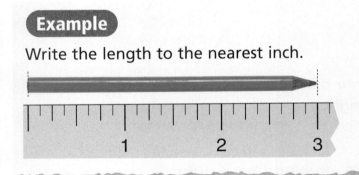

To measure to the nearest inch:

• Line up one end of the object with the left end of the ruler.

• Find the inch mark nearest the other end of the object.

The pencil is 3 inches long to the nearest inch.

▶ Practice

Write the length to the nearest inch.

1.

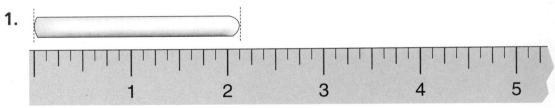

✔ FIND A RULE

You can write a rule to describe a number pattern.

Example

Write a rule for the table. Then complete the table.

Tricycles	1	2	3	4	5	6
Wheels	3	6	9	12	■	■

Pattern: The number of wheels equals the number of tricycles times 3.

Rule: Multiply the number of tricycles by 3.

So, the missing numbers are 15 and 18.

▶ Practice

Write a rule for the table. Then complete the table.

1.
Packs	1	2	3	4	5	6
Cookies	6	12	18	24	■	■

2.
Chairs	3	6	2	5	1	4
Legs	12	24	8	20	■	■

✔ USE A METRIC RULER

A **centimeter (cm)** is a metric unit used to measure length.

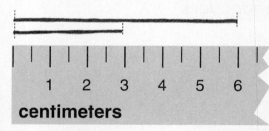

- The blue string is about 6 cm long.
- The red string is about 3 cm long.

▶ Practice

For 1–3, use the drawing below.

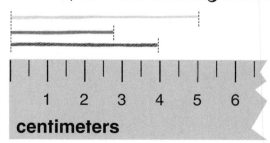

1. Which string is about 5 cm long?

2. Which string is about 1 cm longer than the pink string?

3. Which string is less than 3 cm long?

TROUBLESHOOTING

✓ MEASURE TO THE NEAREST CENTIMETER

To measure to the nearest centimeter:

• Line up one end of the object with the left end of the ruler.

• Find the cm mark nearest the other end of the object.

Example

Write the length to the nearest centimeter.

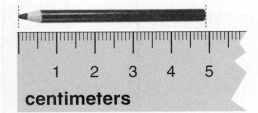

The pencil is 5 centimeters long to the nearest centimeter.

▶ Practice

Write the length to the nearest centimeter.

1.

2.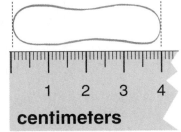

✓ MODEL PARTS OF A WHOLE

This figure has 5 equal parts.

• 3 equal parts are shaded.

• 2 equal parts are *not* shaded.

▶ Practice

Write how many equal parts are in the whole figure.
Then write how many parts are shaded.

1.

2.

3.

✓ MODEL PARTS OF A GROUP

This group has 6 squares.

• 1 square is yellow.

• 5 squares are *not* yellow.

▶ Practice
Write the number in each group. Then write the number that are *not* yellow in each group.

1.   2.   3.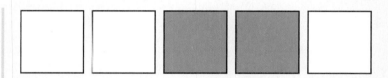

✓ NAME THE FRACTION

A number that names part of a whole or part of a group is called a **fraction**.

• 1 part is shaded.

• There are 3 parts in all.

So, $\frac{1}{3}$ of the circle is shaded.

• 2 squares are blue.

• There are 5 squares in all.

So, $\frac{2}{5}$ of the squares are blue.

▶ Practice
Write the fraction for the part that is shaded.

1.   2.   3.

Write the fraction for the part that is blue.

4.    5.   6.

TROUBLESHOOTING

✔ COMPARE FRACTIONS

As you compare fractions, remember the symbols you have used.

< means *is less than*.　　　　> means *is greater than*.

$1 < 3$　　　　　　　　　　　$3 > 1$

Examples

A Compare $\frac{2}{4}$ and $\frac{3}{4}$.

$\frac{2}{4} < \frac{3}{4}$ or $\frac{3}{4} > \frac{2}{4}$

B Compare $\frac{3}{5}$ and $\frac{4}{10}$.

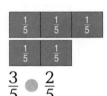

$\frac{3}{5} > \frac{4}{10}$ or $\frac{4}{10} < \frac{3}{5}$

▶ Practice

Compare. Write < or > for each ●.

1.

$\frac{1}{4}$ ● $\frac{2}{4}$

2.

$\frac{3}{5}$ ● $\frac{2}{5}$

3.
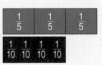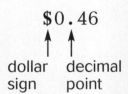

$\frac{1}{4}$ ● $\frac{3}{8}$

✔ USE MONEY NOTATION

Use a **dollar sign** and a **decimal point** to write money amounts.

Example

Count and write the amount.

$0.25　　$0.35　　$0.40　　$0.45　　$0.46

$$\$0.46$$
↑ ↑
dollar　decimal
sign　　point

So, the value of the coins is $0.46.

▶ Practice

Count and write the amount.

1.

2.

✅ NAME THE DECIMAL

You can use decimals to show tenths and hundredths.

Mixed Number: $1\frac{6}{10}$ Fraction: $\frac{40}{100}$

Decimal: 1.6   Decimal: 0.40

▶ Practice
Write each fraction or mixed number as a decimal.

1. $\frac{3}{10}$ 2. $\frac{9}{10}$ 3. $\frac{45}{100}$ 4. $1\frac{5}{10}$

5. $\frac{2}{100}$ 6. $\frac{50}{100}$ 7. $1\frac{7}{10}$ 8. $1\frac{75}{100}$

✅ COMPARE AND ORDER DECIMALS

Use a number line to compare and order decimals.

Example

Write 0.9, 0.5, and 0.8 in order from least to greatest.

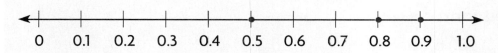

0.5 is to the left of 0.8   0.8 is to the left of 0.9

So, the numbers from least to greatest are 0.5, 0.8, 0.9.

▶ Practice
Use the number line above. Write < or > for each ●.

1. 0.6 ● 0.5   **2.** 0.3 ● 0.6   **3.** 0.7 ● 0.4   **4.** 0.9 ● 1.0

Use the number line to order the decimals from least to greatest.

5. 0.1, 0.7, 0.4   **6.** 1.0, 0.2, 0.1   **7.** 0.9, 0.8, 0.7   **8.** 0.8, 0.4, 1.0

Use the number line to order the decimals from greatest to least.

9. 0.4, 0.9, 0.2   **10.** 0.1, 0.8, 1.0   **11.** 0.7, 0.3, 0.5   **12.** 0.6, 0.2, 0.8

EXTRA PRACTICE

Set A (pp. 4–5)

Write each number in standard form.

1.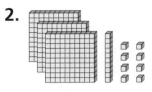

2.

3.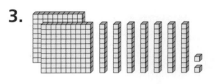

4. 500 + 60 + 6   **5.** 700 + 4   **6.** 800 + 10 + 4

7. four hundred seventy-six   **8.** nine hundred ninety-one

Write the value of the blue digit.

9. 346   **10.** 872   **11.** 13   **12.** 554

Set B (pp. 6–9)

Write in standard form.

1. 1,000 + 900 + 40 + 2   **2.** 5,000 + 700 + 80 + 3

3. 3,000 + 5   **4.** 9,000 + 900 + 10 + 4

5. two thousand, four hundred sixty-seven   **6.** eight thousand, eighteen

Write in expanded form.

7. 5,487   **8.** 6,055   **9.** 6,170   **10.** 7,796

Set C (pp. 10–11)

Write in standard form.

1. 10,000 + 6,000 + 900 + 60 + 5   **2.** 50,000 + 3,000 + 400 + 80 + 6

3. 30,000 + 6,000 + 1   **4.** 90,000 + 5,000 + 900 + 10 + 3

5. fifty-one thousand, four hundred   **6.** twenty-two thousand, eighteen

7. seventy thousand, forty-nine   **8.** ninety-nine thousand, four

Write in expanded form.

9. 65,487   **10.** 76,055   **11.** 36,173   **12.** 47,796

Set A (pp. 18–19)

For 1–2, choose a benchmark of 10, 100, or 1,000 to estimate each.

1. the number of pieces in a large bag of dog food

2. the number of teeth in your mouth

3. There are 25 students in Ken's class. There are 4 third-grade classes. About how many students are in the third grade?

Set B (pp. 20–23)

Compare the numbers. Write <, >, or = for each ●.

1. 400 ● 12
2. 646 ● 600
3. 741 ● 741
4. 57 ● 75
5. 165 ● 164
6. 313 ● 515

Set C (pp. 24–25)

Write the numbers in order from least to greatest.

1. 124; 562; 347
2. 102; 89; 157
3. 702; 546; 212

Write the numbers in order from greatest to least.

4. 1,466; 1,365; 1,988
5. 2,218; 3,010; 2,115
6. 5,010; 5,100; 5,310

Set D (pp. 28–29)

Round to the nearest hundred and the nearest ten.

1. 414
2. 888
3. 502
4. 635
5. 157
6. 733
7. 374
8. 498

Set E (pp. 30–31)

Round to the nearest thousand.

1. 3,345
2. 8,866
3. 5,533
4. 6,500
5. 9,457
6. 1,168
7. 7,662
8. 2,220

Set A (pp. 36–37)

Use the Grouping Property to find the sum.

	1.	**2.**	**3.**	**4.**	**5.**
	7	7	16	12	31
	8	13	14	37	19
	+3	+ 9	+27	+18	+76

6. $6 + 6 + 4 = \blacksquare$ **7.** $27 + 10 + 13 = \blacksquare$ **8.** $3 + 7 + 18 = \blacksquare$

9. Ron bought 8 goldfish, 15 guppies, and 12 minnows. How many fish did he buy?

Set B (pp. 38–39)

Estimate the sum.

	1.	**2.**	**3.**	**4.**	**5.**
	11	24	63	165	298
	+23	+17	+27	+220	+456

	6.	**7.**	**8.**	**9.**	**10.**
	645	584	2,375	560	6,757
	+594	+248	+4,082	+439	+4,446

Set C (pp. 42–43)

Find the sum. Estimate to check.

	1.	**2.**	**3.**	**4.**	**5.**
	465	978	789	$3.12	919
	+521	+200	+123	+$5.29	+453

6. $327 + 246 = \blacksquare$ **7.** $299 + 403 = \blacksquare$ **8.** $544 + 671 = \blacksquare$

Set D (pp. 46–49)

Find the sum. Estimate to check.

	1.	**2.**	**3.**	**4.**	**5.**
	3,120	2,899	6,247	7,231	3,485
	+5,771	+ 901	+2,319	+1,191	+9,856

6. Estimate to decide if the sum of $1,874 + 3,205$ is greater than 4,000. Explain your answer.

7. $\frac{a+b}{c}$ **Algebra** Write the missing addend. $4,000 + \blacksquare = 5,100$

Set A (pp. 54–55)

Estimate the difference.

1. 63 −49	**2.** 54 −14	**3.** 98 −55	**4.** 595 −227	**5.** 873 −475
6. 795 −309	**7.** 7,850 −2,187	**8.** 8,026 −4,826	**9.** 5,575 −4,746	**10.** 2,521 −1,779

Set B (pp. 58–61)

Find the difference. Estimate to check.

1. 205 − 67	**2.** 608 −409	**3.** 500 −165	**4.** 900 −198	**5.** 402 −317

6. (a+b/c) **Algebra** The sum of 885 and another number is 901. What is the other number?

7. MENTAL MATH How might you use mental math to find 500 − 199?

Set C (pp. 62–65)

Find the difference. Estimate to check.

1. 4,561 −4,327	**2.** 8,135 −3,645	**3.** 7,608 −5,810	**4.** 4,005 −3,318	**5.** 8,000 − 997

6. Lindsey paddled 4,033 feet in her canoe. Ronald paddled 2,077 feet in his. How much farther did Lindsey paddle than Ronald?

7. REASONING The library was built 6 years before the post office was built. The post office was built in 1971. How old was the library in 1988?

Set D (pp. 68–69)

Write + or − to make the number sentence true.

1. 8 ● 15 = 23   **2.** 95 ● 16 = 79   **3.** 517 ● 483 = 1,000

4. 35 = 29 ● 6   **5.** 42 ● 9 = 51   **6.** 27 ● 4 = 23

7. Shelby had $18.50. She spent $7.25 at the car wash. Write an expression that shows how much she has left.

Set A (pp. 84–85)

Use < or > to compare the amounts of money.

1. a.

b.

2. a.

b.

3. a.

b.

4. Maria has 3 quarters, 3 dimes, and 1 nickel. She wants to buy a slice of pie for $1.10. Does she have enough money? Explain.

5. Ronnie has 1 quarter, 4 dimes, and 3 nickels. Lydia has 3 quarters. Who has more money? Explain.

Set B (pp. 88–89)

Find the sum or difference. Estimate to check.

1. $3.35 +$2.14	**2.** $8.56 −$4.45	**3.** $7.08 −$5.35	**4.** $6.45 +$5.77	**5.** $4.07 +$1.96
6. $8.00 −$4.44	**7.** $9.05 +$8.96	**8.** $5.40 +$2.95	**9.** $9.42 −$3.58	**10.** $10.00 −$ 5.65

USE DATA For 11–12, use the table.

11. How much more does a pound of boysenberries cost than an apricot?

12. Ezra buys 2 plums and a pound of grapes with a $5 bill. The clerk gives him two $1 bills, 2 quarters, 3 dimes, and 7 pennies. Is the change correct? Explain.

FREIDA'S FRUIT STAND	
Fruit	**Price**
Apricots	$0.65 each
Plums	$0.45 each
Grapes	$1.33 each pound
Apples	$0.55 each
Boysenberries	$1.79 each pound

Set A (pp. 96–97)

Write two ways you can read each time. Then write the time, using A.M. or P.M.

1.

recess

2.

play at park

3.

plant flowers

4.

go to library

Set B (pp. 100–101)

Copy and complete the schedule.

MONDAY NIGHT ON CHANNEL 8		
Program	Time	Elapsed Time
1. Game Show	4:00 P.M. – 5:00 P.M.	▨
2. Evening News	5:00 P.M. – 5:45 P.M.	▨
3. Basketball Game	5:45 P.M. – ▨	2 hours 30 minutes
4. Mystery Theater	▨ – 10:00 P.M.	1 hour 45 minutes
5. Nighttime News	10:00 P.M. – 10:30 P.M.	▨

Set C (pp. 102–103)

Solve. Use the calendars.

November						
Sun	Mon	Tue	Wed	Thu	Fri	Sat
					1	2
3	4	5	6	7	8	9
10	11	12	13	14	15	16
17	18	19	20	21	22	23
24	25	26	27	28	29	30

December						
Sun	Mon	Tue	Wed	Thu	Fri	Sat
1	2	3	4	5	6	7
8	9	10	11	12	13	14
15	16	17	18	19	20	21
22	23	24	25	26	27	28
29	30	31				

1. Mr. Burns went on a trip from November 24 to December 6. How many days was he gone?

2. Enya went on a trip for 3 weeks and 3 days. She returned December 4. When did Enya leave?

3. What date is 2 weeks before November 19?

4. What day of the week is 3 days after December 14?

Set A (pp. 116–117)

Copy and complete.

1.

   a. ■ groups of ■ = ■

   b. ■ + ■ + ■ + ■ + ■ = ■

   c. ■ × ■ = ■

2.

   a. ■ groups of ■ = ■

   b. ■ + ■ + ■ + ■ = ■

   c. ■ × ■ = ■

For 3–8, find how many in all.

3. 5 groups of 3

4. 4 + 4 + 4 + 4

5. 7 + 7 + 7

6. 5 × 4

7. 4 groups of 5

8. 5 × 5

9. Ana bought 6 packages of 5 cards each. How many cards did she buy?

Set B (pp. 118–119)

Find the product.

1. $7 \times 2 = $ ■

2. $5 \times 2 = $ ■

3. $7 \times 5 = $ ■

4. $8 \times 2 = $ ■

5. $4 \times 5 = $ ■

6. $9 \times 5 = $ ■

7. ■ $= 6 \times 5$

8. ■ $= 2 \times 2$

9. ■ $= 5 \times 5$

10. Keith bought 5 packages of toy cars. Each package has 3 cars in it. David has 16 toy cars. Who has more cars? How do you know?

Set C (pp. 122–125)

Find the product.

1. $7 \times 3 = $ ■

2. $5 \times 3 = $ ■

3. ■ $= 3 \times 9$

4. $4 \times 5 = $ ■

5. ■ $= 3 \times 3$

6. $8 \times 5 = $ ■

7. $\begin{array}{r} 6 \\ \times 3 \\ \hline \end{array}$

8. $\begin{array}{r} 5 \\ \times 6 \\ \hline \end{array}$

9. $\begin{array}{r} 3 \\ \times 8 \\ \hline \end{array}$

10. $\begin{array}{r} 3 \\ \times 9 \\ \hline \end{array}$

11. $\begin{array}{r} 5 \\ \times 9 \\ \hline \end{array}$

Set A (pp. 132–133)

Find the product.

1. $0 \times 5 = \blacksquare$ **2.** $3 \times 7 = \blacksquare$ **3.** $1 \times 7 = \blacksquare$ **4.** $4 \times 3 = \blacksquare$

5. $\blacksquare = 6 \times 3$ **6.** $\blacksquare = 8 \times 5$ **7.** $\blacksquare = 0 \times 9$ **8.** $\blacksquare = 1 \times 1$

9. Is the product of 3 and 0 *greater than, less than,* or *equal to* the product of 0 and 6? Explain.

Set B (pp. 134–135)

Find the product.

1. $\begin{array}{r} 8 \\ \times 4 \\ \hline \end{array}$ **2.** $\begin{array}{r} 4 \\ \times 0 \\ \hline \end{array}$ **3.** $\begin{array}{r} 4 \\ \times 6 \\ \hline \end{array}$ **4.** $\begin{array}{r} 4 \\ \times 4 \\ \hline \end{array}$ **5.** $\begin{array}{r} 4 \\ \times 9 \\ \hline \end{array}$

6. $3 \times 4 = \blacksquare$ **7.** $\blacksquare = 6 \times 4$ **8.** $\blacksquare = 4 \times 7$ **9.** $5 \times 4 = \blacksquare$

10. Mario has 4 packs of 8 stickers. He also has 19 loose stickers. How many stickers does he have in all?

Set C (pp. 138–141)

Find the product.

1. $4 \times 6 = \blacksquare$ **2.** $5 \times 3 = \blacksquare$ **3.** $8 \times 0 = \blacksquare$ **4.** $9 \times 5 = \blacksquare$

5. $5 \times 8 = \blacksquare$ **6.** $3 \times 6 = \blacksquare$ **7.** $7 \times 4 = \blacksquare$ **8.** $3 \times 9 = \blacksquare$

9. $8 \times 2 = \blacksquare$ **10.** $3 \times 7 = \blacksquare$ **11.** $6 \times 5 = \blacksquare$ **12.** $4 \times 8 = \blacksquare$

13. A movie is shown 5 times each day. How many times is that movie shown in one week?

Set D (pp. 142–143)

Find the missing factor.

1. $\blacksquare \times 9 = 18$ **2.** $5 \times \blacksquare = 20$ **3.** $\blacksquare \times 1 = 8$ **4.** $\blacksquare \times 9 = 9$

5. $2 \times \blacksquare = 14$ **6.** $4 \times \blacksquare = 16$ **7.** $3 \times \blacksquare = 21$ **8.** $\blacksquare \times 8 = 32$

9. Jill has 9 baskets with an equal number of eggs in each. If she has 36 eggs in all, how many are in each basket?

Set A (pp. 148–149)

Find each product.

1. $5 \times 6 = $ ■
2. $6 \times 2 = $ ■
3. $0 \times 6 = $ ■
4. $6 \times 7 = $ ■
5. $6 \times 4 = $ ■
6. $6 \times 3 = $ ■
7. ■ $= 8 \times 6$
8. $4 \times 7 = $ ■
9. $8 \times 2 = $ ■
10. ■ $= 6 \times 6$
11. $6 \times 1 = $ ■
12. $6 \times 9 = $ ■

Complete.

13. ■ $\times 4 = 16$
14. ■ $\times 5 = 35$
15. $2 \times$ ■ $= 8$
16. $8 \times$ ■ $= 24$

Set B (pp. 150–151)

Find each product.

1. $1 \times 7 = $ ■
2. ■ $= 7 \times 5$
3. $6 \times 7 = $ ■
4. $3 \times 7 = $ ■
5. $2 \times 7 = $ ■
6. $3 \times 3 = $ ■
7. $7 \times 7 = $ ■
8. $4 \times 7 = $ ■
9. $8 \times 1 = $ ■
10. $0 \times 7 = $ ■
11. ■ $= 7 \times 8$
12. $9 \times 7 = $ ■

13. Alex has 8 bags of rocks. Each bag has 7 rocks. Vera takes 3 bags. How many rocks does Alex have left?

Set C (pp. 152–153)

Find each product.

1. $7 \times 8 = $ ■
2. $8 \times 0 = $ ■
3. $8 \times 3 = $ ■
4. $6 \times 8 = $ ■
5. ■ $= 8 \times 4$
6. ■ $= 2 \times 5$
7. $8 \times 1 = $ ■
8. $8 \times 8 = $ ■
9. $8 \times 2 = $ ■
10. $5 \times 8 = $ ■
11. $2 \times 2 = $ ■
12. $8 \times 9 = $ ■

13. Dolores has 8 bags of 4 apples. Ruth has 3 bags of 8 apples. How many more apples does Dolores have?

Set D (pp. 156–159)

Find each product.

1. $8 \times 7 = $ ■
2. $6 \times 7 = $ ■
3. ■ $= 6 \times 4$
4. $7 \times 3 = $ ■
5. $9 \times 8 = $ ■
6. $8 \times 6 = $ ■
7. $7 \times 7 = $ ■
8. $3 \times 5 = $ ■
9. $9 \times 7 = $ ■
10. ■ $= 6 \times 5$
11. $8 \times 8 = $ ■
12. $9 \times 6 = $ ■

Set A (pp. 164–167)

Find the product.

1. $9 \times 7 = $ 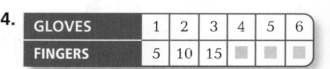   **2.** $6 \times 9 = $ **3.** $6 \times 10 = $ **4.** $10 \times 3 = $

5. $3 \times 9 = $ **6.** $8 \times 9 = $ **7.** $10 \times 7 = $ **8.** $9 \times 5 = $

9. $9 \times 4 = $ **10.** $10 \times 5 = $ **11.** $8 \times 10 = $ **12.** $9 \times 2 = $

Find the missing factor.

13. $\blacksquare \times 9 = 27$ **14.** $90 = 9 \times \blacksquare$ **15.** $7 \times \blacksquare = 56$ **16.** $21 = \blacksquare \times 7$

Set B (pp. 168–169)

Write a rule for the table. Then copy and complete the table.

1.

PACKS	1	2	3	4	5	6
CARDS	4	8	12	■	■	■

2.

BAGS	1	2	3	4	5	6
ORANGES	6	12	18	■	■	■

3.

CANS	1	2	3	4	5	6
TENNIS BALLS	3	6	9	■	■	■

4.

GLOVES	1	2	3	4	5	6
FINGERS	5	10	15	■	■	■

For 5–6, use the table in Exercise 3.

5. If you had 18 tennis balls, how many cans would you have?

6. Suppose you had 7 cans. How many tennis balls would you have?

Set C (pp. 170–171)

Find each product.

1. $(4 \times 1) \times 3 = $ **2.** $(3 \times 2) \times 3 = $ **3.** $(5 \times 1) \times 5 = $

4. $4 \times (2 \times 2) = $ **5.** $10 \times (3 \times 3) = $ **6.** $(6 \times 1) \times 7 = $

Find the missing factor.

7. $(4 \times \blacksquare) \times 1 = 16$ **8.** $5 \times (2 \times \blacksquare) = 20$ **9.** $\blacksquare \times (7 \times 1) = 49$

10. Mason made 2 sandwiches for each of 3 friends. He used 2 slices of bread for each sandwich. How many slices of bread did he use?

Set A (pp. 186–187)

Write a division sentence for each.

1. $\begin{array}{r} 18 \\ -\ 6 \\ \hline 12 \end{array}$ $\begin{array}{r} 12 \\ -\ 6 \\ \hline 6 \end{array}$ $\begin{array}{r} 6 \\ -6 \\ \hline 0 \end{array}$

2. $\begin{array}{r} 36 \\ -\ 9 \\ \hline 27 \end{array}$ $\begin{array}{r} 27 \\ -\ 9 \\ \hline 18 \end{array}$ $\begin{array}{r} 18 \\ -\ 9 \\ \hline 9 \end{array}$ $\begin{array}{r} 9 \\ -9 \\ \hline 0 \end{array}$

Use subtraction to solve.

3. $15 \div 5 = \blacksquare$ 4. $18 \div 3 = \blacksquare$ 5. $12 \div 4 = \blacksquare$ 6. $16 \div 4 = \blacksquare$

7. $7\overline{)14}$ 8. $5\overline{)20}$ 9. $3\overline{)24}$ 10. $8\overline{)40}$

Set B (pp. 188–189)

Complete each number sentence. Draw an array to help.

1. $4 \times \blacksquare = 8$ $8 \div 4 = \blacksquare$
2. $6 \times \blacksquare = 30$ $30 \div 6 = \blacksquare$
3. $8 \times \blacksquare = 32$ $32 \div 8 = \blacksquare$
4. $4 \times \blacksquare = 12$ $12 \div 4 = \blacksquare$
5. $9 \times \blacksquare = 36$ $36 \div 9 = \blacksquare$
6. $2 \times \blacksquare = 12$ $12 \div 2 = \blacksquare$

7. What division sentence could you write for an array that shows $5 \times 8 = 40$?

8. How can you use $5 + 5 + 5 + 5 = 20$ to help you find $20 \div 5$?

Set C (pp. 190–193)

Write the fact family.

1. 2, 3, 6   2. 3, 7, 21   3. 3, 9, 27   4. 3, 6, 18

5. 4, 6, 24   6. 4, 8, 32   7. 5, 5, 25   8. 3, 8, 24

Find the quotient or the missing divisor.

9. $6 \div \blacksquare = 3$ 10. $18 \div 6 = \blacksquare$ 11. $\blacksquare = 12 \div 3$ 12. $20 \div \blacksquare = 4$

13. $30 \div 6 = \blacksquare$ 14. $3 = 15 \div \blacksquare$ 15. $9 \div \blacksquare = 3$ 16. $16 \div 4 = \blacksquare$

17. Jerome made 30 cookies. He ate 3 and divided the rest equally among 3 friends. How many cookies did each friend get?

Set A (pp. 200–201)

Find each missing factor and quotient.

1. $2 \times \blacksquare = 10$ $10 \div 2 = \blacksquare$ **2.** $5 \times \blacksquare = 30$ $30 \div 5 = \blacksquare$

Find each quotient.

3. $15 \div 5 = \blacksquare$ **4.** $\blacksquare = 16 \div 2$ **5.** $\blacksquare = 45 \div 5$ **6.** $10 \div 5 = \blacksquare$

7. $2\overline{)2}$ **8.** $5\overline{)20}$ **9.** $2\overline{)18}$ **10.** $5\overline{)25}$ **11.** $2\overline{)12}$

12. Divide 20 by 2.   **13.** Divide 35 by 5.   **14.** Divide 6 by 2.

Set B (pp. 202–203)

Write the multiplication fact you can use to find the quotient. Then write the quotient.

1. $18 \div 3 = \blacksquare$ **2.** $32 \div 4 = \blacksquare$ **3.** $9 \div 3 = \blacksquare$

Find each quotient.

4. $28 \div 4 = \blacksquare$ **5.** $12 \div 3 = \blacksquare$ **6.** $\blacksquare = 27 \div 3$ **7.** $\blacksquare = 8 \div 4$

8. $4\overline{)16}$ **9.** $3\overline{)15}$ **10.** $4\overline{)24}$ **11.** $3\overline{)21}$ **12.** $4\overline{)12}$

13. Divide 30 by 3.   **14.** Divide 20 by 4.   **15.** Divide 36 by 4.

Set C (pp. 204–205)

Find each quotient.

1. $0 \div 4 = \blacksquare$ **2.** $\blacksquare = 3 \div 3$ **3.** $\blacksquare = 8 \div 1$ **4.** $10 \div 10 = \blacksquare$

5. $7\overline{)7}$ **6.** $8\overline{)0}$ **7.** $1\overline{)4}$ **8.** $3\overline{)0}$ **9.** $9\overline{)9}$

10. Divide 8 by 8.   **11.** Divide 0 by 6.   **12.** Divide 9 by 1.

Set D (pp. 206–207)

Write an expression to describe each problem.

1. Four friends share 28 stickers equally. How many stickers does each friend get?

2. Melinda had $15. She buys slippers for $8. How much money does she have now?

Set A (pp. 214–217)

Find the missing factor and quotient.

1. $8 \times \blacksquare = 32$ $32 \div 8 = \blacksquare$ **2.** $7 \times \blacksquare = 35$ $35 \div 7 = \blacksquare$

Find the quotient.

3. $42 \div 6 = \blacksquare$ **4.** $\blacksquare = 24 \div 4$ **5.** $64 \div 8 = \blacksquare$ **6.** $\blacksquare = 21 \div 7$

7. $7\overline{)49}$ **8.** $2\overline{)2}$ **9.** $6\overline{)36}$ **10.** $5\overline{)40}$ **11.** $8\overline{)48}$

12. Divide 63 by 7.   **13.** Divide 80 by 8.   **14.** Divide 15 by 3.

Set B (pp. 218–219)

Find the quotient.

1. $\blacksquare = 36 \div 9$ **2.** $20 \div 10 = \blacksquare$ **3.** $\blacksquare = 20 \div 5$ **4.** $54 \div 6 = \blacksquare$

5. $3\overline{)12}$ **6.** $10\overline{)70}$ **7.** $9\overline{)72}$ **8.** $4\overline{)16}$ **9.** $9\overline{)27}$

10. Divide 56 by 8.   **11.** Divide 60 by 6.   **12.** Divide 100 by 10.

Set C (pp. 220–223)

Find the quotient.

1. $6 \div 6 = \blacksquare$ **2.** $\blacksquare = 7 \div 1$ **3.** $\blacksquare = 0 \div 5$ **4.** $28 \div 7 = \blacksquare$

5. $9\overline{)81}$ **6.** $3\overline{)18}$ **7.** $6\overline{)24}$ **8.** $2\overline{)20}$ **9.** $7\overline{)0}$

10. Divide 32 by 4.   **11.** Divide 40 by 10.   **12.** Divide 16 by 8.

Set D (pp. 224–225)

USE DATA For 1–4, use the price list at the right to find the cost of each number of items.

PRICE LIST	
Mugs	$4
Aprons	$8

1. 3 aprons   **2.** 5 mugs   **3.** 6 aprons   **4.** 8 mugs

Find the cost of one of each item.

5. 2 pizzas cost $14.   **6.** 4 tapes cost $32.   **7.** 5 books cost $25.

8. 6 pens cost $12.   **9.** 7 balls cost $21.   **10.** 3 shirts cost $27.

Set A (pp. 240–241)

For 1–3, use the tally table.

FAVORITE HOBBY	
Hobby	**Tally**
Collecting stamps	IIII
Collecting sports cards	IIII IIII
Collecting coins	II
Reading	IIII III
Drawing	IIII

1. How many people answered the survey?

2. What is the most popular hobby?

3. How many fewer people chose collecting coins than chose reading?

For 4–6, use the frequency table.

4. How many people answered the survey?

5. Did more people choose peas or carrots?

6. How many more people chose corn than chose broccoli?

FAVORITE VEGETABLE	
Type	**Number**
Carrots	6
Peas	6
Beans	3
Corn	12
Broccoli	1

Set B (pp. 242–243)

For 1–3, use the table.

FAVORITE BREAKFAST FOOD				
	Bacon and Eggs	**French Toast**	**Cereal**	**Muffins**
Boys	9	6	4	1
Girls	4	8	5	3

1. How many girls were surveyed?

2. How many boys liked French toast the best?

3. How many students chose bacon and eggs?

For 4–6, use the table.

4. How many students are wearing blue shoes?

5. How many students are wearing white or black shoes?

6. How many more students are wearing blue shoes than red shoes?

SHOE COLOR				
	White	**Black**	**Red**	**Blue**
Boys	9	4	3	5
Girls	8	3	2	4

EXTRA PRACTICE

Set A (pp. 252–253)

For 1–4, use the bar graph.

1. How many moths are there?

2. What is the total number of insects?

3. How many more butterflies are there than beetles?

4. If Dr. Cooper sold five of his bees, how many bees would be left?

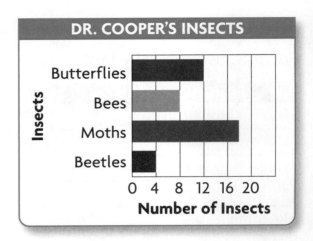

DR. COOPER'S INSECTS

Set B (pp. 256–259)

For 1–3, use the line plot.

1. What is the range of the data?

2. What is the mode for this data? Explain.

3. How many students were surveyed?

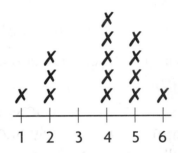

Number of Chores Students Did Last Week

Set C (pp. 260–261)

For 1–6, use the grid. Write the letter of the point named by the ordered pair.

1. (3,1)   2. (6,3)

3. (5,2)   4. (1,3)

5. (4,4)   6. (2,4)

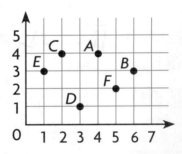

Set D (pp. 262–263)

For 1–2, use the line graph.

1. In what month did Anita's Inn have the most guests? In what month were there the fewest guests?

2. How many guests stayed at Anita's Inn from January through June?

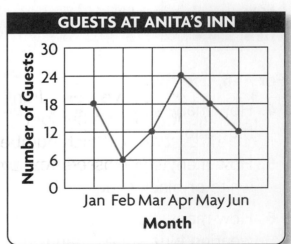

GUESTS AT ANITA'S INN

Set A (pp. 268–269)

Tell whether each event is *certain* or *impossible*.

1. You can choose a nickel from these coins.

2. You can spin red or blue on this spinner.

Set B (pp. 270–271)

1. Name the color that you are most likely to spin.

2. Name the color marble that is least likely to be pulled.

Set C (pp. 274–277)

1. Tenesha pulled animal crackers from the box. She made the graph below of the outcomes. Which animals are equally likely to be pulled from the box?

2. Ned pulled marbles from the bag. He made the graph below of the outcomes. Which color marble is most likely to be pulled from the bag? Which is least likely to be pulled?

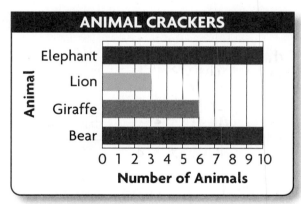

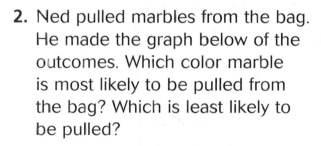

Set D (pp. 278–279)

1. This line plot shows how often each number on the number cube was tossed. Which number is most likely to be tossed next time?

2. This tally table shows the color tiles that were pulled from a bag. Which color tile is most likely to be pulled?

COLOR	TALLIES
green	︎llll lll
blue	l
red	lll
yellow	ll

```
X X X X X X
X X X X X X
X X X X X X
+--+--+--+--+--+--+
1  2  3  4  5  6
```

Set A (pp. 294–297)

Find the product.

1. 15
× 3

2. 27
× 2

Find the product. You may wish to use base-ten blocks.

3. 16
× 2

4. 32
× 3

5. 13
× 6

6. 25
× 5

7. 47
× 4

8. 34
× 8

9. 73
× 2

10. 23
× 7

11. 45
× 5

12. 99
× 3

13. $6 \times 15 = $ ■

14. $9 \times 21 = $ ■

15. $7 \times 33 = $ ■

16. $4 \times 51 = $ ■

17. Jill has 3 boxes of nature magazines. There are 28 magazines in each box. How many nature magazines does Jill have in all?

Set B (pp. 298–299)

Find the product. Tell whether you need to regroup. Write *yes* or *no*.

1. 21
× 4

2. 23
× 9

3. 38
× 5

4. 44
× 3

5. 33
× 3

Find the product.

6. 14
× 6

7. 29
× 5

8. 37
× 7

9. 18
× 2

10. 22
× 3

11. 38
× 4

12. 25
× 8

13. 72
× 7

14. 49
× 8

15. 53
× 9

16. $9 \times 19 = $ ■

17. $4 \times 91 = $ ■

18. $3 \times 12 = $ ■

19. $8 \times 88 = $ ■

20. Angelo sells hot dogs for $3 each. If he sells 56 hot dogs, how much money will he make?

Set A (pp. 306–307)

Copy and complete. Use patterns and mental math.

1. $8 \times 3 = \blacksquare$
$8 \times 30 = \blacksquare$
$8 \times 300 = \blacksquare$
$8 \times 3,000 = \blacksquare$

2. $4 \times 7 = \blacksquare$
$4 \times 70 = \blacksquare$
$4 \times \blacksquare = 2,800$
$4 \times 7,000 = \blacksquare$

3. $\blacksquare \times 6 = 54$
$9 \times \blacksquare = 540$
$9 \times 600 = \blacksquare$
$9 \times \blacksquare = 54,000$

Use mental math and basic multiplication facts to find the product.

4. $2 \times 900 = \blacksquare$ **5.** $5 \times 4,000 = \blacksquare$ **6.** $3 \times 80 = \blacksquare$ **7.** $7 \times 300 = \blacksquare$

Set B (pp. 310–311)

Estimate the product.

1. $\begin{array}{r} 15 \\ \times\ 7 \\ \hline \end{array}$
2. $\begin{array}{r} 46 \\ \times\ 6 \\ \hline \end{array}$
3. $\begin{array}{r} 92 \\ \times\ 9 \\ \hline \end{array}$
4. $\begin{array}{r} 130 \\ \times\ 8 \\ \hline \end{array}$
5. $\begin{array}{r} 683 \\ \times\ 5 \\ \hline \end{array}$

Set C (pp. 312–315)

Find the product.

1. $\begin{array}{r} 141 \\ \times\ 4 \\ \hline \end{array}$
2. $\begin{array}{r} 718 \\ \times\ 2 \\ \hline \end{array}$
3. $\begin{array}{r} 455 \\ \times\ 4 \\ \hline \end{array}$
4. $\begin{array}{r} 257 \\ \times\ 7 \\ \hline \end{array}$
5. $\begin{array}{r} 609 \\ \times\ 6 \\ \hline \end{array}$

6. $3 \times 252 = \blacksquare$ **7.** $5 \times 279 = \blacksquare$ **8.** $9 \times 191 = \blacksquare$ **9.** $8 \times 348 = \blacksquare$

Set D (pp. 316–317)

Find the product in dollars and cents.

1. $\begin{array}{r} \$3.32 \\ \times\ \ \ 5 \\ \hline \end{array}$
2. $\begin{array}{r} \$6.24 \\ \times\ \ \ 4 \\ \hline \end{array}$
3. $\begin{array}{r} \$1.99 \\ \times\ \ \ 8 \\ \hline \end{array}$
4. $\begin{array}{r} \$7.13 \\ \times\ \ \ 6 \\ \hline \end{array}$
5. $\begin{array}{r} \$5.58 \\ \times\ \ \ 2 \\ \hline \end{array}$

6. $7 \times \$1.37 = \blacksquare$ **7.** $3 \times \$8.30 = \blacksquare$ **8.** $9 \times \$4.59 = \blacksquare$ **9.** $4 \times \$2.46 = \blacksquare$

Set E (pp. 318–319)

Find the product. Estimate to check.

1. $\begin{array}{r} 2,481 \\ \times\ \ \ \ 3 \\ \hline \end{array}$
2. $\begin{array}{r} 5,082 \\ \times\ \ \ \ 2 \\ \hline \end{array}$
3. $\begin{array}{r} \$12.39 \\ \times\ \ \ \ \ 7 \\ \hline \end{array}$
4. $\begin{array}{r} 8,469 \\ \times\ \ \ \ 5 \\ \hline \end{array}$
5. $\begin{array}{r} \$35.94 \\ \times\ \ \ \ \ 4 \\ \hline \end{array}$

EXTRA PRACTICE

Set A (pp. 326–327)

Use the model. Write the quotient and remainder.

1.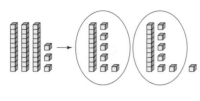

$33 \div 2 = \blacksquare$

2.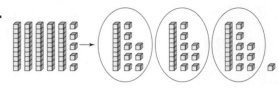

$55 \div 3 = \blacksquare$

Divide. You may use base-ten blocks to help.

3. $29 \div 2 = \blacksquare$ 4. $45 \div 4 = \blacksquare$ 5. $54 \div 3 = \blacksquare$ 6. $68 \div 6 = \blacksquare$

7. $3\overline{)58}$ 8. $2\overline{)23}$ 9. $3\overline{)57}$ 10. $4\overline{)59}$

11. $7\overline{)66}$ 12. $8\overline{)51}$ 13. $4\overline{)63}$ 14. $5\overline{)54}$

Set B (pp. 328–331)

Divide and check.

1. $15 \div 7 = \blacksquare$ 2. $47 \div 5 = \blacksquare$ 3. $68 \div 5 = \blacksquare$ 4. $38 \div 2 = \blacksquare$

5. $9\overline{)16}$ 6. $8\overline{)74}$ 7. $5\overline{)70}$ 8. $4\overline{)89}$

9. $3\overline{)56}$ 10. $9\overline{)32}$ 11. $3\overline{)47}$ 12. $2\overline{)27}$

13. Ms. Payne has 65 crayons. She divides them equally among 9 students. How many crayons does each student get? How many are left over?

14. Jackie has 35 crackers. She divides them equally among 5 plates. How many crackers are on 2 plates?

Set C (pp. 332–333)

Divide.

1. $28 \div 5 = \blacksquare$ 2. $45 \div 7 = \blacksquare$ 3. $15 \div 8 = \blacksquare$ 4. $39 \div 4 = \blacksquare$

5. $52 \div 2 = \blacksquare$ 6. $78 \div 9 = \blacksquare$ 7. $21 \div 6 = \blacksquare$ 8. $54 \div 3 = \blacksquare$

9. $5\overline{)59}$ 10. $2\overline{)25}$ 11. $9\overline{)28}$ 12. $6\overline{)32}$

13. $7\overline{)79}$ 14. $9\overline{)59}$ 15. $8\overline{)67}$ 16. $4\overline{)98}$

Set A (pp. 340–341)

Copy and complete. Use patterns and mental math to help.

1. $9 \div 9 = \blacksquare$

$90 \div 9 = \blacksquare$

$900 \div 9 = \blacksquare$

2. $7 \div 7 = \blacksquare$

$70 \div 7 = \blacksquare$

$700 \div 7 = \blacksquare$

3. $42 \div \blacksquare = 7$

$\blacksquare \div 6 = 70$

$4,200 \div \blacksquare = 700$

Use mental math and a basic fact to find the quotient.

4. $400 \div 5 = \blacksquare$

5. $360 \div 9 = \blacksquare$

6. $1,800 \div 2 = \blacksquare$

7. $5\overline{)450}$

8. $4\overline{)3,600}$

9. $8\overline{)4,800}$

Set B (pp. 342–343)

Estimate each quotient. Write the basic fact you used to find the estimate.

1. $155 \div 3 = \blacksquare$

2. $639 \div 7 = \blacksquare$

3. $374 \div 6 = \blacksquare$

Estimate the quotient.

4. $4\overline{)318}$

5. $5\overline{)212}$

6. $2\overline{)801}$

7. $3\overline{)291}$

8. $8\overline{)653}$

9. $9\overline{)551}$

Set C (pp. 344–345)

Find the quotient.

1. $2\overline{)68}$

2. $3\overline{)108}$

3. $5\overline{)250}$

4. $9\overline{)279}$

5. $6\overline{)552}$

6. $4\overline{)296}$

7. $7\overline{)119}$

8. $8\overline{)624}$

9. Jan has 384 jelly beans. How many can she put in each of 6 bags?

10. Tyrone has 423 cards. How many can he give to each of his 9 friends?

Set D (pp. 346–347)

Find the quotient.

1. $2\overline{)200}$

2. $4\overline{)128}$

3. $6\overline{)672}$

4. $3\overline{)549}$

5. $5\overline{)495}$

6. $9\overline{)369}$

7. $8\overline{)808}$

8. $7\overline{)623}$

9. The product of 7 and a certain number is 238. What is the other number?

10. Marian has 516 sunflower seeds. How many seeds can she put in each of 3 bags?

Set A (pp. 362–365)

Name the solid figure that each object looks like.

1.
2.
3.
4.

Set B (pp. 366–367)

Name the solid figures used to make each object.

1.
2.
3.

Set C (pp. 368–371)

Name each figure.

1.
2.
3.
4.

Use a corner of a piece of paper to tell whether each angle is a *right angle, greater than* a right angle, or *less than* a right angle.

5.
6.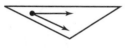
7.

Set D (pp. 372–373)

Describe the lines. Write *parallel* or *intersecting*.

1.
2.
3.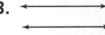

Describe the intersecting lines. Write *form right angles* or *do not form right angles*.

4.
5.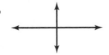
6.

Set A (pp. 382–383)

Tell if each figure is a polygon. Write yes or no.

1.
2.
3.
4.
5.

Write the number of sides and angles each polygon has. Then name the polygon.

6.
7.
8.
9.
10.

11. Melanie was drawing polygons for homework. She decided that squares and rectangles are quadrilaterals. Do you agree? Explain.

Set B (pp. 384–387)

Tell whether the two figures are congruent. Write yes or no.

1.
2.
3.

How many lines of symmetry does each object have?

4.
5.
6.
7.

8. Look at the letters in the word MATH. Which letters have one line of symmetry? Do any of the letters have more than one line of symmetry? Explain.

Set C (pp. 388–389)

Tell if each figure will tessellate. Write yes or no.

1.
2.
3.
4.

5. Use grid paper and pattern blocks to make a design. Repeat your design to make a tessellation.

Set A (pp. 396–397)

Write if each angle is a *right angle, greater than* a
right angle, or *less than* a right angle.

1.
2.
3.
4.
5.

Set B (pp. 398–401)

For 1–3, use the triangles at the right. Write *A, B,* or *C*.

1. Which triangle is equilateral?
2. Which triangle is scalene?
3. Which is a right triangle?

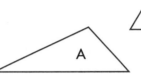

Name each triangle. Write *equilateral, isosceles,* or *scalene*.

4. 5.  6. 7. 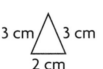 8.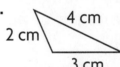

Set C (pp. 402–403)

Describe the angles and sides of each quadrilateral.

1.
2.
3.

Set D (pp. 404–407)

Write the name for each quadrilateral.

1.
2.
3.
4.
5.

Set A (pp. 420–423)

Measure the length to the nearest inch.

1.

2.

Measure the length to the nearest half inch.

3.

4.

5.

Set B (pp. 424–425)

Choose the unit you would use to measure each.
Write *inch, foot, yard,* **or** *mile*.

1. the height of a 2-story house
2. the distance between 2 towns
3. the length of a hockey stick
4. the length of your thumb

Set C (pp. 430–431)

Change the units. Use the Table of Measures on
page 430 to help.

1. ■ cups = 1 quart

16 cups = ■ quarts

2. ■ feet = 1 yard

feet	3	6
yards	1	2

■ feet = 4 yards

Set D (pp. 432–433)

Write the rule and change the units. You may make a
table to help. (2 pints = 1 quart)

1. How many pints are in
   9 quarts?

   ■ pints = 9 quarts

2. How many quarts are in
   6 pints?

   ■ quarts = 6 pints

Set A (pp. 440–443)

Estimate the length in centimeters. Then use a ruler to measure to the nearest centimeter.

1.

2.

3.

4.

5.

Choose the unit you would use to measure each.
Write *cm, m,* or *km*.

6. the length of a marker

7. the height of a two-story building

8. the distance you can kick a ball

9. the distance you can walk in half an hour

Choose the better estimate.

10. Jana's math book is almost 3 _?_ long.

    A meters
    B decimeters

11. Ed walked 1 _?_ to get to Jeff's house.

    A kilometer
    B centimeter

12. The tree is about 6 _?_ high.

    A kilometers
    B meters

13. The ant is 1 _?_ long.

    A decimeter
    B centimeter

Solve.

14. Mathias has a piece of yarn that measures 13 centimeters. Is that more or less than 2 decimeters? Explain.

15. Alice had a poster board 1 m long. She cut 12 cm off of the poster board. How many centimeters long is the poster board now?

Set A (pp. 458–461)

Find the perimeter.

1.

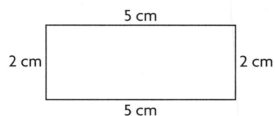

5 cm

2 cm   2 cm

5 cm

2.

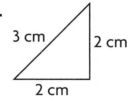

3 cm   2 cm

2 cm

Use your centimeter ruler to find the perimeter.

3.

4.

Set B (pp. 464–465)

Find the total area that covers each solid figure.

1.

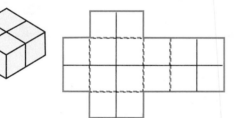

2.

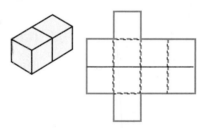

Set C (pp. 468–471)

Find the volume of each solid. Write the volume in cubic units.

1.

2.

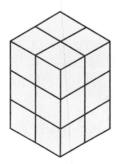

3.

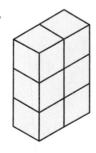

EXTRA PRACTICE

Set A (pp. 482–485)

Write a fraction in numbers and in words that names the shaded part.

1.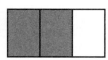

2.

3.

Write the fraction, using numbers.

4. one eighth

5. four out of seven

6. two divided by five

Set B (pp. 486–487)

Write the fraction that names the part of each group that is circled.

1.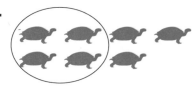

2.

3. Draw 4 nickels. Circle $\frac{3}{4}$ of them.

4. Draw 5 rectangles. Circle $\frac{1}{5}$ of them.

5. Draw 8 triangles. Circle $\frac{1}{2}$ of them.

Set C (pp. 488–491)

Find an equivalent fraction. Use fraction bars.

1.

2.

3.

Set D (pp. 492–495)

Compare. Write <, >, or = for each ●.

1. $\frac{1}{8}$ ● $\frac{3}{10}$

2.  $\frac{4}{6}$ ● $\frac{4}{8}$

3. 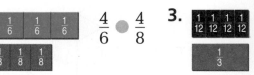 $\frac{4}{12}$ ● $\frac{1}{3}$

4. Use fraction bars to order $\frac{1}{2}$, $\frac{3}{10}$, and $\frac{2}{3}$ from greatest to least.

Set A (pp. 504–507)

**Find the sum. Write the answer in simplest form.
Use fraction bars.**

1. $\frac{2}{4} + \frac{1}{4} = $ ■

2. $\frac{2}{5} + \frac{1}{5} = $ ■

3. $\frac{4}{8} + \frac{1}{8} = $ ■

4. $\frac{4}{10} + \frac{2}{10} = $ ■

5. $\frac{1}{3} + \frac{1}{3} = $ ■

6. $\frac{6}{12} + \frac{1}{12} = $ ■

7. $\frac{3}{6} + \frac{2}{6} = $ ■

8. $\frac{1}{4} + \frac{2}{4} = $ ■

9. $\frac{1}{8} + \frac{2}{8} = $ ■

10. $\frac{6}{10} + \frac{3}{10} = $ ■

11. $\frac{2}{6} + \frac{2}{6} = $ ■

12. $\frac{1}{12} + \frac{1}{12} = $ ■

13. Miles gave $\frac{1}{5}$ of the cake to Molly and $\frac{2}{5}$ to Sarah. What fraction of the cake did Miles give away?

14. Gwen says that the simplest form of $\frac{6}{12}$ is $\frac{3}{6}$. Is she correct? Explain.

Set B (pp. 510–513)

Compare. Find the difference. Write the answer in simplest form.

1.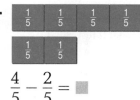

$\frac{4}{5} - \frac{2}{5} = $ ■

2.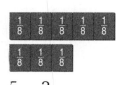

$\frac{5}{8} - \frac{3}{8} = $ ■

3.

$\frac{8}{12} - \frac{2}{12} = $ ■

Find the difference. Write the answer in simplest form. Use fraction bars.

4. $\frac{2}{3} - \frac{1}{3} = $ ■

5. $\frac{5}{6} - \frac{2}{6} = $ ■

6. $\frac{3}{4} - \frac{2}{4} = $ ■

7. $\frac{5}{10} - \frac{3}{10} = $ ■

8. $\frac{3}{4} - \frac{1}{4} = $ ■

9. $\frac{4}{5} - \frac{1}{5} = $ ■

10. $\frac{7}{8} - \frac{5}{8} = $ ■

11. $\frac{10}{12} - \frac{1}{12} = $ ■

Solve.

12. Nate and Sam ate a candy bar. Nate ate $\frac{5}{10}$ of the candy, and Sam ate $\frac{5}{10}$ of the candy. How much of the candy bar did the boys eat in all?

13. Valencia and Dora ate some jelly beans. Valencia ate $\frac{3}{8}$ of them, and Dora ate $\frac{5}{8}$ of the them. What fraction tells how much more of the jelly beans Dora ate than Valencia?

Set A (pp. 520–521)

Write the fraction and decimal for the shaded part.

1. 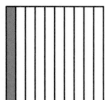 2. 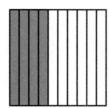 3. 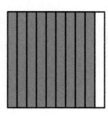 4.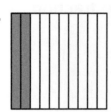

5. What decimal names the amount in Exercise 1 that is **NOT** shaded?

6. How many more parts would you shade in Exercise 2 to show 0.6?

Set B (pp. 526–527)

Write the word form and expanded form for each decimal.

1.
ONES	•	TENTHS	HUNDREDTHS
0	•	1	8

2.
ONES	•	TENTHS	HUNDREDTHS
1	•	3	3

Write the missing number.

3. 0.45 = ■ tenths 5 hundredths

4. 0.16 = 1 tenth ■ hundredths

Set C (pp. 528–529)

Compare. Write < or > for each ●.

1.
ONES	•	TENTHS
2	•	2
2	•	4

2.2 ● 2.4

2.
ONES	•	TENTHS	HUNDREDTHS
1	•	0	9
1	•	0	3

1.09 ● 1.03

3. 2.31 ● 1.32   4. 5.1 ● 1.5   5. 0.09 ● 0.90   6. 1.10 ● 0.10

For 7–10, use the number line to order the decimals from least to greatest.

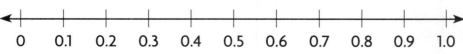

7. 0.2, 0.1, 0.3   8. 0.7, 0.1, 0.4   9. 0.9, 0.6, 0.3   10. 0.4, 0.2, 0.6

Set A (pp. 536–537)

Write the amount of money shown. Then write the amount as a fraction of a dollar.

1.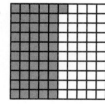

2.

3.

4.

5. Ian had 2 quarters, 2 dimes, and 5 pennies. He gave $0.50 to Lucia. What fraction of a dollar does he have left?

6. Give two examples of how you could model $\frac{1}{4}$ of a dollar using dimes, nickels, and pennies.

Set B (pp. 540–543)

Add or subtract.

1.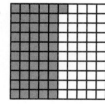

$$\begin{array}{r} 0.30 \\ +0.21 \\ \hline \end{array}$$

2.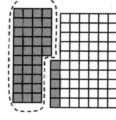

$$\begin{array}{r} 0.40 \\ -0.35 \\ \hline \end{array}$$

3. $\begin{array}{r} 0.1 \\ +0.6 \\ \hline \end{array}$

4. $\begin{array}{r} 3.09 \\ +2.51 \\ \hline \end{array}$

5. $\begin{array}{r} 1.7 \\ -0.2 \\ \hline \end{array}$

6. $\begin{array}{r} \$0.70 \\ -\$0.54 \\ \hline \end{array}$

7. $\begin{array}{r} 0.88 \\ +0.01 \\ \hline \end{array}$

8. $\begin{array}{r} \$4.56 \\ -\$2.38 \\ \hline \end{array}$

9. $\begin{array}{r} 0.5 \\ +0.3 \\ \hline \end{array}$

10. $\begin{array}{r} 8.43 \\ -7.22 \\ \hline \end{array}$

11. Mrs. Saguchi is training for a bike race. She rode 2.5 kilometers on Tuesday, and 2.7 kilometers on Thursday. How far did she ride on those two days?

Tips for Taking Math Tests

Being a good test-taker is like being a good problem solver. When you answer test questions, you are solving problems. Remember to **UNDERSTAND, PLAN, SOLVE,** and **CHECK.**

Read the problem.

- Look for math terms and recall their meanings.
- Reread the problem and think about the question.
- Use the details in the problem and the question.

1. The sum of the digits of a number is 14. Both the digits are odd. The ones digit is 4 less than the tens digit. What is the number?

   A 59   **C** 86

   B 77   **D** 95

TIP! Understand the problem.
Remember the meanings of *sum, digits,* and *odd.* Reread the problem to compare the details to the answer choices. Since all choices have a sum of 14, look for the odd digits. Then look for a ones digit that is 4 less than the tens digit. The answer is **D.**

- Each word is important. Missing a word or reading it incorrectly could cause you to get the wrong answer.
- Pay attention to words that are in **bold** type, all CAPITAL letters, or *italics*.
- Some other words to look for are <u>round</u>, <u>about</u>, <u>only</u>, <u>best</u>, or <u>least to greatest</u>.

2. Mr. Karza drew a diagram of 8 squares. He made 3 squares red, 2 squares blue, 2 squares yellow, and 1 square green. What fraction of the squares was **not** green?

   F $\frac{7}{8}$ **H** $\frac{1}{2}$

   G $\frac{5}{8}$ **J** $\frac{1}{8}$

TIP! Look for important words.
The word **not** is important. Without the word **not,** the answer would be $\frac{1}{8}$. Find the number of squares that were **not** green. The answer is **F.**

Think about how you can solve the problem.

- See if you can solve the problem with the information given.

- Pictures, charts, tables, and graphs may have the information you need.

- You may need to think about information you already know.

- The answer choices may have information you need.

3. Soccer practice started at 12:00. The clock shows the time practice ended. How long did soccer practice last?

A 10 minutes   **C** 35 minutes

B 20 minutes   **D** 50 minutes

TIP! Get the information you need.
Use the clock to find how long soccer practice lasted. You can find out how much time passed by counting by fives. The answer is **D**.

- You may need to write a number sentence and solve it.

- Some problems have two steps or more.

- In some problems you need to look at relationships instead of computing an answer.

- If the path to the solution isn't clear, choose a problem solving strategy and use it to solve the problem.

4. June always has 30 days. Mary takes swimming lessons every three days in June, starting on June 3. How many times will she have lessons?

F 5   **H** 12

G 10   **J** 30

TIP! Decide on a plan.
Lessons every three days sounds like a pattern. Use the strategy *find a pattern*. Count by 3 beginning with June 3 until you reach 30. You need to count 10 numbers, so the answer is **G**.

Follow your plan, working logically and carefully.

- Estimate your answer. Look for unreasonable answer choices.
- Use reasoning to find the most likely choices.
- Solve all steps needed to answer the problem.
- If your answer does not match any answer choice, check your numbers and your computation.

5. The cafeteria served 76 lunches each day for a week. How many lunches were served in 5 days?

A 76   **C** 380

B 353   **D** 1,380

TIP! **Eliminate choices.**
Estimate the product (5 × 80). The only reasonable answers are B and C. Since 5 times the ones digit 6 is 30, the answer must end in zero. If you are still not certain, multiply and check your answer against B and C. The answer is **C**.

- If your answer still does not match, look for another form of the number, such as a decimal instead of a fraction.
- If answer choices are given as pictures, look at each one by itself while you cover the other three.
- If the answer choices include NOT HERE and your answer is not given, make sure your work is correct and then mark NOT HERE.
- Read answer choices that are statements and relate them to the problem one by one.
- If your strategy isn't working, try a different one.

6. Mr. Rodriguez is putting a wallpaper border around a room. The room is 9 feet wide and 12 feet long. How many feet of border does he need?

F 21 ft   **H** 108 ft

G 84 ft   **J** NOT HERE

TIP! **Choose the answer.**
The border goes around all four walls, two that are 9 feet and two that are 12 feet. Answer choices are given using the abbreviation for feet (ft). Add the lengths of the four walls (9 + 9 + 12 + 12). That total is not given. So, mark **J** for NOT HERE.

Take time to catch your mistakes.

- Be sure you answered the question asked.
- Check for important words you might have missed.
- Did you use all the information you needed?
- Check your computation by using a different method.
- Draw a picture when you are unsure of your answer.

7. Katy is buying 3 books. Their prices are $4.95, $3.25, and $7.49. What is the total cost of the books?

A $14.59

C $15.59

B $14.69

D $15.69

TIP! **Check your work.**

To check column addition, write the numbers in a different order. Then you will be adding different basic facts. For example, add $7.49 + $3.25 + $4.95. The answer is **D**.

Don't Forget!

Before the test

- Listen to the teacher's directions and read the instructions.
- Write down the ending time if the test is timed.
- Know where and how to mark your answers.
- Know whether you should write on the test page or use scratch paper.
- Ask any questions you may have before the test begins.

During the test

- Work quickly but carefully. If you are unsure how to answer a question, leave it blank and return to it later.
- If you cannot finish on time, read the questions that are left. Answer the easiest ones first. Then answer the others.
- Fill in each answer space carefully. Erase completely if you change an answer. Erase any stray marks.
- Check that the answer number matches the question number, especially if you skip a question.

ADDITION FACTS TEST

	K	L	M	N	O	P	Q	R
A	3 + 2	0 + 6	2 + 4	5 + 9	6 + 1	2 + 5	3 + 10	4 + 4
B	8 + 9	0 + 7	3 + 5	9 + 6	6 + 7	2 + 8	3 + 3	7 + 10
C	4 + 6	9 + 0	7 + 8	4 + 10	3 + 7	7 + 7	4 + 2	7 + 5
D	5 + 7	3 + 9	8 + 1	9 + 5	10 + 5	9 + 8	2 + 6	8 + 7
E	7 + 4	0 + 8	3 + 6	6 + 10	5 + 3	2 + 7	8 + 2	9 + 9
F	2 + 3	1 + 7	6 + 8	5 + 2	7 + 3	4 + 8	10 + 10	6 + 6
G	8 + 3	7 + 2	7 + 0	8 + 5	9 + 1	4 + 7	8 + 4	10 + 8
H	7 + 9	5 + 6	8 + 10	6 + 5	8 + 6	9 + 4	0 + 9	7 + 1
I	4 + 3	5 + 5	6 + 4	10 + 2	7 + 6	8 + 0	6 + 9	9 + 2
J	5 + 8	1 + 9	5 + 4	8 + 8	6 + 2	6 + 3	9 + 7	9 + 10

SUBTRACTION FACTS TEST

	K	L	M	N	O	P	Q	R
A	9 − 1	10 − 4	7 − 2	6 − 4	20 − 10	7 − 0	8 − 3	13 − 9
B	9 − 9	13 − 4	7 − 1	11 − 5	9 − 7	6 − 3	15 − 10	6 − 2
C	10 − 2	8 − 8	16 − 8	6 − 5	18 − 10	8 − 7	13 − 3	15 − 6
D	11 − 7	9 − 5	12 − 8	8 − 1	15 − 8	18 − 9	14 − 10	9 − 4
E	9 − 2	7 − 7	10 − 3	8 − 5	16 − 9	11 − 9	14 − 8	12 − 6
F	7 − 3	12 − 10	17 − 9	6 − 0	9 − 6	11 − 8	10 − 9	12 − 2
G	15 − 7	8 − 4	13 − 6	7 − 5	11 − 2	12 − 3	14 − 6	11 − 4
H	7 − 6	13 − 5	12 − 9	10 − 5	13 − 8	11 − 3	16 − 10	14 − 7
I	5 − 0	10 − 8	11 − 6	9 − 3	14 − 5	5 − 4	7 − 7	14 − 9
J	15 − 9	9 − 8	13 − 7	8 − 2	7 − 4	13 − 10	10 − 6	16 − 7

MULTIPLICATION FACTS TEST

	K	L	M	N	O	P	Q	R
A	2 × 7	0 × 6	6 × 6	9 × 2	8 × 3	3 × 4	2 × 8	6 × 1
B	7 × 7	5 × 9	2 × 2	7 × 5	2 × 3	10 × 8	4 × 10	8 × 4
C	4 × 5	5 × 1	7 × 0	6 × 3	3 × 5	6 × 8	7 × 3	9 × 9
D	0 × 9	6 × 4	6 × 10	1 × 6	9 × 8	4 × 4	3 × 2	9 × 3
E	0 × 7	9 × 4	1 × 7	9 × 7	2 × 5	7 × 9	5 × 6	5 × 8
F	4 × 3	6 × 9	1 × 9	7 × 6	7 × 10	6 × 0	2 × 9	10 × 3
G	5 × 3	1 × 5	7 × 1	3 × 8	3 × 6	8 × 10	3 × 9	6 × 7
H	7 × 4	7 × 2	3 × 7	2 × 4	7 × 8	4 × 7	5 × 10	8 × 6
I	4 × 6	5 × 5	5 × 7	3 × 3	9 × 6	8 × 0	4 × 9	8 × 8
J	8 × 9	6 × 2	4 × 8	9 × 5	5 × 4	0 × 5	10 × 6	9 × 10

DIVISION FACTS TEST

	K	L	M	N	O	P	Q	R
A	$1\overline{)1}$	$3\overline{)9}$	$2\overline{)6}$	$2\overline{)4}$	$1\overline{)6}$	$3\overline{)12}$	$5\overline{)15}$	$7\overline{)21}$
B	$6\overline{)24}$	$8\overline{)56}$	$5\overline{)40}$	$6\overline{)18}$	$6\overline{)30}$	$7\overline{)42}$	$9\overline{)81}$	$5\overline{)45}$
C	$5\overline{)30}$	$2\overline{)16}$	$3\overline{)21}$	$7\overline{)35}$	$3\overline{)15}$	$9\overline{)9}$	$8\overline{)16}$	$9\overline{)63}$
D	$4\overline{)32}$	$9\overline{)90}$	$4\overline{)8}$	$8\overline{)48}$	$9\overline{)54}$	$3\overline{)18}$	$10\overline{)50}$	$6\overline{)48}$
E	$7\overline{)28}$	$3\overline{)0}$	$5\overline{)20}$	$4\overline{)24}$	$7\overline{)14}$	$3\overline{)6}$	$5\overline{)50}$	$10\overline{)60}$
F	$9\overline{)18}$	$4\overline{)36}$	$5\overline{)25}$	$7\overline{)63}$	$1\overline{)5}$	$8\overline{)32}$	$9\overline{)45}$	$6\overline{)54}$
G	$2\overline{)14}$	$8\overline{)24}$	$4\overline{)4}$	$5\overline{)40}$	$3\overline{)9}$	$4\overline{)12}$	$7\overline{)56}$	$8\overline{)72}$
H	$5\overline{)35}$	$1\overline{)4}$	$8\overline{)64}$	$5\overline{)10}$	$8\overline{)40}$	$2\overline{)12}$	$6\overline{)42}$	$10\overline{)70}$
I	$7\overline{)49}$	$9\overline{)27}$	$10\overline{)90}$	$3\overline{)27}$	$9\overline{)36}$	$4\overline{)20}$	$9\overline{)72}$	$8\overline{)80}$
J	$8\overline{)0}$	$4\overline{)28}$	$2\overline{)10}$	$7\overline{)70}$	$1\overline{)3}$	$10\overline{)80}$	$6\overline{)60}$	$10\overline{)100}$

TABLE OF MEASURES

METRIC | CUSTOMARY

Length

1 decimeter (dm) = 10 centimeters

1 meter (m) = 100 centimeters

1 meter (m) = 10 decimeters

1 kilometer (km) = 1,000 meters

1 foot (ft) = 12 inches (in.)

1 yard (yd) = 3 feet, or 36 inches

1 mile (mi) = 1,760 yards, or 5,280 feet

Mass/Weight

1 kilogram (kg) = 1,000 grams (g)

1 pound (lb) = 16 ounces (oz)

Capacity

1 liter (L) = 1,000 milliliters (mL)

1 pint (pt) = 2 cups (c)

1 quart (qt) = 2 pints

1 gallon (gal) = 4 quarts

TIME

1 minute (min) = 60 seconds (sec)

1 hour (hr) = 60 minutes

1 day = 24 hours

1 week (wk) = 7 days

1 year (yr) = 12 months (mo), or about 52 weeks

1 year = 365 days

1 leap year = 366 days

MONEY

1 penny = 1 cent (¢)

1 nickel = 5 cents

1 dime = 10 cents

1 quarter = 25 cents

1 half dollar = 50 cents

1 dollar ($) = 100 cents

SYMBOLS

< is less than

> is greater than

= is equal to

°F degrees Fahrenheit

°C degrees Celsius

(2,3) ordered pair

By the end of grade three, students deepen their understanding of place value and their understanding of and skill with addition, subtraction, multiplication, and division of whole numbers. Students estimate, measure, and describe objects in space. They use patterns to help solve problems. They represent number relationships and conduct simple probability experiments.

Number Sense

1.0   Students understand the place value of whole numbers:

    1.1   Count, read, and write whole numbers to 10,000.

        What is the smallest whole number you can make using the digits 4, 3, 9, and 1? Use each digit exactly once (Adapted from TIMSS gr. 4, T–2).

    1.2   Compare and order whole numbers to 10,000.

⊶ **1.3**   Identify the place value for each digit in numbers to 10,000.

    1.4   Round off numbers to 10,000 to the nearest ten, hundred, and thousand.

⊶ **1.5**   Use expanded notation to represent numbers (e.g., $3,206 = 3,000 + 200 + 6$).

        True or false?

        $3,102 \times 3 = 9,000 + 300 + 6$

2.0   Students calculate and solve problems involving addition, subtraction, multiplication, and division:

⊶ **2.1**   Find the sum or difference of two whole numbers between 0 and 10,000.

        1. $591 + 87 = ?$

        2. $1,283 + 6,074 = ?$

        3. $3,215 - 2,876 = ?$

        To prepare for recycling on Monday, Michael collected all the bottles in the house. He found 5 dark green ones, 8 clear ones with liquid still in them, 11 brown ones that used to hold root beer, 2 still with the cap on from his parents' cooking needs, and 4 more that were over-sized. How many bottles did Michael collect? (This problem also supports Mathematical Reasoning Standard 1.1.)

⊶ **2.2**   Memorize to automaticity the multiplication table for numbers between 1 and 10.

Note: The sample problems illustrate the standards and are written to help clarify them.

The symbols ⊶ *and* ⬤ *identify the Key Standards for grade three.*

Number Sense (Continued)

2.3 Use the inverse relationship of multiplication and division to compute and check results.

2.4 Solve simple problems involving multiplication of multidigit numbers by one-digit numbers $(3{,}671 \times 3 = \underline{\quad})$.

2.5 Solve division problems in which a multidigit number is evenly divided by a one-digit number $(135 \div 5 = \underline{\quad})$.

2.6 Understand the special properties of 0 and 1 in multiplication and division.

True or false?

1. $24 \times 0 = 24$

2. $19 \div 1 = 19$

3. $63 \times 1 = 63$

4. $0 \div 0 = 1$

2.7 Determine the unit cost when given the total cost and number of units.

2.8 Solve problems that require two or more of the skills mentioned above.

A tree was planted 54 years before 1961. How old is the tree in 1998?

A class of 73 students go on a field trip. The school hires vans, each of which can seat a maximum of 10 students. The school policy is to seat as many students as possible in a van before using the next one. How many vans are needed?

A price list in a store states: pen sets, $3; magnets, $4; sticker sets, $6. How much would it cost to buy 5 pen sets, 7 magnets, and 8 sticker sets?

3.0 Students understand the relationship between whole numbers, simple fractions, and decimals:

3.1 Compare fractions represented by drawings or concrete materials to show equivalency and to add and subtract simple fractions in context (e.g., $\frac{1}{2}$ of a pizza is the same amount as $\frac{2}{4}$ of another pizza that is the same size; show that $\frac{3}{8}$ is larger than $\frac{1}{4}$).

Which is longer, $\frac{1}{3}$ of a foot or 5 inches? $\frac{2}{3}$ of a foot or 9 inches?

Which rectangle is NOT divided into four equal parts? (Adapted from TIMSS gr. 4, K–8)

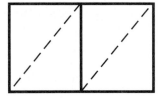

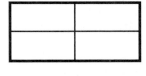

Number Sense (Continued)

3.2 Add and subtract simple fractions (e.g., determine that $\frac{1}{8} + \frac{3}{8}$ is the same as $\frac{1}{2}$).

Find the values:

1. $\frac{1}{6} + \frac{2}{6} = ?$

2. $\frac{7}{8} - \frac{3}{8} = ?$

3.3 Solve problems involving addition, subtraction, multiplication, and division of money amounts in decimal notation and multiply and divide money amounts in decimal notation by using whole-number multipliers and divisors.

Pedro bought 5 pens, 2 erasers and 2 boxes of crayons. The pens cost 65 cents each, the erasers 25 cents each, and a box of crayons $1.10. The prices include tax, and Pedro paid with a ten-dollar bill. How much change did he get back?

3.4 Know and understand that fractions and decimals are two different representations of the same concept (e.g., 50 cents is $\frac{1}{2}$ of a dollar, 75 cents is $\frac{3}{4}$ of a dollar).

Algebra and Functions

1.0 Students select appropriate symbols, operations, and properties to represent, describe, simplify, and solve simple number relationships:

1.1 Represent relationships of quantities in the form of mathematical expressions, equations, or inequalities.

1.2 Solve problems involving numeric equations or inequalities.

1.3 Select appropriate operational and relational symbols to make an expression true (e.g., if 4 __ 3 = 12, what operational symbol goes in the blank?).

1.4 Express simple unit conversions in symbolic form (e.g., __ inches = __ feet $\times$ 12).

If number of feet = number of yards $\times$ 3, and number of inches = number of feet $\times$ 12, how many inches are there in 4 yards?

1.5 Recognize and use the commutative and associative properties of multiplication (e.g., if 5 $\times$ 7 = 35, then what is 7 $\times$ 5? and if 5 $\times$ 7 $\times$ 3 = 105, then what is 7 $\times$ 3 $\times$ 5?).

Algebra and Functions (Continued)

2.0   Students represent simple functional relationships:

○━ꞁ **2.1**   Solve simple problems involving a functional relationship between two quantities (e.g., find the total cost of multiple items given the cost per unit).

John wants to buy a dozen pencils. One store offers pencils at 6 for $1. Another offers them at 4 for 65 cents. Yet another sells pencils at 15 cents each. Where should John purchase his pencils in order to save the most money?

2.2   Extend and recognize a linear pattern by its rules (e.g., the number of legs on a given number of horses may be calculated by counting by 4s or by multiplying the number of horses by 4).

Here is the beginning of a pattern of tiles. Assuming that the pattern continues linearly, how many tiles will be in the sixth figure? (Adapted from TIMSS gr. 4, K–6)

Measurement and Geometry

1.0   Students choose and use appropriate units and measurement tools to quantify the properties of objects:

1.1   Choose the appropriate tools and units (metric and U.S.) and estimate and measure the length, liquid volume, and weight/mass of given objects.

○━ꞁ **1.2**   Estimate or determine the area and volume of solid figures by covering them with squares or by counting the number of cubes that would fill them.

○━ꞁ **1.3**   Find the perimeter of a polygon with integer sides.

1.4   Carry out simple unit conversions within a system of measurement (e.g., centimeters and meters, hours and minutes).

Measurement and Geometry (Continued)

2.0   Students describe and compare the attributes of plane and solid geometric figures and use their understanding to show relationships and solve problems:

2.1 Identify, describe, and classify polygons (including pentagons, hexagons, and octagons).

2.2 Identify attributes of triangles (e.g., two equal sides for the isosceles triangle, three equal sides for the equilateral triangle, right angle for the right triangle).

2.3 Identify attributes of quadrilaterals (e.g., parallel sides for the parallelogram, right angles for the rectangle, equal sides and right angles for the square).

2.4   Identify right angles in geometric figures or in appropriate objects and determine whether other angles are greater or less than a right angle.

Which of the following triangles include an angle that is greater than a right angle?

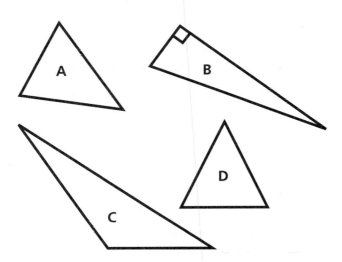

2.5   Identify, describe, and classify common three-dimensional geometric objects (e.g., cube, rectangular solid, sphere, prism, pyramid, cone, cylinder).

2.6   Identify common solid objects that are the components needed to make a more complex solid object.

Statistics, Data Analysis, and Probability

1.0   Students conduct simple probability experiments by determining the number of possible outcomes and make simple predictions:

    1.1   Identify whether common events are certain, likely, unlikely, or improbable.

        Are any of the following certain, likely, unlikely, or impossible?

        1.   Take two cubes each with the numbers 1, 2, 3, 4, 5, 6 written on its six faces. Throw them at random, and the sum of the numbers on the top faces is 12.

        2.   It snows on New Year's day.

        3.   A baseball game is played somewhere in this country on any Sunday in July.

        4.   It is sunny in June.

        5.   Pick any two one-digit numbers, and their sum is 17.

  1.2   Record the possible outcomes for a simple event (e.g., tossing a coin) and systematically keep track of the outcomes when the event is repeated many times.

  1.3   Summarize and display the results of probability experiments in a clear and organized way (e.g., use a bar graph or a line plot).

    1.4   Use the results of probability experiments to predict future events (e.g., use a line plot to predict the temperature forecast for the next day).

Mathematical Reasoning

1.0   Students make decisions about how to approach problems:

    1.1   Analyze problems by identifying relationships, distinguishing relevant from irrelevant information, sequencing and prioritizing information, and observing patterns.

    1.2   Determine when and how to break a problem into simpler parts.

2.0   Students use strategies, skills, and concepts in finding solutions:

    2.1   Use estimation to verify the reasonableness of calculated results.

        Prove or disprove a classmate's claim that 49 is more than 21 because 9 is more than 1.

    2.2   Apply strategies and results from simpler problems to more complex problems.

    2.3   Use a variety of methods, such as words, numbers, symbols, charts, graphs, tables, diagrams, and models, to explain mathematical reasoning.

Mathematical Reasoning (Continued)

2.4   Express the solution clearly and logically by using the appropriate mathematical notation and terms and clear language; support solutions with evidence in both verbal and symbolic work.

2.5   Indicate the relative advantages of exact and approximate solutions to problems and give answers to a specified degree of accuracy.

2.6   Make precise calculations and check the validity of the results from the context of the problem.

3.0   Students move beyond a particular problem by generalizing to other situations:

3.1   Evaluate the reasonableness of the solution in the context of the original situation.

3.2   Note the method of deriving the solution and demonstrate a conceptual understanding of the derivation by solving similar problems.

3.3   Develop generalizations of the results obtained and apply them in other circumstances.

GLOSSARY

Pronunciation Key

a	add, map	f	fit, half	n	nice, tin	p	pit, stop	y$\overline{oo}$	fuse, few
ā	ace, rate	g	go, log	ng	ring, song	r	run, poor	v	vain, eve
â(r)	care, air	h	hope, hate	o	odd, hot	s	see, pass	w	win, away
ä	palm, father	i	it, give	ō	open, so	sh	sure, rush	y	yet, yearn
b	bat, rub	ī	ice, write	ô	order, jaw	t	talk, sit	z	zest, muse
ch	check, catch	j	joy, ledge	oi	oil, boy	th	thin, both	zh	vision,
d	dog, rod	k	cool, take	ou	pout, now	th	this, bathe		pleasure
e	end, pet	l	look, rule	ŏŏ	took, full	u	up, done		
ē	equal, tree	m	move, seem	$\overline{oo}$	pool, food	û(r)	burn, term		

ə   the schwa, an unstressed vowel representing
the sound spelled *a* in **a**bove, *e* in sick**e**n,
i in poss**i**ble, *o* in mel**o**n, *u* in circ**u**s

Other symbols:
• separates words into syllables
′ indicates stress on a syllable

A

acute angle [ə•ky$\overline{oo}$t′ ang′gəl] An angle that has a measure less than a right angle

addend [a′dend] Any of the numbers that are added (p. 36)
Example: 2 + 3 = 5
  ↑ ↑
 addend addend

addition [ə•dish′ən] The process of finding the total number of items when two or more groups of items are joined; the opposite operation of subtraction (p. 36)

A.M. [ā em] Between midnight and noon (p. 96)

angle [ang′gəl] The figure formed when two rays share the same endpoint (p. 369)
Example:

area [âr′ē•ə] The number of square units needed to cover a flat surface (p. 462)
Example:

area = 9 square units

array [ə•rā′] An arrangement of objects in rows and columns (p. 120)
Example:

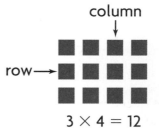

3 × 4 = 12

bar graph [bär graf] A graph that uses bars to show data (p. 252)
Example:

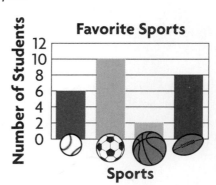

benchmark numbers [bench′märk num′bərz] Numbers that help you estimate the number of objects without counting them, such as 25, 50, 100, 1,000 (p. 18)

C

calendar [ka′lən•dər] A table that shows the days, weeks, and months of a year (p. 102)
Example:

July						
Sun	Mon	Tue	Wed	Thu	Fri	Sat
	1	2	3	4	5	6
7	8	9	10	11	12	13
14	15	16	17	18	19	20
21	22	23	24	25	26	27
28	29	30	31			

capacity [kə•pa′sə•tē] The amount a container can hold (p. 426)

center [sen′tər] A point in the middle of a circle that is the same distance from anywhere on the circle (p. 375)
Example:

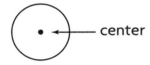
center

centimeter (cm) [sen′tə•mē•tər] A metric unit that is used to measure length (p. 440)
Example:

1 cm

certain [sûr′tən] An event is certain if it will always happen. (p. 268)

classify [kla′sə•fī] To group pieces of data according to how they are the same; for example, you can classify data by size, color, or shape. (p. 242)

closed figure [klōzd fi′•gyər] A shape that begins and ends at the same point (p. 382)
Examples:

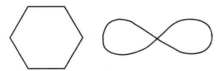

compare [kəm•pâr′] To describe whether numbers are equal to, less than, or greater than each other (p. 20)

cone [kōn] A solid, pointed figure that has a flat, round base (p. 362)
Example:

congruent [kən•grōō′ənt] Figures that have the same size and shape (p. 384)
Example:

counting back [koun′ting bak] A way to find the difference when you subtract 1, 2, or 3
Example: 8 − 3 = ■   Count: 8 . . . 7, 6, 5

counting on [koun′ting on] A way to find the sum when one of the addends is 1, 2, or 3
Example: 5 + 2 = ■   Count: 5 . . . 6, 7

counting up [koun′ting up] A way to find the difference by beginning with the smaller number
Example: 7 − 4 = ■

Count: 4 . . . 5, 6, 7 ← 3 is the difference.

cube [kyōōb] A solid figure with six congruent square faces (p. 362)
Example:

cubic unit [kyōō′bik yōō′nət] A cube with a side length of one unit; used to measure volume (p. 468)

cup (c) [kup] A customary unit used to measure capacity (p. 426)

cylinder [sil′in•dər] A solid or hollow object that is shaped like a can (p. 362)
Example:

D

data [dāʹtə] Information collected about people or things (p. 238)

decimal [deʹsə•məl] A number with one or more digits to the right of the decimal point (p. 520)

decimal point [deʹsə•məl point] A symbol used to separate dollars from cents in money and to separate the ones place from the tenths place in decimals (p. 520)
Example: 4.5
        ↳ decimal point

decimeter (dm) [deʹsə•mē•tər] A metric unit that is used to measure length; 1 decimeter = 10 centimeters (p. 440)

degrees Celsius (°C) [di•grēzʹ selʹsē•əs] A unit for measuring temperature in the metric system (p. 450)

degrees Fahrenheit (°F) [di•grēzʹ farʹən•hīt] A unit for measuring temperature in the customary system (p. 450)

denominator [di•näʹmə•nā•tər] The part of a fraction that tells how many equal parts are in the whole (p. 482)
Example: $\frac{3}{4}$ ←denominator

diameter [dī•aʹmə•tər] A line segment that passes through the center of a circle and whose endpoints are on the circle (p. 375)
Example:

diameter

difference [difʹrən(t)s] The answer in a subtraction problem
Example: 6 − 4 = 2
        ↳ difference

digits [diʹjəts] The symbols 0, 1, 2, 3, 4, 5, 6, 7, 8, and 9 (p. 4)

divide [di•vīdʹ] To separate into equal groups; the opposite operation of multiplication (p. 184)

dividend [diʹvə•dend] The number that is to be divided in a division problem (p. 188)
Example: 35 ÷ 5 = 7
        ↳ dividend

divisor [də•viʹzər] The number that divides the dividend (p. 188)
Example: 35 ÷ 5 = 7
        ↳ divisor

E

edge [ej] A line segment formed where two faces meet (p. 362)
Example:

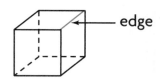
edge

elapsed time [i•lapstʹ tīm] The amount of time that passes from the start of an activity to the end of that activity (p. 98)

equal sign (=) [ēʹkwəl sīn] A symbol used to show that two numbers have the same value (p. 20)
Example: 384 = 384

equally likely [ēʹkwəl•lē līʹklē] Having the same chance of happening (p. 272)

equilateral triangle [ē•kwə•latʹər•əl trīʹang•gəl] A triangle that has all sides equal (p. 399)

equivalent [ē•kwivʹə•lənt] Two or more sets that name the same amount are equivalent. (p. 80)

equivalent fractions [ē•kwivʹə•lənt frakʹshənz] Two or more fractions that name the same amount (p. 488)
Example:

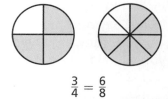
$\frac{3}{4} = \frac{6}{8}$

estimate [esʹtə•māt] To find about how many or how much (p. 38)

even [ēʹvən] A whole number that has a 0, 2, 4, 6, or 8 in the ones place is even (p. 2)

event [i•ventʹ] Something that happens (p. 268)

expanded form [ik•spandʹid fôrm] A way to write numbers by showing the value of each digit (p. 4)
Example: 7,201 = 7,000 + 200 + 1

experiment [ik•sper′ə•mənt] A test that is done in order to find out something (p. 274)

expression [ik•spre′shən] The part of a number sentence that combines numbers and operation signs, but doesn't have an equal sign (p. 68)
Example: 5 × 6

face [fãs] A flat surface of a solid figure (p. 362)
Example:

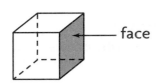

face

fact family [fakt fam′ə•lē] A set of related multiplication and division, or addition and subtraction, number sentences (p. 190)
Example:

4 × 7 = 28	28 ÷ 7 = 4
7 × 4 = 28	28 ÷ 4 = 7

factor [fak′tər] A number that is multiplied by another number to find a product (p. 118)
Example: 3 × 8 = 24
⬆ ⬆
factor factor

fair [fâr] A game is fair if every player has an equal chance to win. (p. 280)

foot (ft) [foŏt] A customary unit used to measure length or distance;
1 foot = 12 inches (p. 424)

fraction [frak′shən] A number that names part of a whole or part of a group (p. 482)
Example:

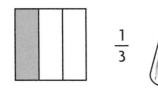

$\frac{1}{3}$

frequency table [frē′kwen•sē tā′bəl] A table that uses numbers to record data (p. 238)

gallon (gal) [ga′lən] A customary unit for measuring capacity; 4 quarts = 1 gallon (p. 427)

gram (g) [gram] A metric unit that is used to measure mass (p. 448)

greater than (>) [grā′tər than] A symbol used to compare two numbers, with the larger number given first (p. 20)
Example: 6 > 4

grid [grid] Horizontal and vertical lines on a map (p. 260)

Grouping Property of Addition [groō′ping prä′pər•tē əv ə•di′shən] A rule stating that you can group addends in different ways and still get the same sum (p. 36)
Example:
4 + (2 + 5) = 11 and
(4 + 2) + 5 = 11

Grouping Property of Multiplication [groō′ping prä′pər•tē əv məl•tə•plə•kā′shən] A rule stating that when the grouping of factors is changed, the product remains the same (p. 170)
Example:
3 × (4 × 1) = 12 and
(3 × 4) × 1 = 12

hexagon [hek′sə•gän] A polygon with six sides and six angles (p. 382)
Example:

horizontal bar graph [hôr•ə•zän′təl bär graf] A bar graph in which the bars go from left to right (p. 252)

hour (hr) [our] A unit used to measure time; in one hour, the hour hand on a clock moves from one number to the next. 1 hour = 60 minutes (p. 94)

hour hand [our hand] The short hand on an analog clock (p. 94)

hundredth [hən′drədth] One of one hundred equal parts (p. 524)
Example:

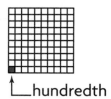

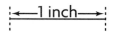
hundredth

impossible [im•pä′sə•bəl] An event is impossible if it will never happen. (p. 268)

inch (in.) [inch] A customary unit used to measure length (p. 420)
Example:

←1 inch→

intersecting lines [in•tər•sek′ting līnz] Lines that cross (p. 372)
Example:

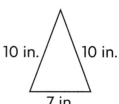

inverse operations [in′vərs ä•pə•rā′shənz] Opposite operations, or operations that undo each other, such as addition and subtraction or multiplication and division (p. 188)

isosceles triangle [ī•sos′ə•lēz trī′ang•gəl] A triangle that has two equal sides (p. 399)
Example:

10 in.　10 in.

7 in.

kilogram (kg) [kil′ə•gram] A metric unit that is used to measure mass;
1 kilogram = 1,000 grams (p. 448)

H82 Glossary

kilometer (km) [kə•lä′mə•tər] A metric unit that is used to measure length and distance;
1 kilometer = 1,000 meters (p. 440)

less than (<) [les than] A symbol used to compare two numbers, with the lesser number given first (p. 20)
Example: 3 < 7

like fractions [līk frak′shənz] Fractions that have the same denominator (p. 502)

likely [līk′lē] Having a good chance of happening (p. 270)

line [līn] A straight path extending in both directions with no endpoints (p. 368)
Example:

←—————→

line graph [līn graf] A graph that uses a line to show how something changes over time (p. 262)
Example:

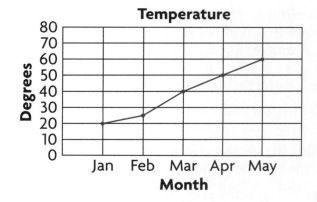

line of symmetry [līn əv sim′ə•trē] An imaginary line that divides a figure into two congruent parts (p. 385)
Example:

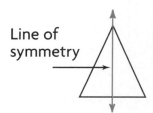
Line of symmetry

line plot [līn plot] A diagram that records each piece of data on a number line (p. 256)
Example:

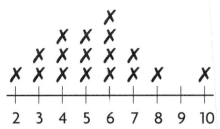

Hours Band Members Practiced

line segment [līn seg′mənt] A part of a line that extends between two points, called endpoints (p. 368)
Example:

liter (L) [lē′tər] A metric unit that is used to measure capacity; 1 liter = 1,000 milliliters (p. 446)

M

mass [mas] The amount of matter in an object (p. 448)

meter (m) [mē′tər] A metric unit that is used to measure length and distance; 1 meter = 100 centimeters (p. 440)

midnight [mid′nīt] 12:00 at night (p. 96)

mile (mi) [mīl] A customary unit used to measure length and distance; 1 mile = 5,280 feet (p. 424)

milliliter (mL) [mi′lə·lē·tər] A metric unit that is used to measure capacity (p. 446)

minute (min) [mi′nət] A unit used to measure short amounts of time; in one minute, the minute hand moves from one mark to the next. (p. 94)

minute hand [mi′nət hand] The long hand on an analog clock (p. 94)

mixed number [mikst nəm′bər] A number represented by a whole number and a fraction (p. 495)
Example: $4\frac{1}{2}$

mode [mōd] The number found most often in a set of data (p. 256)

multiply [mul′tə·plī] When you combine equal groups, you can multiply to find how many in all; the opposite operation of division. (p. 116)

multistep problem [məl′tē·step prä′bləm] A problem with more than one step (p. 172)

N

noon [nōōn] 12:00 in the day (p. 96)

number sentence [num′bər sen′təns] A sentence that includes numbers, operation symbols, and a greater than or less than symbol or an equal sign (p. 68)
Example:

5 + 3 = 8 is a number sentence.

numerator [nōō′mə·rā·tər] The part of a fraction above the line, which tells how many parts are being counted (p. 482)
Example: $\frac{3}{4}$ ←numerator

O

obtuse angle [əb·t(y)ōōs′ ang′gəl] An angle that has a measure greater than a right angle

octagon [äk′tə·gän] A polygon with eight sides and eight angles (p. 382)
Example:

odd [od] A whole number that has a 1, 3, 5, 7, or 9 in the ones place is odd (p. 2)

Order Property of Addition [ôr′dər prä′pər·tē əv ə·dish′ən] A rule stating that you can add two numbers in any order and get the same sum

Order Property of Multiplication [ôr′dər prä′pər·tē əv məl·tə·plə·kā′shən] A rule stating that you can multiply two factors in any order and get the same product (p. 121)
Example: 4 × 2 = 8
2 × 4 = 8

ordered pair [ôr′dərd pâr] A pair of numbers that names a point on a grid (p. 260)
Example: **(3,4)**

ounce (oz) [ouns] A customary unit used to measure weight (p. 428)

outcome [out'kum'] A possible result of an experiment (p. 270)

parallel lines [par'ə•lel līnz] Lines that never cross (p. 372)
Example:

parallelogram [par•ə•lel'ə•gram] A quadrilateral with 2 pairs of parallel sides and 2 pairs of equal sides (p. 405)
Example:

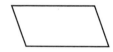

pentagon [pen'tə•gän] A polygon with five sides and five angles (p. 382)
Example:

perimeter [pə•ri'mə•tər] The distance around a figure (p. 456)
Example:

pictograph [pik'tə•graf] A graph that uses pictures to show and compare information. (p. 250)
Example:

HOW WE GET TO SCHOOL	
Walk	✹ ✹ ✹
Ride a Bike	✹ ✹ ✹ ✹
Ride a Bus	✹ ✹ ✹ ✹ ✹ ✹
Ride in a Car	✹ ✹

Key: Each ✹ = 10 students.

pint (pt) [pīnt] A customary unit for measuring capacity; 1 pint = 2 cups (p. 426)

place value [plās val'yoo] The value of each digit in a number, based on the location of the digit (p. 4)

P.M. [pē em] Between noon and midnight (p. 96)

point [point] An exact position or location (p. 368)

polygon [pol'ē•gän] A closed plane figure with straight sides; each side is a line segment. (p. 382)
Examples:

possible outcome [pos'ə•bəl out'kəm] Something that has a chance of happening (p. 272)

pound (lb) [pound] A customary unit used to measure weight; 1 pound = 16 ounces (p. 428)

predict [pri•dikt'] To make a reasonable guess about what will happen (p. 272)

product [prä'dəkt] The answer in a multiplication problem (p. 118)
Example: 3 × 8 = 24
⌐ product

pyramid [pir'ə•mid] A solid, pointed figure with a flat base that is a polygon (p. 362)
Example:

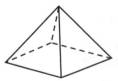

Q

quadrilateral [kwa•drə•lat'ər•əl] A polygon with four sides and four angles (p. 382)
Example:

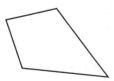

quart (qt) [kwôrt] A customary unit for measuring capacity; 1 quart = 2 pints (p. 426)

quotient [kwō′shənt] The number, not including the remainder, that results from dividing (p. 188)
Example: 8 ÷ 4 = 2
          ∟ quotient

radius [rā′dē•əs] A line segment whose endpoints are the center of a circle and any point on the circle (p. 375)
Example:

range [rānj] The difference between the greatest number and the least number in a set of data (p. 256)

ray [rā] A part of a line, with one endpoint, that is straight and continues in one direction (p. 369)
Example:

rectangle [rek′tang•gəl] A quadrilateral with 2 pairs of parallel sides, 2 pairs of equal sides, and 4 right angles (p. 405)
Example:

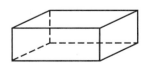

rectangular prism [rek•tan′gyə•lər pri′zəm] A solid figure with six faces that are all rectangles (p. 362)
Example:

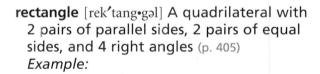

regroup [rē•grōōp′] To exchange amounts of equal value to rename a number (p. 40)
Example: 5 + 8 = 13 ones or 1 ten 3 ones

remainder [ri•mān′dər] The amount left over when a number cannot be divided evenly (p. 325)

results [ri•zults′] The answers from a survey (p. 240)

rhombus [räm′bəs] A quadrilateral with 2 pairs of parallel sides and 4 equal sides (p. 405)
Example:

right angle [rīt ang′gəl] A special angle that forms a square corner (p. 369)
Example:

right triangle [rīt trī′ang•gəl] A triangle with one right angle (p. 399)
Example:

rounding [roun′ding] One way to estimate (p. 28)

scale [skāl] The numbers on a bar graph that help you read the number each bar shows (p. 252)

scalene triangle [skā′lēn trī′ang•gəl] A triangle in which no sides are equal (p. 399)
Example:

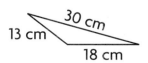

schedule [ske′•jо̄о̄l] A table that lists activities or events and the times they happen (p. 100)

sequence [sē′kwəns] To write events in order (p. 104)

simplest form [sim′pləst fôrm] When a fraction is modeled with the largest fraction bar or bars possible (p. 504)

sphere [sfir] A solid figure that has the shape of a round ball (p. 362)
Example:

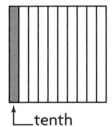

square [skwâr] A quadrilateral with 2 pairs of parallel sides, 4 equal sides, and 4 right angles (p. 405)
Example:

square number [skwâr nəm′bər] A product of two factors that are the same (p. 159)
Example: 4 × 4 = 16;
16 is a square number.

square unit [skwâr yо̄о̄′nət] A square with a side length of one unit; used to measure area (p. 462)

standard form [stan′dərd fôrm] A way to write numbers by using the digits 0–9, with each digit having a place value (p. 4)
Example: 345 ← standard form

subtraction [səb•trak′shən] The process of finding how many are left when a number of items are taken away from a group of items; the process of finding the difference when two groups are compared; the opposite operation of addition (p. 54)

sum [səm] The answer to an addition problem (p. 36)

survey [sər′vā] A question or set of questions that a group of people are asked (p. 240)

symmetry [sim′mə•trē] When one half of a figure looks like the mirror image of the other half (p. 384)

tally table [ta′lē tā′bəl] A table that uses tally marks to record data (p. 238)

tenth [tenth] One of ten equal parts (p. 520)
Example:

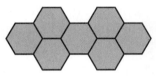
tenth

tesselate [tes′ə•lāt] To combine plane figures so they cover a surface without overlapping or leaving any space between them (p. 388)

tessellation [te•sə•lā′shən] A repeating pattern of closed figures that covers a surface with no gaps and no overlaps (p. 388)
Example:

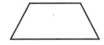

time line [tīm līn] A schedule of events, or an ordered list of historic moments (p. 109)

trapezoid [trap′ə•zoid] A quadrilateral with one pair of parallel sides (p. 390)
Example:

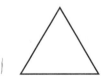

triangle [trī′ang•gəl] A polygon with three sides and three angles (p. 382)
Example:

unlikely [ən•lī′klē] An event is unlikely if it does not have a good chance of happening. (p. 270)

H86 Glossary

V

vertex [vûr′teks] In a solid figure, a corner where three or more edges meet (p. 362)
Example:

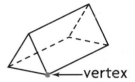
—vertex

vertical bar graph [vûr′ti•kəl bär graf] A bar graph in which the bars go up from bottom to top (p. 252)

volume [väl′yəm] The amount of space a solid figure takes up (p. 468)

 W

whole number [hōl nəm′bər] One of the numbers 0, 1, 2, 3, 4, ... The set of whole numbers goes on without end.

word form [wûrd form] A way to write numbers by using words (p. 4)
Example: The word form of 212 is two hundred twelve.

 Y

yard (yd) [yärd] A customary unit used to measure length or distance; 1 yard = 3 feet (p. 424)

Z

Zero Property of Addition [zē′rō prä′pər•tē əv ə•di′shən] A rule stating that when you add zero to, or subtract zero from, a number, the result is that number

Zero Property of Multiplication [zē′rō prä′pər•tē əv məl•tə•plə•kā′shən] A rule stating that the product of zero and any number is zero (p. 132)

INDEX

A

A.M., 96–97
Addends, 36, 37, 75
　missing, 43, 48, 60
Addition
　to check subtraction, 60, 62, 64
　of columns, 36–37, 115, 237, 455
　estimating sums, 38–39, 42–43, 75
　four-digit numbers, 46–48, 75
　fractions, 504–505
　greater numbers, 46–48
　grouping addends, 36
　Grouping Property of, 36–37
　of hundred thousands, 73
　mental math and, 35
　modeling for
　　simple fractions, 502
　　whole numbers, 40
　of money, 42, 88–89, 111, 540–543
　of more than two addends, 36–37
　and multiplication, 116–117
　of one-digit numbers, 36–37, 75, 131
　order of addends, 36
　Order Property of, 131
　properties of, 36, 131
　regrouping, 40–43, 46–49
　repeated, in multiplication, 116–117
　three-digit numbers, 40–43, 75
　three or more addends, 36–37, 75
　two-digit numbers, 35, 291
Addition strategies
　predicting and testing, 44–45
Algebra
　add/subtract money, 88–89
　connecting addition and multiplication, 116–117
　expressions, 68–69, 206–207
　　evaluating, 158
　　writing, 68–69, 206–207
　extending linear pattern by its rules, 136–137, 164–165, 168–169, 201, 203, 432–433
　fact families, 190–193
　find a rule, 168–169
　finding linear patterns, 390–391, 484
　functional relationships, 81, 216, 224–225
　inequalities, 22, 25–27, 151, 205, 222, 299
　inverse operations, 188
　　multiplication and division, 188–192, 196
　missing addend, 43, 48, 60, 216
　missing digit, 216, 327
　missing factors, 142–143, 151, 153, 187, 189, 192, 201, 203, 216, 296, 319, 341, 460, 470, 542
　missing operation symbol, 219
　multiply with three factors, 170–171
　number sentences, 68–69, 330, 333
　patterns in, 168–169, 306–307, 340–341
　practice the facts, 156–159
　recording division of two-digit numbers, 328–331
　relating multiplication and division, 188–189
　rules for changing units, 432–433

solving unit cost problems, 224–225
subtraction of fractions, 508–513
subtraction of greater numbers, 62–65
Angles, 368–369
　greater than a right, 369–371, 395–399, 414
　less than a right, 369–371, 395–399, 414
　in polygons, 382–383
　in quadrilaterals, 402
　right, 368–373, 396–409
　　identify in polygons, 370, 396–409
　sorting triangles by, 398, 399
Area, 462–463
　plane figures, 462–463, 466
　solid figures, 464–465
　surface area of solids, 477
Arrays, 120–121, 292, 293
　dividing through tens, 220–222, 232
　in division, 188–189, 215, 216
　to find products, 138–140, 148–151, 152, 154, 156, 159, 170, 178, 188–189
　to find quotients, 188–189, 215–216
　and multiplication, 120–121, 178, 180, 188–189
Assessment
　Chapter Review/Test, 14, 32, 50, 70, 90, 106, 128, 144, 160, 174, 196, 210, 228, 246, 264, 282, 302, 320, 336, 352, 378, 392, 410, 436, 452, 472, 498, 516, 532, 546
　Cumulative Review, 15, 33, 51, 71, 91, 107, 129, 145, 161, 175, 197, 211, 229, 247, 265, 283, 303, 321, 337, 353, 379, 393, 411, 437, 453, 473, 499, 517, 533, 547
　Mixed Review and Test Prep, 3, 5, 9, 11, 19, 23, 25, 29, 31, 37, 39, 41, 43, 49, 55, 57, 61, 65, 69, 81, 85, 87, 89, 95, 97, 99, 101, 103, 117, 119, 121, 125, 133, 135, 141, 143, 149, 151, 153, 159, 167, 169, 171, 185, 187, 189, 193, 201, 203, 205, 207, 217, 219, 223, 225, 239, 241, 243, 253, 255, 259, 261, 263, 269, 271, 273, 277, 279, 293, 297, 299, 307, 311, 315, 317, 319, 325, 327, 331, 333, 341, 343, 345, 347, 349, 365, 367, 371, 373, 375, 383, 387, 389, 397, 401, 403, 407, 423, 425, 427, 429, 431, 433, 443, 447, 449, 451, 457, 461, 463, 465, 471, 485, 487, 491, 495, 503, 507, 509, 513, 521, 523, 525, 527, 529, 537, 539, 543
　Study Guide and Review, 74–75, 110–111, 178–179, 232–233, 286–287, 356–357, 414–415, 476–477, 550–551
Associative Property
　See Grouping Property

B

Bar graph, 60, 124, 155, 218, 252, 253, 254, 255, 259, 264, 274, 276, 286, 297, 309, 320, 331, 377, 401, 425, 434, 435, 436, 461, 494, 531
　horizontal, 252, 253, 254, 255
　making, 254–255
　reading, 252–253
　vertical, 252, 253, 255, 259
Base-ten blocks, 1, 4–8, 17, 20, 21, 40, 41, 56, 57, 59
　in division, 326–328, 340
　in multiplication, 292–296, 298
Basic facts
　dividing by multiples of ten, hundred, and thousand, 340, 341, 342, 343
　multiplication, 306–307, 308–309, 315, 455
　multiplication table through ten by ten, 221
　multiplying multidigit numbers by one-digit numbers, 312–315
　multiplying multiples of ten, 298, 306

multiplying multiples of hundred, 298, 306
multiplying multiples of thousand, 306
related addition and subtraction facts, 62, 63
related multiplication and division facts, 188–189, 200–201, 214, 339
Be a Good Problem Solver, xxiv–xxv
Benchmark numbers, 18
in estimation, 18–19
in measurement, 18–19
Bills, counting, 79–81, 86–87
Breaks problems into simpler parts strategy, 350–351, 376–377, 544–545

C

Calendar, 93, 102–103, 104–105, 106
use of, 112
California Connections, 76–77, 112–113, 180–181, 234–235, 288–289, 358–359, 416–417, 478–479, 552–553
California Mathematics Content Standards, H71–H77
Capacity
customary, 426–427, 476
estimating, 426–427
metric, 446–447
Celsius temperature, 450–451
Center (of circle), 374, 375
Centimeter, 439, 440–445, 476, 477
Certain events, 268, 284
Challenge, 73, 109, 177, 231, 285, 355, 413, 475, 549
Change, making, 86–87, 110
Chapter Review/Test, 14, 32, 50, 70, 90, 106, 128, 144, 160, 174, 196, 210, 228, 246, 264, 282, 302, 320, 336, 352, 378, 392, 410, 436, 452, 472, 498, 516, 532, 546
Check What You Know, 1, 17, 35, 53, 79, 93, 115, 131, 147, 163, 183, 199, 213, 237, 249, 267, 291, 305, 323, 339, 361, 381, 395, 419, 439, 455, 481, 501, 519, 535
Choosing the operation, 208–209, 300–301
Circle, 361, 374–375, 381, 385, 389
Circumference, 475
Classify, 242
polygons, 382–383, 408
quadrilaterals, 382–383, 408
triangles, 382, 398–401
Clocks
analog, 93, 94, 95, 96, 97, 106, 217, 345, 407
digital, 94, 95, 97, 106
Closed figures, 382
Coins, 108
counting, 79–81, 86–87
dimes, 79–81, 86–87
half dollars, 108, 536
nickels, 79–81, 86–87
pennies, 79–81, 86–87
quarters, 79–81, 86–87
Column addition, 36–37, 115, 237, 291, 455
Commutative Property
of Multiplication, 120–125
See also Order Property of Multiplication
Compare, 20, 23
and contrast, 471
Comparing
data, 259
decimals, 528–529
fractions, 492–495, 501, 511
graphs, 252–253

money amounts, 84
numbers, 17, 20–23, 74
on number line, 20, 21
parts of a whole, 267
Composite numbers, 355
Computer Software
E-Lab, 3, 40, 57, 81, 95, 98, 116, 150, 188, 240, 293, 325, 363, 384, 385, 441, 463, 488, 525
Mighty Math Calculating Crew, 47, 63, 87, 317, 347
Mighty Math Carnival Countdown, 124, 139, 459
Mighty Math Number Heroes, 268, 399, 506, 520
Cone, 361, 362, 363, 366
Congruent figures, 384–387, 395
Corners, 381
Counting
change, 86–87
money, 84–85
skip-counting, 115, 249
backward, 186
by fives, 3, 115
by tens, 3, 163, 249
by threes, 2
by twos, 2, 115
Cross Curricular Connections
Art, 401
Reading, 23, 167, 217, 259, 297, 407, 471, 543
Science, 61, 277, 387, 491
Social Studies, 223, 331, 461, 485, 513
Cube, 361, 362–365, 366
and rectangular prism, 365
surface area of, 464
Cubic unit, 468–471, 477
Cumulative Review, 15, 33, 51, 71, 91, 107, 129, 145, 161, 175, 197, 211, 229, 247, 265, 283, 303, 321, 337, 353, 379, 393, 411, 437, 453, 473, 499, 517, 533, 547
Cup, 426–427
Customary units
of capacity, liquid measure
cups, 426–427
gallons, 426–427, 476
pints, 426–427, 476
quarts, 426–427, 476
changing units in, 430–431, 432–433
of length
feet, 424–425
inches, 420–423, 424–425, 476
miles, 424–425, 478
yards, 424–425, 478
of weight
ounces, 428–429
pounds, 428–429
Cylinder, 361, 362, 363, 366–367

D

Data, 238, 246
analyzing charts and tables, 4, 23, 26, 27, 29, 31, 32, 37, 41, 45, 48, 64, 67, 76, 83, 89, 127, 158, 180, 246, 286, 309, 314, 335, 351, 552
bar graph, 60, 124, 155, 253, 254, 255, 274, 276, 286, 320, 331, 341, 425, 434, 435, 436, 494, 531
classifying, 242–243
collecting and organizing, 238–239
diagrams, 400, 479
in identifying relationships, 27

line graphs, 262–263
line plots, 278–279, 409
ordering numbers, 24–27
pictographs, 8, 49, 103, 141, 167, 173, 195, 209, 250, 264, 391
rounding, 28–31
schedules, 100, 101, 106
survey, 257, 286
tally table, 238–241, 244
understanding, 240–241
Days, 102, 103
Decimal point, 520, 526, 551
Decimals
adding, 540–543, 551
comparing, 528–529, 535, 552
defined, 520
equivalent fractions and, 549
forms of, 526–527
fractions and, 520–521, 535, 551
greater than one, 535
hundredths, 524–525
mixed numbers as, 535
money amounts, 536–537
money and, 538–539, 540–543
number line, 528–529, 535, 549
ordering, 528–529, 535
place value in, 520
reading and writing, 526–527
subtracting, 540–543, 551
tenths, 520–523
Decimeter, 440–442
Degrees
Celsius, 450–451
Fahrenheit, 450–451
Denominator, 482–485, 501
Diameter, 374–375
Differences. *See* Subtraction
Digits, 4
place in quotient, 344–345
place value of, 18–27, 74
Dime, 79–81, 86, 108
Distance, measuring, 358, 478
Divide, 184
Dividend, 188, 200, 323, 324
three-digit, 346–347
two-digit, 220–223
Division, 184, 196
basic facts, 323, 339
by one and zero, 204–205
by two and five, 200–201
by three and four, 202–203
by six, seven, and eight, 214–217
by nine and ten, 218–219
by eleven and twelve, 231
as connected to multiplication, 230, 233, 357
divisors, 188, 200, 323, 324
estimate quotients, 342–343
fact families, 190–193
facts through five, 213
facts through ten, 220–223
finding cost of one item, 233
as inverse of multiplication, 184, 188–189, 344
meaning of, 184–185
mental math and, 340–342
model of two-digit numbers, 326–327
of money amounts, 348–349
patterns in, 340–341
placing first digit in quotient, 344–345
practice, 332–333
procedures, 220–221
properties, 184
quotients, 188, 220–221
relating to subtraction, 186–187
with remainders, 324–325, 334–335
repeated subtraction and, 186–187
rules for, 204
of three-digit numbers, 346–347, 357
of two-digit numbers, 357
writing a number sentence, 194–195
zeros in, 204–205
Divisor, 188
Dollar, 80–85, 86–87
Doubles (doubling), 179
even factors, 156, 157
to find products, 138–139
using to multiply, 148, 156, 157
Draw a Picture strategy, 154–155
Drawing conclusions, 280–281

E

Edge, 362
E-Lab. *See* Technology Link
Elapsed time, 98–99, 102, 106, 111
Endpoints, 368–369
Equal groups, 115–117, 138, 147, 184
dividing, 204
to find quotient, 215, 221
Equal to, 20
symbol for, 20
Equally likely events, 272–273, 284
Equilateral triangle, 398–401
Equivalent fractions, 488–491, 501, 550, 553
decimals and, 549
Equivalent sets, 80–81, 90, 110
Estimate, 38
Estimate or Exact Answer, 66–67
Estimation, 50
add/subtract money, 89
benchmark numbers, 18–19
of capacity, 427
to check reasonableness of answer, 42, 62, 63, 312, 313, 318
of differences, 54–55, 75
greater numbers, 46
of measurement, 420, 477
perimeter, 458–461
products, 310–311
quotients, 342–343
and rounding, 28–29, 54
of sums, 38–39
three-digit numbers, 42–43
use of calendar, 103
using a number line, 28–29
volume, 468–471
weight, 428
Even/odd numbers, 2–3, 74
Event, 268
Events, sequencing, 104–105
Expanded notation, 4, 7, 8, 10, 11, 73, 74, 526, 527

Experiments, in probability, 274–277
Expression, 68–69, 206, 207
   evaluating, 68, 69
   expressing simple unit conversions in symbolic form,
     430–435
   writing, 206–207, 232
Extra Practice, H32–H61

F

Faces, 362, 414
   of solid figures, 464
   tracing and naming, 363
Fact family, 190–193, 232
   defined, 190
   dividing through, 10, 220, 221
   multiplication, 190, 191
   multiplication and division, 190–191, 220, 221
Factors, 118, 134
   in arrays, 120–121, 220–221
   missing, 142–143, 178, 179, 213, 230, 355
Facts test, H66–H69
   multiplying by
     zero, 132–133
     one, 132–133, 178–179
     two, 118–119
     three, 122–125
     four, 134–135
     five, 118–119
     six, 148–149
     seven, 150–151
     eight, 152–153
     nine, 164–167
     ten, 164–167
Fahrenheit temperature, 450–451
Fair, defined, 280
Feet, 424–425, 478
Figures
   closed, 382
   congruent, 384, 395
   open, 382
   patterns with, 381
   plane, 462–463
   solid, 464–465
   that tessellate, 388–389
Find a Pattern strategy, 136–137, 308–309, 390–391
Foot, 424–425
Fraction bars, 482, 498
   and equivalent fractions, 489, 490
   model making, 496
   to order fractions, 493–494
   and parts of a whole, 492
Fractions
   adding, 502–503, 504–507, 550
   comparing, 492–495, 501, 519, 550
   concept of, 482–485
   decimals and, 520–521, 535, 551
   denominator, 502
   equivalent, 488–491, 550, 553
   like, 502
   in measurements, 421–422
   in mixed numbers, 495
   modeling
     with fraction bars, 482, 489, 490, 492, 493, 496, 501, 504,
      505, 510
     on number line, 483
     as part of group, 481, 486–487
     as part of whole, 482–485
     with tiles, 492
   and money, 536–537
   naming, 501, 519
   on number line, 549
   numerator, 502
   ordering, 492–495
   simplest form of, 504–506, 516, 550
   subtracting, 508–513, 550
Frequency table, 238, 241, 255, 289

G

Gallon, 426, 436, 476
Geometry
   angles, 368–373
     classifying, 372–373
     in polygons, 382–383
     in quadrilaterals, 402
   area, concept of, 462–463
   classifying
     angles, 372–373
     figures, 382–383
     polygons, 382–383, 408
     solid figures, 416, 464–465
     triangles, 398–399
   closed figures, 382
   congruent figures, 384–387, 417
   curves in plane figures, 382–383
   faces in solids, 464
   line, 368–371
     angle relationships and, 382–383, 395–401, 404
     parallel, 402
     ray, 368–369
     segment, 368–369
   line symmetry, 384–387, 417
   making complex solid forms from simpler solids, 366–367
   one-dimensional figures, 368, 408
   open figures, 382
   patterns in, 381
   perimeter of polygons, 456–457
   plane figures, 361
     circles, 374–375, 381, 385, 389
     combine, 388–389
     hexagons, 382–383, 388, 390
     octagons, 382–383
     parallelograms, 408
     pentagons, 382–383
     polygons, 382–383, 408
     quadrilaterals, 382–383, 408
     rectangles, 408, 463
     squares, 376, 381, 390, 391, 408
     triangles, 381, 382–383, 388, 390, 391, 396–401
   point, 368
   polygons, 382–391, 396, 402–409
   quadrilaterals, 382–383, 402–409
   solid figures, 362–363
     combine, 366–367
     cones, 361, 362, 363, 366
     cubes, 361–366
     cylinders, 361–366
     edges, 362, 416
     faces, 362, 416, 464

pyramids, 361, 362–367
rectangular prisms, 361–366
spheres, 361–364, 381
vertices, 362, 414, 416
symmetry, 385–387, 417
tessellation, 388–389
three-dimensional figures, 361–367
triangle
equilateral, 398–401
isosceles, 398–401
right, 398–401
scalene, 398–401
two-dimensional figures, 381, 382–383, 408
volume, 468–471
Gram, 448–449
Graphic aids, 259, 407
Graphs
bar, 155, 252–255, 259, 309, 341, 377, 434–435
choosing, 250–257
data labels on, 254
and frequency table, 255
identifying parts of, 250
key of, 209, 250–251
kinds of, 250–255
line, 262–263
line plots, 256–259
making, 254–255, 289, 309
pictographs, 137, 167, 195, 209, 249, 250
predictions and, 543
as problem solving skill, 434–435
as problem solving strategy, 250–251
scale and, 252, 254–255
Greater than, 17, 89
symbol for, 20
Greater than likely event, 284
Grid, 260
to find products, 315
locating points on, 260–261, 287
ordered pairs, 260–261
Grid paper, in multiplication, 293
Grouping Property
of Addition, 36, 50, 65
of Multiplication, 170–171, 178

Half dollar, 108, 536
Hands On
add fractions, 502–503
add three-digit numbers, 40–41
area of plane figures, 462–463
arrays, 120–121
capacity, 426–427
liters and milliliters, 446–447
circles, 374–375
collect and organize data, 238–239
decimals and money, 538–539
divide with remainders, 324–325
dividing amounts of money, 348–349
elapsed time, 98–99
hundredths, 524–525
make bar graphs, 254–255
make change, 86–87
make equivalent sets, 80–81
mass: grams and kilograms, 448–449

meaning of division, 184–185
measure temperature, 450–451
multiply two-digit numbers, 292–293
number patterns on a hundred chart, 2–3
perimeter, 456–457
possible outcomes, 272–273
relate decimals and money, 538–539
subtract fractions, 508–509
subtract three-digit numbers, 56–57
tenths, 522–523
time to the minute, 94–95
weight, 428–429
Harcourt Math Newsroom Video, 21, 100, 165, 214, 262, 295, 369, 421, 542
Hexagons, 382–383, 388, 390, 413
Highly likely event, 284
Horizontal bar graph, 252
Hour, 93, 94–99
Hundred chart, 12–13, 72
patterns in, 2–3
Hundred thousands, 73
Hundreds, place value and, 1
Hundredths, 524–525
subtracting, 541

Identify relationships, 26–27, 408–409
Impossible event, 268, 284
Inch(es), 420–423, 424–425, 476
Inequalities
represent relationships of qualities in, 85, 205, 222, 299
solve problems involving, 22, 25–27, 151, 153, 158, 205, 222, 299
Information, problem solving with too much/too little, 126–127
Interpret remainder, 334–335
Intersecting lines, 372
Inverse operations, 188
multiplication and division, 190–191, 196
Isosceles triangle, 398–401

Key, 250–251
Kilogram, 448–449
Kilometer, 440–442, 445, 479

L

Lattice multiplication, 315
Length. *See* Measurement
Less than, 17, 89
symbol for, 20
Like fractions, 502–503
addition of, 502–507, 516
subtraction of, 508–513
Likely events, 270–271
Lines, 368
intersecting, 372–373, 402, 404
parallel, 372–373, 402–409
types of, 372–373

Line graphs, 262, 264, 287
  reading, 262–263
Line of symmetry, 384–387
Line plots, 256–259, 264, 278, 409
Line segments, 368–371, 412, 416
Linkup
  Art, 401
  Geography, 513
  Math History, 315
  Reading Strategies
    analyze information, 167
    choose important information, 217, 297
    compare, 23, 471
    make predictions, 543
    use graphic aids, 259, 407
  Science, 61, 277, 387, 491
  Social Studies, 223, 331, 461, 485
Liters, 446–447

M

Make Generalizations, 466–467
Make a Model strategy, 496–497
Make a Table strategy, 82–83, 244–245, 444–445
Manipulatives and visual aids
  arrays, 120, 121, 220, 221
  balance, 428, 448
  base-ten blocks, 1, 4–8, 17, 20, 21, 40, 41, 56, 57, 59, 292–296, 298, 326–328, 340
  centimeter ruler, 460
  clock dials, 93–95, 97–99
  counters, 118, 184, 185, 215, 216, 220–222, 324, 325
  decimal models, 522–525, 540, 541
  fact cards, 190, 193
  fraction bars, 482, 489, 494, 496, 497, 501–506, 508–512, 516, 519, 521
  geometric wood solids, 363–364
  grid paper, 6, 7, 230, 406, 462, 474
  multiplication tables, 221, 231
  number cards, 9, 49
  number cube, 443
  number lines, 122–124, 528, 529, 535
  pattern blocks, 390–391, 413
  place-value chart, 7, 10, 21, 73, 528, 529
  play money, 79–81, 84–87, 348, 535–538
  ruler, 459, 460, 475
  spinners, 244, 268, 270, 273, 274–276, 280–281, 282, 284
  string, 475
  tiles, 120–121, 138–140, 148, 150–152, 156, 159, 170, 188, 189, 215, 216, 220–222, 275, 276, 456, 462
  unit cubes, 468–471
  yardstick, 424
Mass, 448–449
Math Detective, 72, 108, 176, 230, 284, 354, 412, 474, 548
Mean, 285
Measurement
  area, 462–463, 464–465
  capacity, 426–427, 476
  Celsius degrees, 450–451
  changing units, 430–433, 476, 477
  choosing an appropriate measuring tool, 420, 430–431, 440, 441, 446–449, 476
  choosing a reasonable unit, 430–433, 440, 441, 446–449
  customary units
    cups, 426–427

    feet, 421–425, 478
    gallons, 426–427, 476
    inches, 419–425, 476
    miles, 424–425, 478
    ounces, 428–429
    pints, 426–427
    pounds, 428–429
    quarts, 426–427
    yards, 424–425, 478
  distance, 358, 478–479
  estimating
    capacity, 446–447
    length, 422–423, 477
    volume, 468
    weight/mass, 428–429, 448–449
  Fahrenheit degrees, 450–451
  half inch, 421
  length, 420–425, 440–443, 476
  liquid volume, 426–427, 446–447, 476
  mass, 448–449
  metric system
    centimeters, 439, 440, 477
    decimeters, 440, 477
    grams, 448–449
    kilograms, 448–449
    kilometers, 440, 479
    liters, 446–447
    meters, 440
    milliliters, 446–447
  to nearest half inch, 421
  to nearest inch, 419–423
  perimeter of polygon, 477
  square units, 462, 464–465, 477
  Table of Measures, 424, 430
  temperature, 450–451, 478
  time
    days, 102, 103
    hour, 94–99
    minutes, 94–99
    months, 102, 103
    weeks, 102, 103
    year, 102
  unit conversions, 430–435, 444–445
  using rulers, 419–422, 439, 475
  volume of solid figure, 468–471
  weight, 428–429
Median, 285
Mental math
  adding two-digit numbers, 35
  division and, 333, 340–341
  multiplication and, 306–307
  patterns in, 306–307, 340–341
  subtracting two-digit numbers, 53
Meter, 440–443, 477
Metric system
  capacity, 446–447
    liters, 446–447
    milliliters, 446–447
  changing units in, 444–445
  customary units compared to, 450
  length, 440–443
    centimeters, 439, 440–445, 477, 479
    decimeters, 440–445, 477
    kilometers, 440–445, 479
    meters, 440–445, 477, 479

mass, 448–449
  grams, 448–449
  kilograms, 448–449
  relationship of units, 441
Midnight, 96–97
Mighty Math Calculating Crew
  Nick Knack, 347
  Superhero Superstore, 47, 63, 87, 317
Mighty Math Carnival Countdown
  Pattern Block Roundup, 459
  Snap Clowns, 124, 139
Mighty Math Number Heroes
  Fraction Fireworks, 506, 520
  Geoboard, 398
  Probability, 268
Mile, 424–425
Milliliter, 446–447
Minute, 94–99
Mixed Applications, 27, 67, 105, 127, 173, 209, 281, 301, 335, 409, 435, 467, 515, 531
Mixed numbers, 495
  decimals and, 535
  fractions and, 535
Mixed Review and Test Prep, 3, 5, 9, 11, 19, 23, 25, 29, 31, 37, 39, 41, 43, 49, 55, 57, 61, 65, 69, 81, 85, 87, 89, 95, 97, 99, 101, 103, 117, 119, 121, 125, 133, 135, 141, 143, 149, 151, 153, 159, 167, 169, 171, 185, 187, 189, 193, 201, 203, 205, 207, 217, 219, 223, 225, 239, 241, 243, 253, 255, 259, 261, 263, 269, 271, 273, 277, 279, 293, 297, 299, 307, 311, 315, 317, 319, 325, 327, 331, 333, 341, 343, 345, 347, 349, 365, 367, 371, 373, 375, 383, 387, 389, 397, 401, 403, 407, 423, 425, 427, 429, 431, 433, 443, 447, 449, 451, 457, 461, 463, 465, 471, 485, 487, 491, 495, 503, 507, 509, 513, 521, 523, 525, 527, 529, 537, 539, 543
Mixed Strategy Practice, 13, 45, 83, 137, 155, 195, 227, 245, 251, 309, 351, 377, 391, 445, 497, 545
Mode, 256, 264
Models
  area, 464–467
  decimal models, 520–527
  division, 348
  division of two-digit numbers, 326–327
  to find sum, 40–41
  fractions, 501, 502–506
  hundredths, 524–527
  multiplication, 292–296
  parts of a group, 486
  parts of a whole, 482
  problem solving strategy, 496–497
  simple fractions, 488–490
  solid figures, 362
  subtraction, 56–59
Money
  adding, 88–89, 111, 113, 540–543
  adding and subtracting, 88–89
  comparing amounts of, 84–85, 110
  counting bills and coins, 79–81, 86–87
  and decimal notation, 535–545
  dividing amounts of, 348–349
  equivalent sets, 80–81, 110
  estimating with, 111
  find the cost, 224–225
  finding products using, 316–317
  fractions and, 536–537
  making change, 86–87, 110, 113
  subtracting, 88–89, 111, 540–543

Multiplication
  and addition, 116–117
  Associative Property of, 170–171
  basic facts, 307
  checking division with, 200–201
  Commutative Property of, 121–124, 138, 139, 156–157
  and division, 230, 233
  factors, 118, 134
    zero, 132–133
    one, 132–133
    two, 118–119
    three, 122–125
    four, 134–135
    five, 118–119
    six, 148–149, 179
    seven, 150–151, 179
    eight, 152–153, 179
    nine, 164–167
    ten, 164–167
  to find product, 118–119, 134, 139
  four-digit numbers, 318
  Grouping Property of, 170–171, 178
  as inverse of division, 188–189, 344
  of money amounts, 316–319
  multiples and, 233
  Order Property of, 120–125, 138, 139, 156, 157, 178, 179, 213
  patterns in, 179, 306–307
  practice, 138–141, 298–299, 318–319
  product, 118–119
  recording, 294–297
  skip-counting in, 164
  three-digit numbers, 312–315, 356
  with three factors, 170–171
  two-digit numbers, 292–293, 305, 356
  uses of, 224
  writing sentences, 120, 132
Multiplication facts, 131, 179, 291, 339, 455
  through eight, 163
  through five, 147
  through ten, 183, 199
Multiplication table, 178
  to find quotient, 231
  to five, 200
  to nine, 134, 139, 142, 149, 156, 202
  patterns in, 231
  to six, 191
  to ten, 164, 221
Multiply, 116–117
Multistep problems, 39, 60, 61, 64, 65, 89, 90, 117, 135, 140, 141, 143, 144, 148, 149, 151–153, 155, 156, 164–167, 172–173, 187, 190–192, 200–203, 208, 214–225, 227, 300–301, 329–331, 428, 435, 443, 445, 447, 463, 466, 522, 523, 531

N

Nickel, 108
  counting, 79–81, 86–87
Noon, 96
Number
  benchmark, 18–19
  comparing, 17, 20–23, 74
  composite, 355
  even or odd, 2–3, 74
  expanded form, 4–5, 7, 8, 10–11, 73, 74, 526, 527

fractions, 484
hundred thousands, 73
mixed, 495
ordering, 17, 20, 24–25, 74
prime, 355
reading and writing, 1, 116, 484
rounding, 74
size of, 18–19
square, 159, 177
standard form, 4–5, 7–8, 10–11, 73, 74
to ten thousand, 6–11
word form, 4–5, 7–8, 10–11, 73, 74
Number line
comparing numbers on, 20, 21, 535
decimals on, 549
to find product, 124
finding a pattern on, 136
fractions on, 549
multiplication, 122–123
ordering numbers, 24–25, 483, 484, 535
patterns on, 136
in rounding, 28, 30
skip-counting on, 138, 164
Number Sense
adding
decimals, 540–545
fractions, 502–503, 504–507
multidigit whole numbers, 36–49, 55–58, 60–69
benchmarks, 18–19
checking division with multiplication, 200, 218, 344
comparing and ordering
fractions, 492–497
whole numbers, 20–25, 26–27, 48, 257–259, 263
connecting addition and multiplication, 116–117
counting whole numbers, 2–11
dividing multidigit numbers, 324–327, 328–333, 340–341, 344–348
estimation, 38–39, 54–55, 66–67, 310–311, 342–343
Grouping Property of Addition, 36–37
identify place value, 4–5, 7–11, 24, 25, 29, 31
inverse, 143, 188–192, 200–203, 214–219, 221, 222, 224, 228, 329, 330, 332–333
memorize division facts, 200–205, 214–224
memorize multiplication facts, 116–125, 132–135, 138–142, 148–154, 156–159, 164–167, 170–171, 214–226
modeling fractions, 482–485, 486–491
money, 80–89, 316–317, 540–545
multiply multidigit numbers, 292–301, 306–307, 308–311, 312–315, 316–319
Order Property of Multiplication, 120–121, 156
Property of One, 132–133, 158, 204, 205
reading and writing
decimals, 520–530
fractions, 482–485, 486
whole numbers, 2–9
relate multiplication and division, 188–190
relating subtraction and division, 186–187
relating whole numbers, fractions, and decimals, 482–497, 520–527, 529–530, 536–539
rounding, 28–31, 38–39, 310–311
subtracting
decimals, 540–545
fractions, 508–513
multidigit whole numbers, 56–65
unit cost, 224–225

using expanded notations, 4–5, 7, 8–11, 73–74, 526–527
Zero Property, 132–133, 158, 204, 205
Number sentence, 199, 210, 354
expressions in, 206
as problem solving strategy, 194–195
true/false, 68, 69
writing, 194–195
Numerator, 482–485, 501

O

Octagons, 382–383
Odd numbers, 2–3, 74
One-digit number
dividing by, 200–201, 214–219
multiplying with, 118–119, 122–125, 132–135, 138–141, 148–159, 164–167
Ones
division by, 204–205
multiplying by, 132–133, 144
place value and, 1
Open figure, 382
Operation
choosing, 208–209, 300–301
symbols for, 206, 354
Ordered pair, 260, 261
Ordering
decimals, 528–529
fractions, 493–495
on number line, 528–529, 535
numbers, 17, 20, 24–25, 74
Order Property of Addition, 131
Order Property of Multiplication, 120–125, 138, 139, 156, 178, 179, 213
Ordinal numbers, 1
Ounce, 428–429, 436
Outcomes, 270
possible, 272–273
predicting, 278–279
recording, 272–273, 274–276

P

P.M., 96–97
Paper clip, as unit of measure, 420, 424
Parallel lines, 372–373, 402–409
Parallelogram, 404–407, 408
Parts of a group
counting, 486, 487
model for, 481
Parts of a whole, 548
counting, 482–485
identifying, 267
model for, 481
Pattern finding strategy, 136–137
Patterns
in division, 340–341
even/odd numbers, 2
finding, 136–137, 308–309, 390–391
in geometry, 331
in a hundred chart, 2–3
of hundred thousands, 73
hundredths, 524–525

with multiples, 148, 152, 164, 306–307
with number line, 136
with pattern blocks, 390–391, 413
place value, 306
plane figures, 388–389, 413
in problem solving, 308–309
square numbers, 159, 177
in tenths, 522–523
and tessellation, 388–389
Pennies, counting, 79–81, 86–87
Pentagons, 382–383
Perimeter, 456–457, 472, 474, 475
estimate and find, 458–461
Pictographs, 8, 49, 103, 141, 167, 173, 195, 209, 249, 250, 264, 391
Pint, 426–427
Place value
chart, 7, 10, 21–22
comparing numbers, 22
decimal use of, 520–521
in division, 344–347, 349
in hundred thousands, 73
hundreds, 1
multiply nine and ten, 164–167
ones, 1
ordering numbers, 24–25
patterns, 306
ten thousands, 10
tens, 1
thousands, 6–7
understanding, 4–5, 74
Plane figures
area of, 462–463
combine, 388–389
identifying, 361, 374, 382–383
patterns with, 413
Points, 368
end point, 368
on grid, 260–261, 287
Polygons, 382–391, 408–409
classifying, 370, 382, 383, 398–401, 402–409
describing, 382, 383, 384–389, 396–397, 402–409
hexagons, 382–388, 413
identify right angles in, 370, 396–409
identifying, 376, 377, 382, 383, 390–391, 415
octagons, 382–383, 415
pentagons, 382–383, 415
quadrilaterals, 382–383, 402–409
rectangles, 391, 408
rectangular prism, 361–362, 364, 381
rhombus, 390, 391, 404, 405, 408–409
squares, 361, 381, 390, 391, 405, 408
trapezoid, 413
triangles, 381, 388, 396–401, 413, 414, 415, 416
Possible outcome, 272–273
Pound, 428–429, 436
Predict, 272
Predict and test strategy, 44–45
Predictions, 272
of certain events, 268–269
experiments in, 274–277, 287
of impossible events, 268–269
of likely or equally likely events, 270–271, 272–276
of outcomes, 278–279, 287, 543
of possible outcomes, 272–273, 274–279, 543
of unlikely events, 270, 271, 275, 282

Prime numbers, 355
Prisms, rectangular, 381
Probability
certain events, 268–269, 271, 282, 284
equally likely or likely, 272–273, 284
greater than likely, 284
highly likely, 284
impossible, 268–269, 271, 282, 284
likely events, 270–271
possible events, 272–273, 543
predictions, 543
recording possible outcomes in an experiment, 543
slight chance, 284
slightly greater chance, 284
summarizing and displaying results of experiments, 274–277
unlikely events, 270, 271, 282
Problem solving skills
choose the operation, 208–209, 300–301
draw conclusions, 280–281
estimate or exact answer, 66–67
identify relationships, 26–27, 408–409
interpret remainder, 334–335
make generalizations, 466–467
multistep problems, 39, 60, 61, 64, 65, 89, 90, 117, 135, 140, 141, 143, 144, 148, 149, 151–153, 155, 156, 164–167, 172–173, 187, 190–192, 200–203, 208, 214–225, 227, 300–301, 329–331, 428, 435, 443, 445, 447, 463, 466, 522, 523, 531
observing patterns, 2, 3, 169
reasonable answers, 514–515, 530–531
sequence events, 104–105
too much/too little information, 126–127
use a graph, 434–435
Problem solving strategies
break problems into simpler parts, 376–377, 544–545
draw a picture, 154–155
find a pattern, 136–137, 308–309, 390–391
make a graph, 250–251
make a model, 496–497
make a table, 82–83, 244–245, 444–445
predict and test, 44–45
solve a simpler problem 350–351
use logical reasoning, 12–13, 384–387
work backward, 226–227
write a number sentence, 194–195
Products, 118–119. *See* Multiplication
Properties
Associative, or Grouping, Property of Multiplication, 170–171
Commutative, or Order, Property of Multiplication, 121–124, 138–139, 156–157
Grouping Property of Addition, 36–37
Order Property of Addition, 131
Order Property of Multiplication, 120–125, 138–139, 156–157, 178, 179, 213
Zero Property of Multiplication, 132
Pyramids, 361, 362–367
square, 362–367, 414, 416

Q

Quadrilaterals, 382, 402, 409
attributes of, 382, 383, 402–409
classifying, 415

identify right angles in, 402–409
naming, 405
parallelograms, 404, 408–409
rectangles, 361, 408–409
rhombus, 404, 405, 409
sorting, 404–407
squares, 381, 390, 391, 405, 408
Quart, 426–427
Quarter, 79–81, 86–87
Quotient, 188, 196, 200–201, 323–325, 339–341
estimating, 342–343
finding, 202, 203, 221
placing first digit in, 344–345
use of multiplication table to find, 200, 202, 221

R

Radius, 374, 375
Range, 256
Ray, 368, 369
Reading strategies
analyze information, 167
choose important information, 217, 297
compare, 23, 471
make predictions, 543
use graphic aids, 259, 407
Reasonable answers, 514–515, 530–531
Reasoning
applying strategies from simpler problem to solve more
complex problem, 81, 119, 383, 509, 512
breaking problem into simpler parts, 87, 133, 187, 219, 245,
281, 330, 343, 347, 351, 506, 513
in Challenge, 73, 109, 177, 231, 285, 355, 413, 475, 549
checking results of precise calculations, 442
choosing problem solving strategy, 101, 207, 208–209, 542
estimating, 39, 67, 311
evaluate reasonableness, 12–13, 42–43, 46, 62–63, 88, 154,
318, 344, 514–515, 530–531, 544
explaining in
bar graphs, 155, 461
diagrams, 400, 406, 460, 462
graphs, 253, 262–263
grids, 261
models, 450–451, 496–497
numbers, 450–451
symbols, 244, 249
tables, 29, 246, 249
words, 122, 244, 271, 279, 280, 417
generalizing beyond a particular problem to other
situations, 123, 171, 202, 406, 463, 466–467
identifying relationships among numbers, 25, 26–27, 116,
117, 132, 151, 153, 190, 191, 192, 201, 215, 333, 447, 484,
521, 537
in Math Detective, 72, 108, 176, 230, 284, 354, 412, 474, 548
note method of finding solution, 26, 66, 157, 186, 188, 202,
204, 215, 292, 332, 444, 466, 544
observing patterns, 2, 3, 137, 169, 225, 269, 408–409,
486–487
recognizing relevant and irrelevant information, 126–127,
217
sequencing and prioritizing, 126–127
in Thinker's Corner, 9, 49, 65, 125, 141, 159, 193, 365, 371,
423, 443, 495, 507
use of estimation to verify reasonableness of an answer, 42,
62–63, 312–313, 318–319, 442, 551

using mathematical notation, 20–23, 26–27, 40–43, 54–57,
68–69, 80–89, 100–103, 148–155, 168–169, 172–173,
194–195, 206–209, 224–227, 273, 278–279, 292–297,
300–301, 308–309, 312–317, 324–325, 328–331, 362–377,
408–409, 420–435, 440–451, 458–471, 520–529, 536–545
*Opportunities to explain reasoning are contained in every
exercise set. Some examples are:* 2, 3, 5, 8, 25, 29, 39, 57,
60, 67, 81, 84, 87, 95, 101, 103, 116–117, 119, 122, 124,
132, 133, 134, 140, 151, 153, 158, 169, 171, 185, 187, 189,
190, 191, 192, 201, 203, 204, 205, 208, 216, 219, 222, 225,
253, 261, 269, 271, 276, 279, 280, 292, 293, 296, 299, 307,
311, 330, 333, 343, 347, 349, 351, 362, 363, 364, 368, 371,
372, 373, 375, 422, 429, 433, 442, 447, 484, 490, 497, 506,
512, 537
Recording
data, 238
division of two-digit numbers, 328–331
multiplication, 294–297
outcomes of experiments, 272, 274–276
Rectangle, 361, 405, 408–409
Rectangular prism, 361, 362, 364, 381
combining, 366
vs. cube, 365
Regrouping, 40–41, 42, 56–57, 58–59, 62–63, 312–313
in division, 348
greater numbers, 46
hundreds, 323, 339
ones and tens, 291, 305
practice multiplication, 298–299
rules for, 47, 63
tens, 323, 339
Remainder, 324–325, 336
division with, 324–333
interpreting, 334–335
use of, 334
Repeated subtraction, 220, 221, 232
Results of probability experiments
display and summarize, 274–276, 278–279
Results of a survey, 240
Review Test, *See* Chapter Review Test
Rhombus, 390, 391, 404–407, 409
Right angles, 368–371, 395, 396
defined, 369
in quadrilaterals, 402
Right triangle, 398–401, 415
Rounding, 28, 310, 318
to nearest hundred, 28–29, 74
to nearest ten, 28–29, 54
to nearest thousand, 30–31, 74
rules for, 30, 310
using number line, 28–29
Rule
finding, 168–169, 419
for changing units, 432–433
Rulers
centimeter, 439, 441
using customary, 419–423
using metric, 439, 441

S

Scale, 252–255
Scalene triangle, 398, 399
Schedules, 100–101, 106
Sequence, 104–105

Sharpen Your Test-Taking Skills, H62–H65
Sides, 381
  and angles, 382–383
  in quadrilaterals, 382–383, 402–409
  sorting triangles by, 398, 399
Simplest form of fraction, 504–507
Skip-count, 2–3, 115, 139, 249
  backward, 186
  by fives, 3, 115
  by tens, 3, 163
  by threes, 2
  by two hundred, 444
  by twos, 2, 115
  on number line, 138
Solid figures, 362–367, 414
  combining, 366–367
  identifying, 361
  surface area of, 464–465
Solve simpler problem strategy, 350–351, 376–377, 544–545
Sphere, 361–364
Square number, 159, 177
Square pyramid, 362–365, 366, 377
Square unit, 462–463, 476, 477
Squares, 361, 381, 390, 391, 405, 408
Standard form, 4, 7, 8, 10, 11, 73, 74, 526, 527
Statistics
  bar graph, 218, 434
  line graph, 262–263
  line plot, 256–259
  pictograph, 167, 173, 195, 209, 249–251
  survey, 240, 246, 257
Stretch Your Thinking, 72, 109, 230, 284, 354, 412, 474, 548
Student Handbook, H1–H99
Study Guide and Review, 74–75, 110–111, 178–179, 232–233, 286–287, 356–357, 414–415, 476–477, 550–551
Subtraction, 199
  addition and, 60, 62, 64
  basic facts, 323
  decimals, 540–543
  differences, 54, 75
  and division, 186–187
  estimation and, 54, 75
  facts, 53, 323
  four-digit numbers, 62–65, 75
  fractions, 508–513, 550
  greater numbers, 62–65
  with money, 88–89, 111
  relating to division, 186–187
  repeated, 186, 220, 221, 232
  three-digit numbers, 56–61, 75
  two-digit numbers, 53
  across zeros, 59, 63
Sum, *See* Addition
Survey, 246, 257
  defined, 240
Symbols, 20–21
  equal to, 20–21
  finding missing operation symbol, 68–69, 187, 207, 219
  greater than, 20–21
  less than, 20–21
  select relational symbol, 85, 205, 222, 299
Symmetry
  lines of, 384–387
  patterns with, 388–389

T

Table of Measures, 424, 430, H70
Tables and charts
  analyzing data from, 4, 37, 41, 45, 48, 54, 55, 64, 66, 67, 127, 158, 237, 251, 277, 309, 314, 408, 420, 447, 449, 471
  bar graphs from, 255
  choosing the operation, 300–301
  classifying data from, 243
  completing, 218
  division, 201, 203, 215, 216, 218, 219
  estimating from, 426
  frequency, 238, 241, 246
  grouping data in, 424, 440, 458, 538
  line segments and angles, 370
  making, 215, 216, 242, 419, 430
    as problem solving strategy, 82–83, 244–245, 444–445
  multiplication, 185
  schedules, 100–101, 106
  tally, Refer to entry **Tally Table** below.
  using pattern, 487
  writing rules, 168, 169
Tally mark, 240, 244
  grouping of, 238
Tally table, 238, 239, 240, 241, 244, 249, 256, 257, 267, 272, 274, 279
Technology Link
  E-Lab, 3, 40, 57, 81, 95, 98, 116, 150, 188, 240, 293, 325, 363, 384, 385, 441, 463, 488, 525
  Harcourt Math Newsroom Video, 21, 100, 165, 214, 262, 295, 369, 421, 542
  Mighty Math Calculating Crew, 47, 63, 87, 317, 347
  Mighty Math Carnival Countdown, 124, 139, 203, 459
  Mighty Math Number Heroes, 268, 399, 506, 520
Temperature
  changing, 450
  measuring, 450–451
  on line plot, 258
Ten thousand
  numbers to, 6–11
  understanding, 10–11
Tens, place value and, 1
Tenths, 520–521
  hands on, 522–523
Tessellate, 388–389
Tessellation, 388–389
Thinker's Corner, 9, 49, 65, 125, 141, 159, 193, 365, 371, 423, 443, 495, 507
Thousands
  comparing, 21
  multiples of, 6–10
  numbers to, 6–11
  place value of, 7, 10
  rounding, 30–31
Three-digit numbers
  adding, 40–43, 75
  division of, 346–347
  estimating with, 54–55
  multiplying, 312–315
  rounding, 28
  subtracting, 56–59
Three-dimensional figures, 361–367
Time
  A.M., 96–97

analog clocks, 93, 94, 95, 96, 97
calendars, 93, 102–103
century, 109
days, 102, 103
decade, 109
digital clocks, 94, 95, 97
elapsed, 98–99, 102–103, 111
hour, 94, 111
midnight, 96
minute, 94, 111
months, 102
noon, 96
P.M., 96–97
schedules, 100–101, 111
telling, 93, 94–95, 111
units of, 102, 109
weeks, 102–103
years, 102, 109
Time line, 109
Too Much/Too Little Information, 126–127
Trapezoid, 390
Triangles, 361, 381, 382, 388, 389, 390, 391
attributes of, 396–397, 398–401
classifying, 398–401, 415
equilateral, 398–401, 415
as face of solid figure, 362–365
isosceles, 398–401, 415
naming, 399
right, 398, 399, 415
scalene, 398, 399, 415
sorting, 398–401
Two-digit numbers
addition, 35
comparing, 17
model division of, 325–327
multiplication, 292–293
ordering, 17
record division of, 328–331
subtraction, 53

Unfair games, 281
Unit cost, finding, 224–225
Units, changing, 430–431, 432–433, 434–435, 476
Unlikely events, 270, 271, 274–277, 282
Use a Graph, 434–435
Use Logical Reasoning Strategy, 12–13

Vertex (vertices), 362
Vertical bar graph, 252
Volume, 468
estimate and find, 468–471

Weeks, 102, 103
Weight
customary units, 428–429
metric units, 448–449
What's the Error?, 11, 25, 43, 48, 60, 85, 87, 89, 97, 119, 124, 137, 140, 153, 166, 193, 201, 227, 241, 261, 279, 296, 311, 314, 327, 330, 343, 347, 349, 375, 386, 403, 431, 443, 457, 470, 484, 512, 525, 527, 542
What's the Question?, 11, 22, 43, 64, 67, 99, 127, 143, 153, 173, 189, 209, 216, 245, 251, 253, 271, 301, 307, 314, 325, 345, 373, 391, 400, 422, 425, 447, 463, 487, 506, 523, 539, 545
Whole numbers
benchmark, 18
comparing, 20–23
expanded form of, 4–5, 7–11, 73, 74
ordering, 24–25
place value of, 4–5
reading and writing, 4–5
rounding, 28–31
standard form, 4–5, 7–11, 73, 74
word form, 4–5, 7–11, 73
Word form, 4–5, 7–11, 73, 74, 526–527
Work backward strategy, 226–227
Write About It, 8, 22, 27, 31, 37, 83, 97, 99, 101, 121, 133, 149, 166, 185, 203, 219, 241, 258, 259, 276, 281, 293, 299, 319, 325, 347, 364, 370, 373, 383, 387, 389, 397, 406, 409, 425, 442, 449, 460, 465, 470, 490, 515, 521, 525, 537, 539
Write a Number Sentence Strategy, 194–195
Write a problem, 13, 19, 29, 30, 37, 39, 41, 45, 55, 64, 105, 124, 133, 137, 158, 173, 181, 195, 207, 222, 239, 241, 263, 289, 296, 315, 317, 331, 345, 365, 386, 403, 429, 467, 487, 506, 529, 542
Writing in Math. *See* What's the Error? *and* What's the Question *and* Write About It *and* Write a problem.

Yard, 424–425
Year, 102, 109

Zero, 144
in division, 204–205, 340–341
multiply with, 132–133
with subtraction, 59, 63

PHOTO CREDITS